Proceedings of the 25th Pan-American Conference of Naval Engineering—COPINAVAL

Adán Vega Sáenz · Newton Narciso Pereira
Luis Manuel Carral Couce
José Angel Fraguela Formoso
Editors

Proceedings of the 25th Pan-American Conference of Naval Engineering—COPINAVAL

Springer

Editors
Adán Vega Sáenz
Technological University of Panama
Panama City
Panama

Newton Narciso Pereira
Federal Fluminense University
Volta Redonda, Rio de Janeiro
Brazil

Luis Manuel Carral Couce
Escola Politécnica Superior
Universidade da Coruña
Ferrol, A Coruña
Spain

José Angel Fraguela Formoso
Escola Politécnica Superior
Universidade da Coruña
Ferrol, A Coruña
Spain

ISBN 978-3-030-07865-2 ISBN 978-3-319-89812-4 (eBook)
https://doi.org/10.1007/978-3-319-89812-4

Softcover re-print of the Hardcover 1st edition 2019

Printed on acid-free paper

This Springer imprint is published by the registered company Springer International Publishing AG part of Springer Nature
The registered company address is: Gewerbestrasse 11, 6330 Cham, Switzerland

To all those visionaries who, more than 50 years ago, founded the Pan-American Institute of Naval Engineering (IPIN). They are the source of our inspiration to continue working to keep alive the flame of naval engineering through the continuous generation of knowledge.

Adán Vega Sáenz
Newton Narciso Pereira
Luis Manuel Carral Couce
José Angel Fraguela Formoso

Foreword I

The Pan American Conference of Naval Engineering, Maritime Transport and Port Engineering (COPINAVAL) is a long-standing prestigious conference. It has been held in practically every American country, from the United States in the north, to Chile and Argentina in the south of the continent. Strangely enough, until 2017, it was never before held in Panama, a bioceanic country whose importance to world maritime activities is simply too great to need an explanation.

It is only fair to acknowledge that COPINAVAL 2017 put things right. The conference was outstanding in organization, venue, and scientific quality of the papers. The outcome, in the form of this book, shows that a small resolute team can achieve results second to none.

Valdivia, Chile

Dr. Marcos Salas,
Universidad Austral de Chile

The publication of XXV COPINAVAL papers by SPRINGER shows the progress and academic success of the Congress and rewards the authors and the organization of the event for the excellent work developed. The 40 selected papers in the areas of ship design, ship maintenance, sustainability, maritime transportation, corrosion, legal aspects, and education, demonstrate high quality and the ability to produce good research with interesting themes that was achieved by the Pan-American countries.

Sao Paulo, Brazil

Dr. Rui Carlos Botter
Naval and Ocean Engineering Department
University of São Paulo, Brazil

The Pan-American Conference of Naval Engineering, Maritime Transportation and Ports Engineering (COPINAVAL) has been an excellent opportunity for the meeting of researchers, professors, and users related to the maritime sector. The importance of the exchange of experiences of different technological centers and universities is of crucial importance in the new era of this fourth revolution, also called Industry 4.0, in which we are immersed. The observation of reality shows that R&D investment generates economic return, quality employment, and great benefits for society. Reference institutions such as the United Nations or the OECD confirm that sustainable growth in research is the engine for social improvement.

The papers presented are a good example of research and represent an excellent manual to learn about innovation in different aspects in shipbuilding, maritime transport, and naval engineering. In this new era of the optimization of processes, whose characteristics are to make intensive use of process simulation and to use other cutting-edge technologies, the primary goal is to develop industrial plants with much better interconnected production chains and obviously much more competiveness. The naval field is not outside this practice and the presentations, like those included in this book, are good evidence that innovation in this field produces benefits that sooner or later will reach the desired social improvement.

A Coruña, Spain

Salvador Naya
Vicerector for Science Policy, Research and Transfer
Universidade da Coruña, Spain

Foreword II

Over the years the Pan-American Conference of Naval Engineering, Maritime Transportation and Port Engineering—COPINAVAL, has been constituted in the ideal space for the development and strengthening of technological and scientific capabilities of academics and professionals of the area, those who with huge effort, discipline, perseverance, and motivation have worked for the progress of our brotherly nations.

The XXV COPINAVAL, followed with a five-decade trajectory, has been developed in a week that allowed the assistants, speakers, and master lecturers, to share experiences, good practices, and new knowledge that surely will generate impact in the solving of today's problems linked to the tasks of naval engineers and related professionals, people who with hard work and determination assume the responsibility of generating development and social welfare, taking the great ocean as the referent of new opportunities.

Panama, with commitment and dedication, was the manager of all intercultural and scientific exchange in October of 2017. Proof of this was shown through the excellent academic agenda that considered a wide range of issues that considered aspects regarding the 4.0 industry, corrosion, maritime trade, legal aspects of the naval world, offshore, and ship recycling as an opportunity to shipyard crises, among other topics of no less importance.

This book is an acknowledgment of all the authors who contributed to the continued enrichment of the naval profession in the last COPINAVAL. All the works considered here, that were elaborated with excellent quality, are the reflection of the innovative and creative spirit of the professionals in Pan-American nations, all committed to the consolidation of naval engineering. Surely for our authors, the ocean has pointed to the horizon as a life project for self-realization.

With this motivation, and recognizing the success of the XXV COPINAVAL Congress, we invite all authors, professionals, and related researchers to continue in the generation of knowledge and development, using science and technology as the main tools for continuous improvement of our nations.

Colombia accepts with great enthusiasm and happiness assuming the challenge that the Pan-American Institute of Naval Engineering—IPEN has given to us. Our Panamanian kin and their labor will be a reference point for accomplishing the assigned duty, so that in the next March of 2019 the XXVI COPINAVAL Congress in union with the VI International Ship Design and Naval Engineering Congress (ISDNEC) and the VIII International Fair of Naval Industry—Colombiamar, all the Pan-American countries and invited nations will continue to contribute to the development of society through knowledge.

To all of you, thank you for the effort. We will be waiting for you in Cartagena in order to continue with this noble purpose.

Bogota, Colombia

Vice Admiral Jorge Enrique Carreño Moreno
President
Pan-American Institute of Naval Engineering

Acknowledgements

This book would not have been possible without the support received from each of the authors who cordially accepted our invitation. The great success achieved in the XXV COPINAVAL was based on the hard work of the organizing committee, the editors, the reviewers, and of course, thanks to each of the organizations that believed in the project. To the Maritime Authority of Panama, Asmar Shipyard, Bureau Veritas, IBS Class, Copa Airlines, COTECMAR, International Marine Expert, MEC Shipyard, Marine Jet Power, OMCS Class, SENACYT, CILIP, International MarConsult, Talleres Industriales, and the International Maritime University of Panama, our special gratitude.

Special thanks also to the students of the School of Mechanical Engineering of the Technological University of Panama, to the students and professors of the International Maritime University of Panama, and to Barcelo Congress, for all their support.

Last but not least important, we want to thank the board of directors of the Pan-American Institute of Naval Engineers for giving us the opportunity to organize the XXV COPINAVAL in Panama.

Dr. Adán Vega Sáenz
President
Pan-American Institute of Naval Engineers (2015–2017)

Contents

Editors and Contributors

About the Editors

Dr. Adán Vega Sáenz is a mechanical engineer with a Master's degree in materials science and another in plant engineering from the Technological University of Panama. He holds a Ph.D. in marine engineering from Osaka University, Japan. The author of more than 50 scientific articles and a textbook, Dr. Vega's professional experience includes, among others, being a university professor for 15 years at the Technological University of Panama, and visiting professor at the Technological University of Bolivar and UNINORTE, both in Colombia. He is also academic vice president and research vice president at the International Maritime University of Panama and a consultant to several national and international companies.

Dr. Newton Narciso Pereira joined the Fluminense Federal University (UFF), Rio de Janeiro—Brazil, School of Metallurgical Industrial Engineering of Volta Redonda (EEIMVR) in 2016. In 2014 he completed his postdoctorate in Naval Architecture and Ocean Engineering focusing on ballast water management and control in Brazilian ports. He holds a Ph.D. and Master's degree in Naval Architecture and Ocean Engineering from the Polytechnic School of University of São Paulo. He is a graduate in Naval Technology from the Faculdade de Tecnologia de Jahu—UNESP and Production Engineering from Guarulhos University. Dr. Pereira is the general coordinator of the Center for Sustainable System Studies (CESS-UFF) and the graduate coordinator of the Production Engineering course at (EEIMVR).

Dr. Luis Manuel Carral Couce holds a Ph.D. in Naval Architecture and Marine Engineering from the Universidad Politécnica de Madrid (Technical University of Madrid) and a Master's degree in Maritime Law and Port Administration from the University of La Coruna (Spain). He previously worked at the former Spanish Navy shipyard of Bazan (now Navantia) prior to joining the faculty as a professor at the

EPS-Ferrol, which is part of the University of La Coruna (Spain), where he continues to teach and conduct research. As a faculty specialist in Maritime Transport and Ship Auxiliary Systems he is part of the university's Multidisciplinary Research Group (GEM). He has developed several postgraduate degree programs at the university, including Design, Production and Inspection of Sailing and Recreational Watercraft and Maritime Logistics and Transport. In addition, Dr. Couce has been the Chairman of the Spanish Association of Naval Architects in the region of Galicia for several years.

Dr. José Angel Fraguela Formoso holding a Ph.D. in Naval Architecture and Marine Engineering, Dr. Fraguela is a professor at the University of A Coruña (Spain) and the director, organizer, and speaker in numerous masters', expert courses, specialization courses, and technical conferences in Europe and Latin America and the director of numerous Ph.D. and Final Degree projects. He is a member of scientific committees, and peer reviewer and editor in international and national congresses as well as president of the IPIN-SPAIN (Pan-American Institute of Naval Engineering) and a member of the Consultative Council of IPIN-AMERICA. Dr. Fraguela is also the author of many books and technical articles in JCR journals and other databases.

Contributors

Natasha Águila Valdés Empresa de Proyectos, Construcciones y Servicios Navales CEPRONA, La Habana, Cuba

Jose Carlos Alvarez-Feal Universidade da Coruña, A Coruña, Spain

Asunción López Arranz Universidade da Coruña, Campus de Esteiro, Ferrol, A Coruña, Spain

Johnny Bogle Universidad Marítima Internacional de Panamá, Panama City, Panama

Alan Borges Federal University of Pará, Belém, Brazil

Rui Carlos Botter USP—Universidade de São Paulo, Cidade Universitária, São Paulo, SP, Brazil

Yasmin C. Brey Fornaguera Empresa de Proyectos, Construcciones y Servicios Navales CEPRONA, La Habana, Cuba

Vitor Hugo Macedo Cardoso Federal University of Pará, Belém, PA, Brazil

Raul Villa Caro Universidade da Coruña, Campus de Esteiro, A Coruña, Spain

Luis Manuel Carral Couce Universidade da Coruña, A Coruña, Spain

Laura Castro-Santos Universidade da Coruña, A Coruña, Spain

C. Cifuentes Instituto de Ciencias Navales y Marítimas, Universidad Austral de Chile, Valdivia, Chile

Olga Cifuentes Universidad Tecnológica Nacional FRBB, Bahía Blanca, Argentina

F. J. Correa Cantabria University, Santander, Spain

Juan Carlos Carral Couce Carral Design Engineering Solutions, A Coruña, Spain

Fredy Armando Cuervo Lara Universidad Cooperativa de Colombia, Santa Marta, Colombia; Universidad Internacional Iberoamericana de Puerto Rico, Puerto Rico, USA; Universidad Sergio Arboleda, Santa Marta, Colombia; Universidad del Magdalena, Santa Marta, Colombia; Universidad Pedagógica y Tecnológica de Colombia, Tunja, Colombia; Universidad Jorge Tadeo Lozano, Bogotá, Colombia

Fernando Costa da Cruz Federal University of Pará, Belém, PA, Brazil

Breno Farias da Silva Federal University of Pará, Belém, PA, Brazil

Rubens C. da Silva Numerical Offshore Tank (TPN), University of São Paulo, São Paulo, Brazil; Hydrodynamics Laboratory (LABHIDRO), Brazilian Navy Nuclear Development Directorate (DDNM), São Paulo, Brazil

Caio M. da Silva Anastácio Department of Production Engineering, School of Industrial, Metallurgical Engineering at Volta Redonda—EEIMVR, Federal Fluminense University, Vila Sta. Cecília, Volta Redonda, RJ, Brazil

Severino Fonseca Da Silva Neto Programa de Engenharia Oceânica, Centro de Tecnologia, Universidade Federal do Rio de Janeiro, Cidade Universitária, Ilha do Fundão, Rio de Janeiro, Brazil

Vanessa de Almeida Guimarães Federal Center for Technological Education Celso Suckow da Fonseca, Rio de Janeiro, Brazil

Davi Goulart de Andrade Federal University of ABC, Santo André, São Paulo, Brazil

Luis De Gracia Osaka University, Osaka, Japan

Delmo Alves de Moura Federal University of ABC, Santo André, São Paulo, Brazil

Joyce A. Oliveira de Sousa Department of Production Engineering, School of Industrial, Metallurgical Engineering at Volta Redonda—EEIMVR, Federal Fluminense University, Vila Sta. Cecília, Volta Redonda, RJ, Brazil

Natalia de Souza Ribeiro Department of Production Engineering, School of Industrial and Metallurgical Engineering at Volta Redonda - EEIMVR, Federal Fluminense University, Volta Redonda, Rio de Janeiro, Brazil; Center for

Sustainable Systems Studies - CSSE(Eng.) - CESS(Port), Vila Sta. Cecília, Volta Redonda, RJ, Brazil

José de Troya Calatayud Universidade da Coruña, Campus de Esteiro, Ferrol, A Coruña, Spain

José Manuel Domínguez-Esquivel Instituto Mexicano del Petróleo, Ciudad de México, CDMX, Mexico

Franklin Dominguez Ruiz Escuela Superior Politécnica del Litoral, Guayaquil, Ecuador; Universidade da Coruña, A Coruña, Spain

Thadeu L. C. dos Santos University of São Paulo (USP), São Paulo, Brazil

Daniela Escudero Facultad Regional Bahía Blanca, Universidad Tecnológica Nacional, Bahía Blanca, República Argentina

V. Fernández Cantabria University, Santander, Spain

José Angel Fraguela Formoso Universidade da Coruña, Campus de Esteiro, Ferrol, A Coruña, Spain

João Luiz Francisco USP—Universidade de São Paulo, Cidade Universitária, São Paulo, SP, Brazil

J. Freiria Facultad de Ingeniería, Universidad de la República, Montevideo, Uruguay

Toichi Fukasawa National Maritime Research Institute, Tokyo, Japan

Sarah Fiona Gahlen Lebuhn & Puchta, Hamburg, Germany

M. Isabel Lamas Galdo Universidade da Coruña, A Coruña, Spain

S. García Cantabria University, Santander, Spain

Stephen Girvin Centre for Maritime Law, National University of Singapore, Singapore, Singapore

Misael Goicoechea Universidad Politécnica de Madrid, ETSI Navales, Madrid, Spain

Roberto González González School of Economics, State University of Guayaquil, University Citadel, Guayaquil, Ecuador

Ángel Gómez-González Instituto Mexicano del Petróleo, Ciudad de México, CDMX, Mexico

Yuri Victor Remígio Guedes Federal University of Pará, Belém, PA, Brazil

María Jesús Rodríguez Guerreiro Universidade da Coruña, A Coruña, Spain

Ricardo H. R. Gutiérrez Programa de Engenharia Oceânica, Centro de Tecnologia, Universidade Federal do Rio de Janeiro, Cidade Universitária, Ilha do Fundão, Rio de Janeiro, Brazil

Ruiz Hector Department of Naval Architecture and Ocean Engineering, Osaka University, Suita, Osaka, Japan

Murakawa Hidekazu Joining and Welding Research Institute, Osaka University, Osaka, Japan

Isabel Lamas-Galdo Universidade da Coruña, A Coruña, Spain

P. La Paz Facultad de Ingeniería, Universidad de la República, Montevideo, Uruguay

F. Lastra Cantabria University, Santander, Spain

Ilton Curty Leal Junior Federal Fluminense University - Administration Postgraduate Program - PPGA, Volta Redonda, Brazil

Serguei Lonin Research Group of Oceanology, "Almirante Padilla" Naval Cadet School, Cartagena, Colombia

E. López ILUNION, Training Center in Firefighting and Emergencies, Madrid, Spain

Asunción López-Arranz Universidade da Coruña, A Coruña, Spain

Francisco López-Villarreal Instituto Mexicano del Petróleo, Ciudad de México, CDMX, Mexico; Consejo Nacional de Ciencia y Tecnología, Ciudad de México, CDMX, Mexico

Cassia Lima Machado Federal Fluminense University - Administration Postgraduate Program - PPGA, Volta Redonda, Brazil

E. Madariaga Cantabria University, Santander, Spain; Naval Research Center (CINAV), Naval Academy of Marinha Portuguesa, Alfeite, Almada, Portugal

Harlysson Wheiny Silva Maia Federal University of Pará, Belém, PA, Brazil

Adib Maués Federal University of Pará, Belém, Brazil

Silvina Medus Facultad Regional Bahía Blanca, Universidad Tecnológica Nacional, Bahía Blanca, República Argentina

Nadia Mercedes Mendieta Villalba Universidad Politécnica Salesiana, Guayaquil, Ecuador

Marcia E. Ojeda Morales Biotechnology Laboratory, Academic Division of Engineering and Architecture, Juarez Autonomous University of Tabasco, Cunduacán, Tabasco, Mexico

Ulisses A. Monteiro Programa de Engenharia Oceânica, Centro de Tecnologia, Universidade Federal do Rio de Janeiro, Cidade Universitária, Ilha do Fundão, Rio de Janeiro, Brazil

Hito Moraes Federal University of Pará, Belém, Brazil

Rita Moraes Federal University of Pará, Belém, Brazil

Salvador Naya MODES, CITIC and ITMATI Groups, Department of Mathematics, Escuela Politécnica Superior, Universidade da Coruña, A Coruña, Spain

Kazuo Nishimoto University of São Paulo (USP), São Paulo, Brazil; Numerical Offshore Tank (TPN), São Paulo, Brazil

Rodrigo F. Nunes University of São Paulo (USP), São Paulo, Brazil

Euler Sanchez Ocampo Department of Production Engineering, School of Industrial and Metallurgical Engineering at Volta Redonda - EEIMVR, Federal Fluminense University, Volta Redonda, Rio de Janeiro, Brazil; Center for Sustainable Systems Studies - CSSE(Eng.) - CESS(Port), Vila Sta. Cecília, Volta Redonda, RJ, Brazil

Adriana Oliva-Remola Universidad Politécnica de Madrid, ETSI Navales, Madrid, Spain

Naoki Osawa Department of Naval Architecture and Ocean Engineering, Osaka University, Suita, Osaka, Japan

Mayra Agustina Pantoja-Castro Universidad Juárez Autónoma de Tabasco, Villahermosa, TAB, Mexico

Newton Narciso Pereira Department of Production Engineering, School of Industrial and Metallurgical Engineering at Volta Redonda - EEIMVR, Federal Fluminense University, Volta Redonda, Rio de Janeiro, Brazil; Center for Sustainable Systems Studies - CSSE(Eng.) - CESS(Port), Vila Sta. Cecília, Volta Redonda, RJ, Brazil

Luis Pérez-Rojas Universidad Politécnica de Madrid, ETSI Navales, Madrid, Spain

Luiz Vaz Pinto Programa de Engenharia Oceânica, Centro de Tecnologia, Universidade Federal do Rio de Janeiro, Cidade Universitária, Ilha do Fundão, Rio de Janeiro, Brazil

Ricardo Aurélio Quinhões Pinto UFSC—Universidade Federal de Santa Catarina, Santo Antônio Joinville, SC, Brazil

Benjamín Portales-Martínez Instituto Mexicano del Petróleo, Ciudad de México, CDMX, Mexico; Consejo Nacional de Ciencia y Tecnología, Ciudad de México, CDMX, Mexico

David A. Plaza Mendoza Faculty of Maritime Engineering, Biological, Oceanic and Natural Resources, Higher Polytechnic School of the Litoral, University Campus, Guayaquil, Ecuador

Rafael A. Plaza Perdomo Central University of Ecuador, Research and Postgraduate Institute, University Citadel, Quito, Ecuador

Andrés Prieto ITMATI, Santiago de Compostela, A Coruña, Spain; CITIC, A Coruña, Spain; Grupo M2NICA, Departamento de Matemáticas, Facultade de Informática, Universidade da Coruña, A Coruña, Spain

Miguel A. Hernández Rivera Biotechnology Laboratory, Academic Division of Engineering and Architecture, Juarez Autonomous University of Tabasco, Cunduacán, Tabasco, Mexico

Rubén A. Rodriguez COTECMAR - Science and Technology Corporation for Naval, Maritime and Riverine Industry Development, Cartagena de Indias, Colombia

Adán Vega Sáenz Universidad Marítima Internacional de Panamá, Panama City, Panama

Rodolfo Sabonge Universidad Marítima Internacional de Panamá, Panama City, Panama

M. Salas Instituto de Ciencias Navales y Marítimas, Universidad Austral de Chile, Valdivia, Chile

Noela Sánchez Carnero Centro para el Estudio de Sistemas Marinos (CESIMAR), Centro Nacional Patagónico (CENPAT-CONICET), Puerto Madryn, Argentina; Grupo de Oceanografía Física, Universidad de Vigo, Vigo, Spain

Laura Castro Santos Universidade da Coruña, A Coruña, Spain

Rodrigo A. Schiller Numerical Offshore Tank (TPN), University of São Paulo, São Paulo, Brazil

Rashed Sherif Joining and Welding Research Institute, Osaka University, Osaka, Japan

Rodolpho Soares Federal University of Pará, Belém, Brazil

Toshi-Ichi Tachibana Polytechnic School of the University of São Paulo, São Paulo, SP, Brazil

Hitoi Tamaru Tokyo Maritime Science and Technology, Tokyo, Japan

G. Tampier Instituto de Ciencias Navales y Marítimas, Universidad Austral de Chile, Valdivia, Chile

Javier Tarrío Saavedra Grupo MODES, Departamento de Matemáticas, Escola Politécnica Superior, Universidade da Coruña, A Coruña, Spain; ITMATI, Santiago de Compostela, A Coruña, Spain; CITIC, A Coruña, Spain

C. Troncoso Instituto de Ciencias Navales y Marítimas, Universidad Austral de Chile, Valdivia, Chile

A. Trueba Cantabria University, Santander, Spain

Jose Marcio Vasconcellos Naval Architecture and Ocean Engineering Department, COPPE, Federal University of Rio de Janeiro, Rio de Janeiro, Brazil

L. M. Vega Cantabria University, Santander, Spain

Pedro Martínez Villa IPIN Cuba, Lawton, Habana, Cuba

Oscar E. Viteri Escuela Superior Politécnica del Litoral, Guayaquil, Ecuador

Almudena Filgueira Vizoso Universidade da Coruña, A Coruña, Spain

Part I
Naval Engineering

Structural Parametric Model of an Ecological and Efficient Shuttle Tanker for Operations at the Brazilian Pre-salt Region

Rodrigo A. Schiller, Rubens C. da Silva and Kazuo Nishimoto

Abstract The aim of this paper is to present the development of an ecological and efficient shuttle tanker structural parametric model with optimized capacity that was designed to operate at the Brazilian pre-salt region. Such a model estimates the weights and centers of the bare hull, as well as the structural weight, the center of gravity of the ship, and the moments of inertia at different dimensional, geometric, and capacity features of the ship from structural elements of the parallel middle body. These elements were automatically dimensioned at a MATLAB® environment. After an exhaustive search method through several dimensioning cases of these elements, the lowest structural weight was obtained, which satisfies the rules of classification societies. Finally, the results are shown as a response surface, which was built by means of artificial neural networks. These allow assessing the behavior of weights and centers characteristics of the ship, based on the variation of design parameters.

Keywords Pre-salt region · Shuttle tanker · Structural parametric model
Weights and centers

1 Introduction

Structural weight reductions are quite important in design and construction cost savings and have a reasonable effect on decreasing fuel consumption and gas emissions. For large cargo vessels (displacement hulls) a lower structural weight increases the available deadweight for a ship of the same size, thereby improving transport efficiency.

R. A. Schiller (✉) · R. C. da Silva · K. Nishimoto
Numerical Offshore Tank (TPN), University of São Paulo, São Paulo, Brazil
e-mail: digao.schiller@gmail.com

R. C. da Silva
Hydrodynamics Laboratory (LABHIDRO), Brazilian Navy Nuclear Development Directorate (DDNM), São Paulo, Brazil

A. Vega Sáenz et al. (eds.), *Proceedings of the 25th Pan-American Conference of Naval Engineering—COPINAVAL*, https://doi.org/10.1007/978-3-319-89812-4_1

To gain an understanding of the impact of decreased light ship steel weight on fuel consumption, according to [2], a 1% reduction in hull steel weight for each ship in a set of standard designs leads to decreasing fuel consumption by 0.16% for Suezmax tankers (approximately 0.11 ton/day fuel savings).

The discoveries made in the Brazilian pre-salt are among the world's most important in the past decade. The pre-salt region comprises large accumulations of excellent quality, high commercial value light oil, a reality that puts Brazil in a strategic position to meet the great global demand for energy. Daily oil output at the pre-salt progressed from the average of approximately 41,000 barrels per day in 2010, to 1 million barrels per day in mid-2016 [3]. This overall growth of Brazilian oil production generates a demand for a more modern tanker fleet, capable of efficiently transporting the whole production of offshore oil rigs to terminals and following international environmental regulations. Such vessels may present unique design features that can lead to a hull with unconventional dimensions, specific to operate at certain areas with a desirable efficiency.

The purpose of this paper is to present a numerical methodology to estimate the weights and centers of a shuttle tanker hull based on a midsection whose structural components have the lighter possible weight and are in accordance with the International Association of Classification Societies (IACS) Common Structural Rules (CSR) for Double Hull Oil Tankers.

2 Methodology

2.1 General Considerations

The structural model can estimate the weights and centers of a double hull shuttle tanker with a payload (Δ_{load}) greater than 100,000 ton, for different combinations of length overall (L_{OA}), breadth (B), block coefficient (C_B), depth (D), and draft (T). The structural arrangement of the ship's parallel middle body is sized in accordance with the International Association of Classification Societies (IACS) Common Structural Rules (CSR) for Double Hull Oil Tankers with Length 150 m and Above and by comparisons with existing similar ships.

As shown in Fig. 1a, the obtained midship section has a typical double hull oil tanker layout, with a double side structure and one cargo tank symmetrically arranged on each board. The tank arrangement and D were parameterized with L_{OA} and B from existing ships, particularly the NORDIC Rio Shuttle Tanker (Fig. 1b).

The structural model can be used, as demonstrated in the results section, as an analysis tool of variation of weights and centers of the bare hull in preliminary stages design, in which different combinations of main dimensions of the ship are evaluated.

In the specific case of the application example presented, a vessel with a fixed Δ_{load} = 170,000 ton that will operate at the Brazilian pre-salt region was

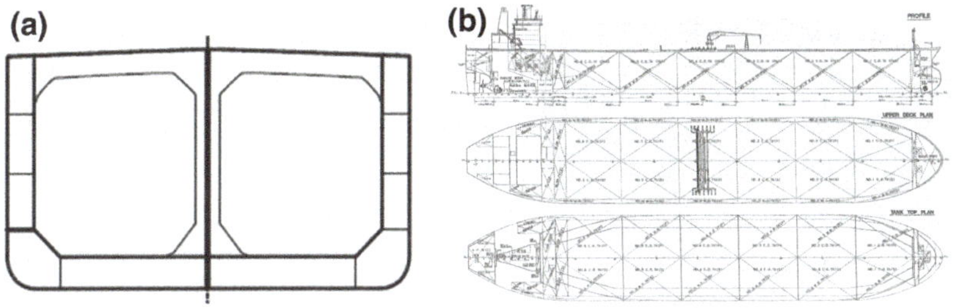

Fig. 1 **a** Typical midship section for double hull tankers [1], **b** NORDIC Rio general arrangement [4]

considered. The results were evaluated by means of a response surface that allowed observing the weights and centers variations in function of a combination of L_{OA} and B.

2.2 *Sequence of Structural Model Operations*

The sequence of the structural synthesis model operations (steps A to I) is shown in Fig. 2. Each step is explained below.

- **Step A**: The model establishes for each (L_{OA}, B) combination the possible values of spacing among primary support members (S), stiffeners (r), and frames (l) (Fig. 3a), based on IACS requirements, and the variation of plating and stiffener thickness. As a simplification, C_B and Δ_{load} were considered constant, and D was estimated by cargo tank height of similar ships. In addition, r, S, and l remained constant around the whole midship section in each case. However,

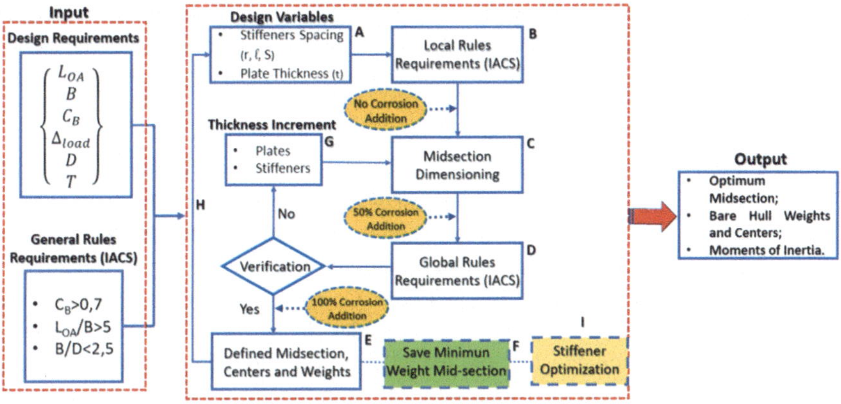

Fig. 2 Structural model flowchart

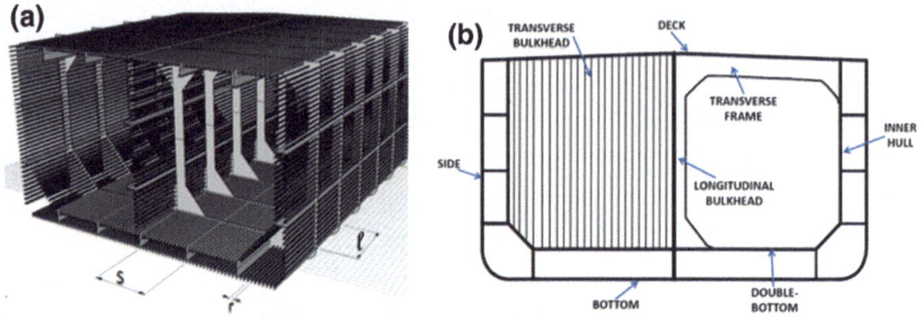

Fig. 3 **a** Spacings considered in the model; **b** different element thickness regions

based on IACS requirements, the plating and stiffeners thickness vary according to their location (Fig. 3b).

- **Step B**: For a given combination of (r, S, l) the minimum thickness and dimensions of plating and stiffeners are calculated according to the IACS local loads and dimensioning criteria, considering the position of each structural element and the static and dynamic pressure combinations on each region for different conditions of draft, loads, and damage. Moreover, thickness values were obtained from manufacturers' tables.
- **Steps C, D**: After meeting the local requirements, the structural elements are placed and the midship section is defined. Hereafter, the section modulus and hull girder inertia are calculated and compared with the minimum global requirements.
- **Steps E, F**: If the midship section meets the global requirements, the weights and centers are calculated. At each cycle defined by a given combination of (r, S, l), the data (longitudinal and transversal elements) of the midsection with the minimum structural weight (considering the linear weight of the parallel middle body) are stored.
- **Step G**: If the global requirements are not met, the structural elements undergo a new thickness increment and a new midship section is defined for further evaluation. The thickness increment is performed to maximize the increase of the section modulus and the hull girder inertia with the smallest possible increasing weight.
- **Step H**: The process is repeated for all (r, S, l) combinations.
- **Step I**: Each midship section stored in **Step F** initially has flat bars as stiffeners. In this step, the stiffener profile is optimized to an equivalent T-bar profile with lower weight and the structural requirements are compared. Finally, the weights and centers of the new midship section are calculated.
- **Corrosion Additions**: The IACS Local Corrosion Additions (LCA) were implemented in the model. The LCA consider the environment that the plates and stiffeners are immersed in and the adjacent ones to increment the thickness due to corrosion. For dimensioning the structural elements, IACS defines three different conditions for the application of the corrosion additions:

- **No Corrosion Addition** (For dimensioning the local elements, corrosion additions are not considered and the local requirements are met for the most critical situation of a lower thickness).
- **50% Corrosion Addition** (The calculation of the section modulus and hull girder inertia is made considering half the corrosion addition and then, the global requirements are assessed).
- **100% Corrosion Addition** (For the final arrangement of the structural elements and for the calculation of the structural weight, the full corrosion addition is considered).

2.3 *Geometric Optimization of Stiffeners*

For each minimum weight midship section stored in Step F, a series of T-bar stiffeners with different flange and web length ratios (L_{flange}/L_{web}) were dimensioned (flange and web have the same thickness). T-Profile was chosen because of its symmetry and it will not be prone to skew bending and is favorable for fatigue strength. The main constraints in this case are the minimum section modulus, inertia, and thickness defined by IACS in a certain position and previously calculated for the flat-bar stiffeners. T-bar stiffeners with L_{flange}/L_{web} around 35–40% were chosen because they presented the higher sectional area reduction (A_T/A_I) if compared with the flat-bar at the same position. Particularly, Fig. 4 shows a result of a stiffener located at the double-bottom region that presented a sectional area reduction of almost 24%.

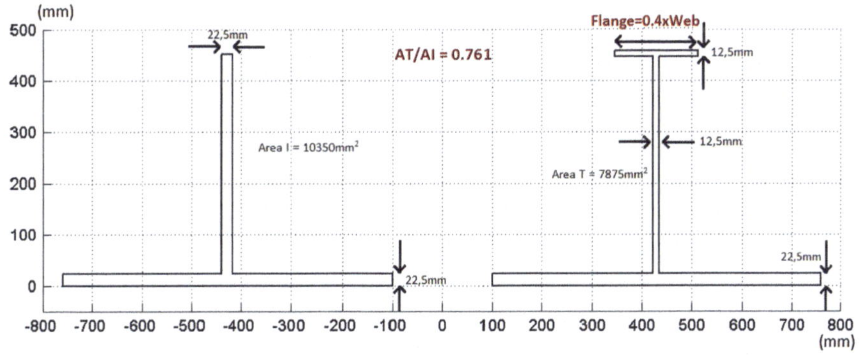

Fig. 4 Sectional area reduction (A_T/A_I) of a stiffener located at double-bottom

2.4 Bare Hull and Lightship Weight and Centers Estimation

The estimation was based on the method of Watson [5], which uses a bare hull steel-weight distribution curve along the length between perpendiculars obtained by analyzing existing vessels with parallel middle body and displacement hull (Fig. 5). This curve is determined by three parameters (r, p, e), where r is the distance between the after perpendicular and the parallel middle body, p is the length of the parallel middle body, and e is the distance between the parallel middle body and the forward perpendicular.

At this preliminary stage in which the hull geometry is not yet defined, the three parameters were determined by the arrangement of the similar vessel NORDIC Rio (Fig. 1b). The distribution of the frames' weight is simple, because its geometry and spacing (l) are defined in the structural model. The number and spacing of transverse bulkheads were also defined by the tank arrangement of the similar vessel. The linear weight of the midship section was defined as the sum of the linear weights of the longitudinal and transverse structural elements. Therefore, by the presented methodology the bare hull weight was estimated by adding a margin of 5% referring to welding and painting material.

The longitudinal position of the center of gravity (LCG) was based on the longitudinal position of the center mass of the steel-weight distribution curve (Fig. 5). The vertical position of the center of gravity (VCG) was approached by the VCG of the parallel middle body, neglecting the stern and bow regions.

For bare hull moments of inertia estimation, the moments of inertia of each structural element "i" of the midship section was calculated according to equations:

$$I_{xx}(i) = \iiint (y^2 + z^2)\rho_{eq}\mathrm{d}v(i) \quad (1)$$

$$I_{yy}(i) = \iiint (x^2 + z^2)\rho_{eq}\mathrm{d}v(i) \quad (2)$$

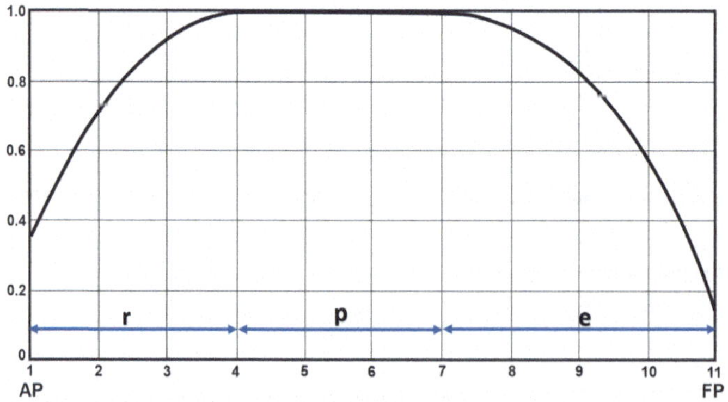

Fig. 5 Steel-weight distribution for ships with parallel middle body [5]

$$I_{zz}(i) = \iiint \left(x^2 + y^2\right)\rho_{eq} \mathrm{d}v(i) \tag{3}$$

where $\rho_{eq} = f(x)\rho_{steel}$ is the equivalent density and $f(x)$ is associated with the steel-weight distribution curve (Fig. 5) to estimate the mass properties of the stern and bow regions. This approach can present significant errors by not considering the geometry of the stern and bow of the ship, especially for the I_{xx}. However, it is satisfactory for the preliminary stages of the project. Furthermore, the parallel axis theorem was applied to calculate the moment of inertia of the whole midship section considering the position of the bare hull center of gravity.

The lightship (W_{LS}) was considered to consist of four components: bare hull (W_{Hull}), superstructure (W_{SS}), machinery (W_M), and outfit (W_O). The weight of each component, except W_{Hull} (calculated by the structural model) was estimated from formulations that consider parameters such as L_{OA}, B, D, and C_B. Such formulations, which are not detailed, were based on the studies of [6–8].

3 Results

In this section, the results of the structural model for a specific case of preliminary stage design of a shuttle tanker made to operate at the Brazilian pre-salt region are presented. The following design requirements were considered.

- Maximum draft ($T_{máx}$) = 18 m.
- Lightship limited to $0.22\Delta_{load.}$
- Δ_{load} = 170,000 ton and $C_B = 0.82$.
- $L_{OA}/B > 5$ and $B/D < 2.5$ (IACS).

As previously explained, the input variables are the main dimensions of the ship, of which just L_{OA} and B vary, and for each combination (L_{OA}, B) also the spacings (r, S, l) of the midship section components vary.

The range of values (r, S, l) were based on the analysis of similar Suezmax tankers, with a small margin on the lower and upper values. Table 1 shows the values of the variables, and it is possible to evaluate the number of simulated cases.

Table 1 Range of values assumed by the variables

Variable	Range	Number of cases
L_{OA}	263:2:357	48
B	44:1:56	13
S	$\frac{B}{[4:2:12]}$	5
r	$\frac{S}{[5:1:20]}$	16
l	$\frac{L_{OA}}{[50:10:150]}$	11
Total number of cases		549,120

3.1 Midship Section and Parallel Middle Body

From the presented variables, the structural model establishes the elements of the midship section, such as the side plating, double hull, main deck, longitudinal and transverse bulkheads, frames, and longitudinal stiffeners.

The mechanism chosen to define the lower weight midship section for each combination of structural variables was the exhaustive search. This is a simple method that will always define a result. However, the computational time in this case is much higher than using optimization algorithms.

As a result, Fig. 6 shows examples of a lower structural weight midship section and frame (Fig. 6a) for a specific case of $L_{OA} = 297$ m and $B = 53$ m and the three-dimensional model of the parallel middle body (Fig. 6b).

3.2 Response Surface

The response surface for each estimated parameter was constructed with artificial neural networks (ANN) because they can present accurate results for simple meshes (with a few simulated cases), depending on the parameter variation. The ANN topology that best fits the characteristics of the response surface with fast convergence and high precision was the one with a 10-neuron layer and back propagation training algorithm (Fig. 7a). According to [9], back propagation is the most widely used ANN algorithm for general engineering applications.

The ANN inputs were L_{OA} and B values and the desired outputs were the estimated parameters from the structural model (one per ANN). The training subset, composed of 60–90% of random samples from the complete set, were used essentially in the ANN learning (training) process. The ANN results and the values obtained by simulations of the structural model were compared. The obtained mean error was almost 2%, and the number of simulated cases was considered sufficient to obtain a good response surface. Figure 7a shows the comparison of the results for the lightship (W_{LS}).

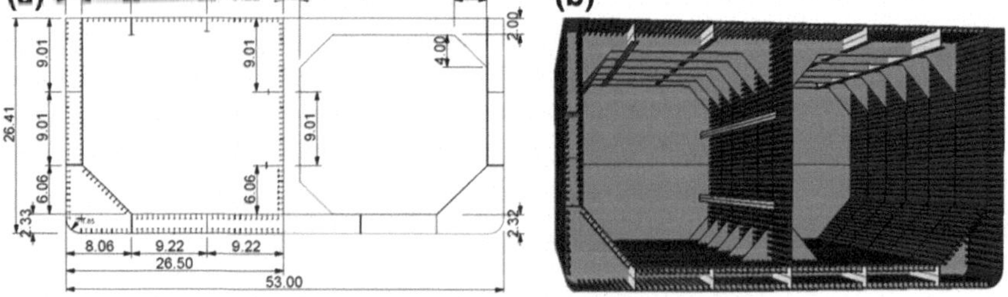

Fig. 6 Lower structural weight midship section (**a**) and parallel middle body (**b**)

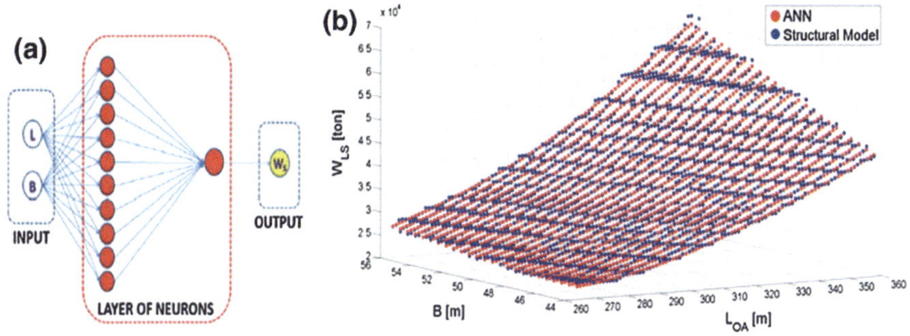

Fig. 7 ANN topology (**a**) and comparison between the W_{LS} results and ANN (**b**)

3.3 General Results

This section presents the results of weights and centers of the lightship according to the variation of L_{OA} and B. Figures 8, 9 and 10 show some of the main results obtained from the structural model to a constant $\Delta_{load} = 170{,}000$ ton.

Applying the design constraints relative to lightweight (limited to $0.22\Delta_{load}$) and maximum draft ($T_{máx} = 18$ m) to the results, a region with the feasible cases was defined (Fig. 11).

To validate the model results, the range of W_{LS} values was compared with the IHS–Fairplay data for Suezmax tankers with 160,000 ton $\leq \Delta_{load} \leq$ 170,000 ton and the same range values of L_{OA} and B from [10], through the lightweight coefficient ($C_{WLS} = W_{LS}/L_{PP}xBxD$). Table 2 shows the comparison between the results from the structural model and the IHS–Fairplay data, and the adherence among the values can be noticed.

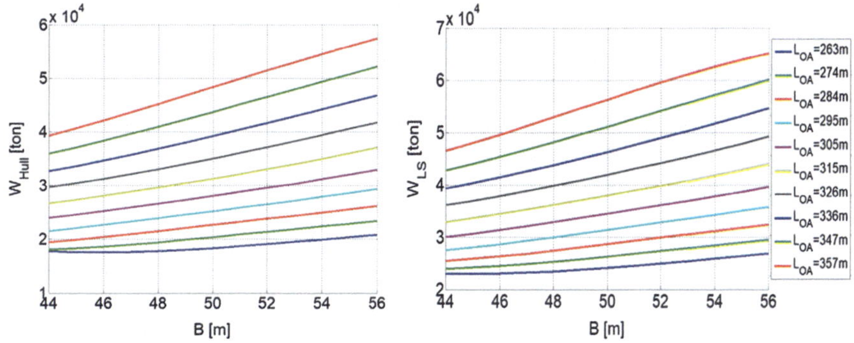

Fig. 8 W_{Hull} and W_{LS} results

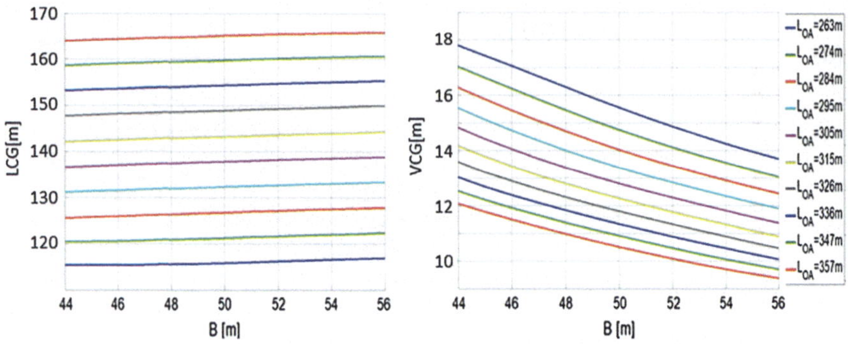

Fig. 9 LCG and VCG results

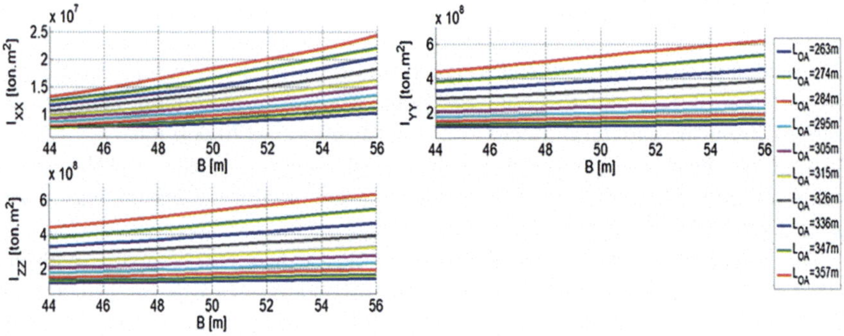

Fig. 10 Inertia results

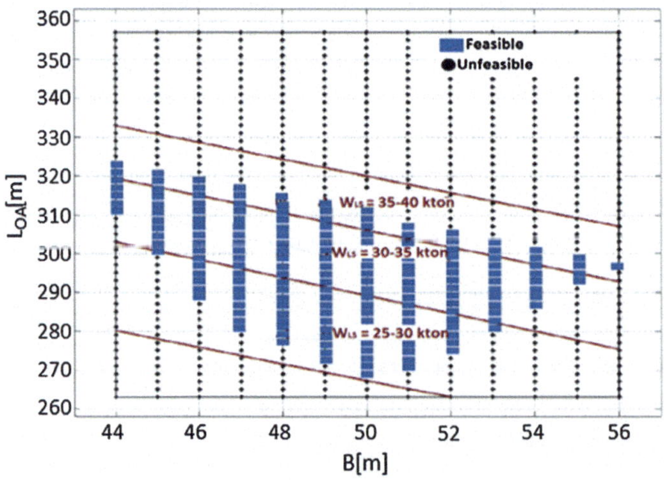

Fig. 11 Feasible region defined by the design constraints

Table 2 Values of C_{WLS} from structural model and IHS–fairplay data

	C_{WLS} [ton/m^3]
IHS–fairplay	0.078–0.09
Structural model	0.079–0.091

4 Conclusion

The structural model generated results that can be quite important at the early stages of the design of a vessel with specific dimensions and requirements as an assistance to decision making. The behavior of the output parameters according to the variation of some variables (in this case, L_{OA} and B) has a great value as a sensitivity analysis of the design requirements. However, an optimization algorithm must be added to the model to make it more efficient, mainly regarding computational time. Moreover, the bow and stern geometry must be considered in more detail to increase the accuracy of the inertia and weights results. These aspects will be improved in the next works.

Acknowledgements The authors gratefully acknowledge PETROBRAS for the financial support provided for this study under a broader R&D project NAVES—Efficient and Ecological Ship.

References

1. DNVGL Homepage: https://rules.dnvgl.com/docs/pdf/DNV/rulesship/2012-07/ts801.pdf, last accessed 12 Mar 2017
2. ABS Homepage: https://ww2.eagle.org/content/dam/eagle/advisories-and-debriefs/ABS_Energy_Efficiency_Advisory.pdf, last accessed 12 Mar 2017
3. PETROBRAS Homepage: http://www.petrobras.com.br/en/our-activities/performance-areas/oil-and-gas-exploration-and-production/pre-salt, last accessed 12 Mar 2017
4. TRANSPETRO Homepage: http://www.transpetro.com.br/pt_br/areas-de-negocios/transporte-maritimo/frota-transpetro.html, last accessed 12 Mar 2017
5. Watson, D.G.: Practical Ship Design, vol. 1. Elsevier, Amsterdam (1998)
6. Schneekluth, H., Bertram, E.V.: Ship Design for Efficiency and Economy. Butterworth Heinemann, United States (1998)
7. Parson, M.G.: Parametric Design, Ch. 11. Ship Design and Construction, SNAME (2003)
8. Ventura, M.: Estimation Methods for Basic Ship Design (2010)
9. Ray, T., Gokarn, R.P., Sha, O.P.: Neural network applications in naval architecture and marine engineering. Artif. Intell. Eng., 213–226 (1995) (Elsevier, India)
10. Kristensen, H.O.: Determination of Regression Formulas for Main Dimensions of Tankers and Bulk Carriers based on IHS Fairplay data. Technical University of Denmark (2012)

Effect of Encountered Wave Condition on Fatigue Life Prediction of Ship Structures

Luis De Gracia, Hitoi Tamaru, Naoki Osawa and Toichi Fukasawa

Abstract Ship weather routing develops an optimum track for ocean voyages based on a forecast of weather, sea conditions, and a ship's individual characteristics for a particular transit. In these days, most ships follow weather routing, and those ships never experience extremely severe seas. In classification society rules, ship structure fatigue assessment is performed without consideration of weather routing. In these assessments, the occurrence probability of severe seas is overestimated and their recurrence interval is underestimated. This might lead to deterioration in fatigue assessment precision. In this study, S–N-based fatigue assessments of a welded joint in a container ship that follows weather routing are performed. This ship sails on a North Atlantic Ocean route. Fatigue lives are evaluated assuming different encountered wave conditions: for a planned route, "Great Circle Route," and a weather routing, "Minimum Time Route." Short sea sequences are generated by a storm model using hindcast data. The storm profiles are determined by using the cumulative frequency of shot seas which is experienced on the MTR routes. Based on these results, the effect of encountered wave conditions on cumulative fatigue damage is discussed.

Keywords Fatigue · Cumulative fatigue damage · Routing · Storm model
Wave load

L. De Gracia (✉) · N. Osawa
Osaka University, Osaka, Japan
e-mail: Luis_DE_Gracia@naoe.eng.osaka-u.ac.jp

H. Tamaru
Tokyo Maritime Science and Technology, Tokyo, Japan

T. Fukasawa
National Maritime Research Institute, Tokyo, Japan

A. Vega Sáenz et al. (eds.), *Proceedings of the 25th Pan-American Conference of Naval Engineering—COPINAVAL*, https://doi.org/10.1007/978-3-319-89812-4_2

1 Introduction

A fatigue assessment is a mandatory assessment in major Classification Society Rules. The fatigue assessment is performed based on the S–N approach (Palmgren–Miners rule). The fatigue damage is caused primarily by the variation of the wave loading acting on ships, resulting in the variation of the stress amplitude [1]. The effectiveness of the S–N-based assessment falls in the reliable description of the encounter wave conditions experienced by ocean-going ships. Tomita et al. [2–4] proposed a "storm model" that can simulate the wave load sequence experienced by ocean-going ships. Kawabe [5] and Prasetyo [6] modified Tomita's model to improve the emulation capability of a real sea state sequence. Recently, De Gracia et al. [7] proposed a modified model that considered the stochastic nature of the wave direction.

In this paper, fatigue assessments of a 6000 TEU (Twenty-foot Equivalent Unit) container ship's welded joint were performed. The target ship is assumed to face two different encountered wave conditions in a North Atlantic Ocean route, following a weather routing, called minimum time route (MTR), and a planned route, great circle route (GCR). Short sea sequences are generated by using Japan Weather Association (JWA) hindcast data, and those for MTR are simulated by adopting Tamaru's weather routing algorithm [8]. SN-based fatigue assessments are performed for MTR and GCR sequences, and the effect of the encountered wave condition on fatigue damage is examined.

2 Encountered Wave Conditions

2.1 Weather Routing Algorithm

The benefits of ship weather routing are primarily in time and cost reductions and increased crew and structural safety. The reduction in transit time, fuel consumption, extreme weather encounters, and hull damage is directly related to saving in operating cost reductions. A weather routing algorithm that can judge the minimum time route from a spatiotemporal distribution of sea states (significant wave height and wave direction) was proposed by Tamaru [8]. The relationship between significant wave height, ship speed loss, and the relative heading angle is considered in the analysis. The ship route is optimized by analysis of the isochrones and the spatiotemporal sea state data was generated from JWA's hindcast data.

2.2 Sea State Data

A North Atlantic route between Boston and Bishop is chosen as the shipping route. The weather routing algorithm explained in Sect. 2.1 is adopted in this study, and the GCRs and MTRs are determined by Tamaru. Figure 1 shows an example of the MTR for the assumed route.

The target ship is a 6000 TEU container ship. It is considered that she sailed on the North Atlantic Ocean for a period of 10 years. The ship experiences the sea state (significant wave height H_S, mean period T_S, and wave direction θ) sequence determined by those at the nearest JWA hindcast data grid point.

2.3 Real Headings Model and Wave Statistics

Consider θ, α, and χ as the wave direction, ship's heading angle, and relative heading angle. The stress in the conventional fatigue design procedure is evaluated assuming that χ is given by a uniform random number. However, throughout the ship's service life, she meets each new wave at a particular relative angle. In this paper, let "real headings model" be the model in which the stress response is calculated by considering the χ's occurrence probability f_χ. θ is determined by random number selection considering θ's occurrence probability f_θ, and χ is calculated by Eq. 1 each time.

$$\chi = \alpha - \theta \tag{1}$$

The averaged f_θ for the North Atlantic Ocean is determined from JWA hindcast data. Figure 2 shows the determined f_θ. It is shown that θ is predominant between 240° and 330°.

Fig. 1 An example of minimum time routes (MTR) for the United States/United Kingdom

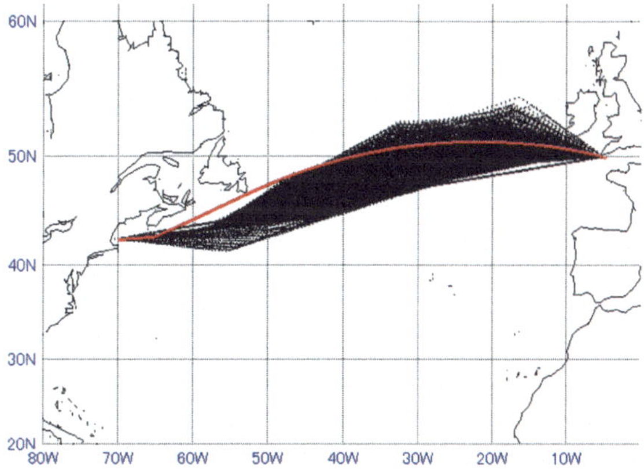

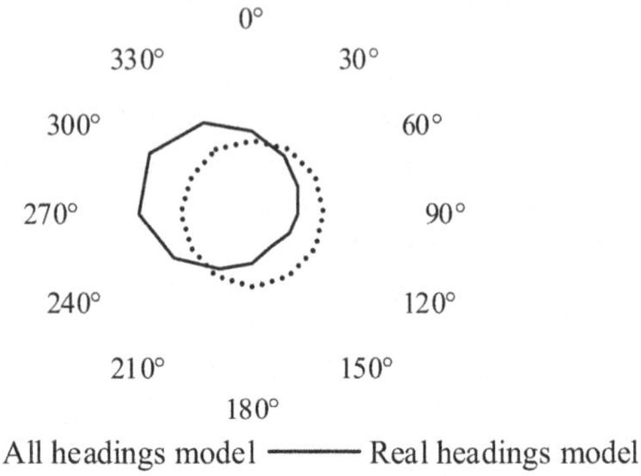

Fig. 2 Wave direction's occurrence probability distribution f_θ, determined from JWA hindcast data in the North Atlantic Ocean

The time history of individual wave height is generated assuming that the individual wave height obeys the Rayleigh distribution, whose energy spectrum can be defined by the ISSC (International Ship and Offshore Structures Congress) spectrum as Eq. (2).

$$\frac{S(\omega)}{H_S^2} = 0.11\left(\frac{\omega T_S}{2\pi}\right)^{-5} \exp\left\{-0.44\left(\frac{\omega T_S}{2\pi}\right)^{-4}\right\} \tag{2}$$

where ω is the wave frequency (rad/s), $S(\omega)$ the specified wave spectrum (m^2/s), and T_S the peak period (s). Let "as-simulated sea sequence" be the sea state sequence directly determined from the GCR or MTR ship position sequence and JWA hindcast data's spatiotemporal wave data, and the "storm sea sequence" be that generated from a storm model simulation. These spatiotemporal wave data are fitted by the log-normal distribution proposed by Wan and Shinkai [9] due to rounding errors that might be found in the histograms. Figure 3 shows the comparison of H_S exceedance probability P_{EX,H_S}, of an as-simulated sea sequence for GCR and MTR routes. It is observed that the difference becomes larger for waves larger than 5 m, whereas the difference increases with H_S. Additionally, differences

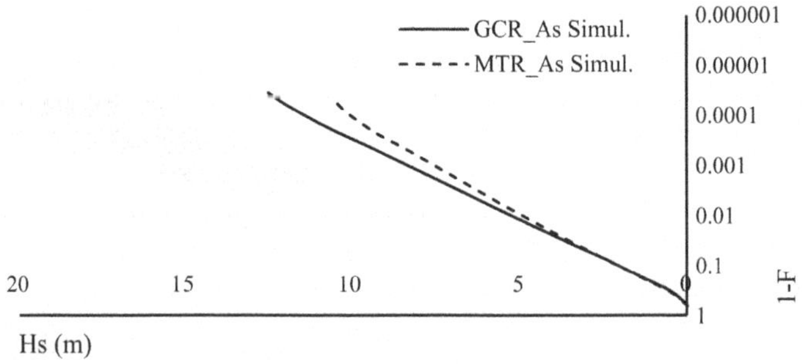

Fig. 3 The comparison of significant wave height's exceedance probability $P_{EX,Hs}$, for as-simulated sea sequence for MTR and GCR routes

in the maximum significant wave height $H_{S,max}$ with MTR having less severe encounter wave conditions than those found in the GCR route are observed.

3 Stress Response

3.1 Stress Statistics

In the as-simulated sea sequences, sequences of (H_S,T_S,θ) are given. The ship course α, is determined by drawing a line segment joining previous and current ship positions. The relative heading angle χ can be determined from θ and α. In this way, the sequence of (H_S, T_S, θ) is determined. ΔS denotes the hotspot stress range. Once (H_S, T_S, θ) is given, the ΔS sequence can be generated by following linear spectrum analysis. In these analyses, ISSC's wave spectrum is adopted as the R parameter. Let this ΔS sequence be the "as-simulated stress sequence." Let $P_{EX,\Delta S}$ be ΔS's exceedance probability. Let $P_{EX,\Delta S|GCR}$ and $P_{EX,\Delta S|MTR}$ be $P_{EX,\Delta S}$ of as-simulated stress sequences for GCR and MTR routes. A comparison of $P_{EX,\Delta S|GCR}$ and $P_{EX,\Delta S|MTR}$ is presented in Fig. 4. In the as-simulated sequence, the difference in the encountered wave condition shows that the difference in the stress exceedance probability becomes evident for $\Delta S > 100$ MPa, and the difference becomes slightly larger with the increase in ΔS. It is considered that these differences are due to χ's randomness and the variation in stress response amplitude operator (RAO) associated with χ. Additionally, the differences in the stress exceedance probability is associated with the encountered wave condition differences between the two routes.

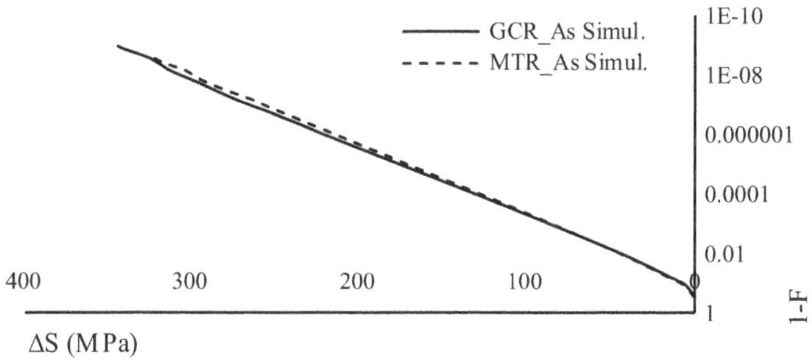

Fig. 4 Comparison of the stress range's exceedance probability $P_{EX,\Delta S}$ for as-simulated stress sequence for MTR and GCR routes

4 Storm Model

4.1 Wave Scatter Diagrams

The joint frequency distributions of (H_S,T_S), known as a wave scatter diagram, are generated by counting sea states recorded in as-simulated sea sequences for both encountered wave conditions: GCR and MTR routes. It is considered that these sea histograms include round errors. These errors are corrected by using the correcting method proposed by Wan and Shinkai [9]. In this method, histograms are fitted with the conditional log-normal distribution $p(T_S|H_S)$ given by Eq. (3).

$$p(T_S|H_S) = \exp\left\{-\frac{[\ln(T_S-\mu)]^2}{2\sigma^2}\right\}\alpha(T_S,H_S),$$
$$\alpha(T_S,H_S) = \frac{\sqrt{2\pi}}{2\pi T_S\sigma}, \mu = E\{\ln(T_S(H_S))\},$$
$$\sigma^2 = \mathrm{var}\{\ln(T_S(H_S))\} \tag{3}$$

H_S's marginal probability distribution $p(H_S)$ is determined in Sect. 2.3. The joint probability distribution $p(T_S, H_S)$ is calculated by Eq. (4).

$$p(H_S, T_S) = p(H_S)p(T_S|H_S) \tag{4}$$

Furthermore, it is recognized that the long-term probability distribution of H_S can be approximated by the Weibull distribution [10]. The characteristic of the Weibull distribution is described as that of $F(H_S)$ in Eq. 5:

$$F(H_S) = 1 - \exp\left[-\left(\frac{H_S}{\lambda}\right)^k\right] \tag{5}$$

and, its p.d.f. is given as

$$\left(\frac{kH_S^{k-1}}{\lambda^k}\right)\exp\left[-\left(\frac{H_S}{\lambda}\right)^k\right] \tag{6}$$

where k and λ are the Weibull's shape and its scale parameters.

In Figs. 5 and 6 are presented the Weibull plot of $F(H_S)$ considering all seasons on the North Atlantic wave scatter diagram for the GCR and MTR cases. The relation between $\ln(H_S)$ and $\ln(\ln(1/1 - F(H_S)))$ can be represented by a straight line. The shape and scale of the Weibull parameters can be identified by using the least square method in conjunction with the correlation of natural logarithms on the

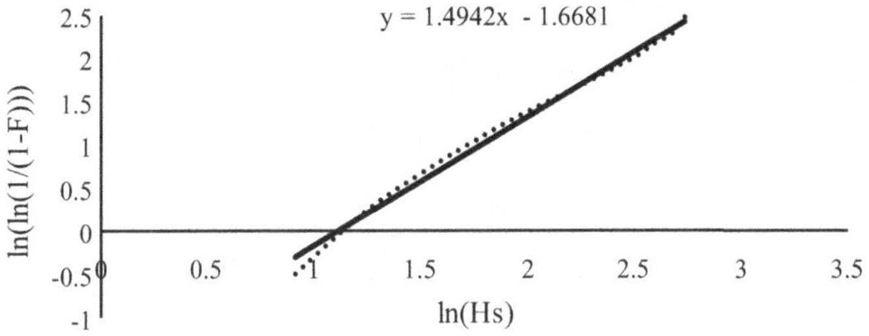

Fig. 5 Weibull plot considering all seasons in the North Atlantic Ocean for GCR cases

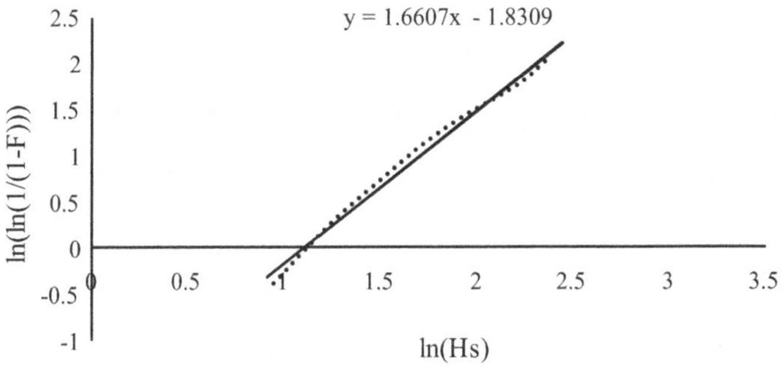

Fig. 6 Weibull plot considering all seasons in the North Atlantic Ocean for MTR cases

left- and right-hand sides of Eq. (5). In this case, for the North Atlantic wave scatter diagram, $F(H_S)$ are determined by performing Weibull fitting from all H_S ranges.

4.2 Storm Models

Tomita et al. [2] studied the time history of wave occurrence experienced by a ship during voyages on the North Pacific Ocean, and demonstrated that the wave-induced load in a ship hull can be divided into two groups: calm sea condition and storm condition. The wave histories are described in calm conditions as time-independent waveforms, whereas in the storm condition the waves can be modeled as time-dependent crescendo–decrescendo waveforms, and they appear randomly. Figure 7 shows an example of wave history generated by the storm model. The "storm model" consists of a "storm profile" and H_S's probability distribution in calm seas. The storm profile consists of a series of storm waveforms and the occurrence probability of storms. In this paper, storm profiles are determined by adopting the "3G storm model" proposed in [6], which can take into account variation of storm duration.

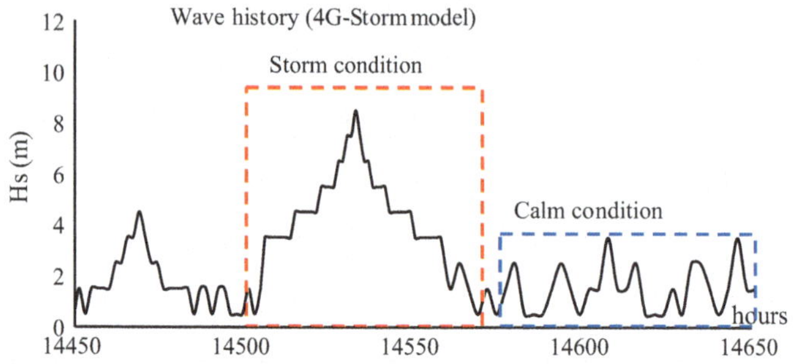

Fig. 7 An example of wave load history generated by the storm model

Once the storm model is established, sea sequences (H_S, T_S, θ) are generated from the storm model. After these sea sequences are generated, stress sequences are generated by adopting real headings or all-headings models. Let $P_{EX,\Delta S,\text{storm}}$ be ΔS's exceedance probability of a storm model's stress sequence. $P_{EX,\Delta S,\text{storm,RH}}$ represent $P_{EX,\Delta S,\text{storm}}$ calculated for a real headings model. A storm sea sequence generated by a storm model with a real heading model emulates the occurrence probability of sea state and relative heading angle.

Figures 8 and 9 show comparisons of $P_{EX,\Delta S,\text{storm,RH}}$ and as-simulated $P_{EX,\Delta S}$ for GCR and MTR routes. It is shown that the differences in $P_{EX,\Delta S}$ are satisfactorily small for both routes. Furthermore, these results demonstrate that the storm models have the emulation capability of generating stress sequences experienced by ocean-going ships. These results are presented for cases where ships follow a weather routing or not, under the conditions chosen.

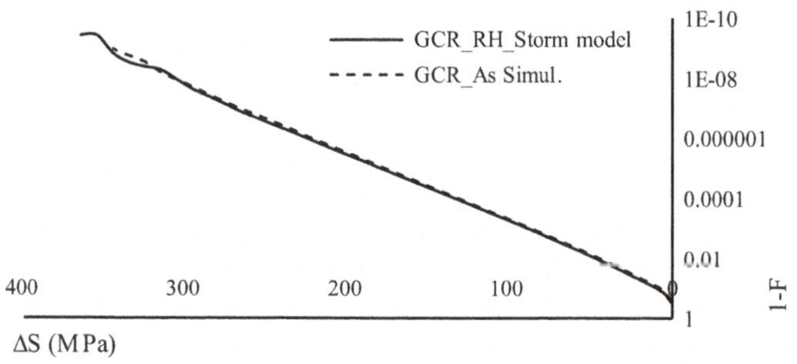

Fig. 8 The comparison of the storm model and as-simulated for GCR route

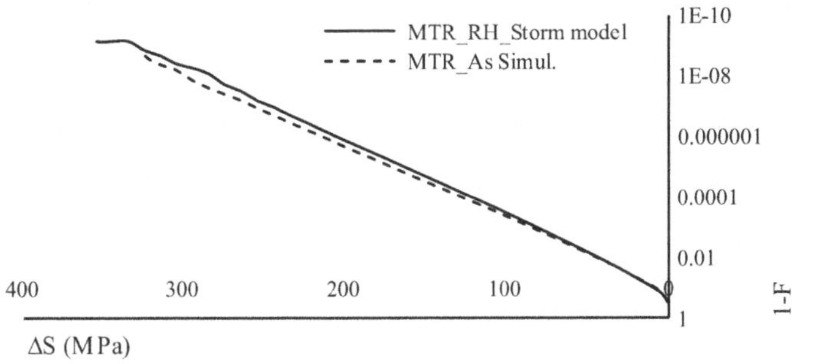

Fig. 9 The comparison of the storm model and as-simulated for MTR route

5 Fatigue Assessment

5.1 Cumulative Fatigue Damage

The cumulative fatigue damage is evaluated by the Palmgren–Miner rule for all classification society rules. It says that the total damage experienced by the structure may be expressed by the accumulated damage from individual load cycles at different stress levels. In this paper, the cumulative damage is calculated over 10 years $D_{10\text{years}}$. The cumulative fatigue damages of the target welded joint $D_{10\text{years}}$ for a given ΔS sequence is calculated by Eq. (7):

$$D_{10\text{years}} = \sum \frac{n_i}{N_i} \tag{7}$$

where n_i is the number of stress cycles in ith stress range block ΔS_i, and N_i the number of cycles to failure for ΔS_i, which is determined using DnV CN.30.7's curve I (for welded joints) [11]. The thickness effect is not considered and the mean stress is assumed to be zero.

5.2 Fatigue Damage Results

In this section, the comparison of the fatigue damage due to differences in the encountered wave conditions is presented. It is assumed that the ship sails in the North Atlantic Ocean. Additionally, the effectiveness of the storm model is examined, comparing the fatigue damage results with those obtained in the as-simulated sequences for both routes, the GCR and MTR. Table 1 shows the cumulative fatigue damage results over 10 years $D_{10\text{years}}$. It is observed that the differences in statistical properties of $D_{10\text{years}}$ are about 15% smaller for ships that encountered wave conditions following the minimum time route, compared to those that sail in the great circle route. These results clearly suggest the effect of the

Table 1 Comparison of the statistic L_f calculated by DnV CN. 30.7

Sequence model	Storm model		As simulated	
Route	GCR	MTR	GCR	MTR
$D_{10\mathrm{YR}}$	0.3694	0.3223	0.3408	0.3402
L_f (year)	27.07	31.03	29.345	29.399

encounter wave condition for vessels that follow a weather routing, is to extend the service life of the structure. These results are expected due to less severe wave conditions encountered in a weather routing compared to those structures experiencing wave loadings in a great circle route (see Figs. 3 and 4).

Moreover, it is recognized that the storm model for weather routing can successfully emulate the (H_S, T_m, χ) sequences experienced by a vessel in the GCR and MTR routes. Under the condition chosen, the storm model results tend to be slightly conservative compared to those obtained in the as-simulated sequence. The differences in statistical properties of $D_{10\mathrm{years}}$ are at most about 8% larger than those obtained in the as-simulated sequence. However, the effects of the high-frequency loading (whipping/springing vibrations) were not considered in this paper, and additional studies are needed to clarify their effect on the statistical properties of the fatigue damage of ship structural members.

6 Conclusions

In this paper, fatigue assessments of the welded joint in the 6000 TEU container ship are performed. Two different encounter wave conditions in the North Atlantic Ocean are considered: great circle (GCR) and minimum time (MTR) routes. Stress sequences are generated by the storm model, with a real heading model that emulates the occurrence probability of sea state and relative heading angle. SN analyses are based on DnV CN 30.7. The following points are obtained from the results.

- To generate the stress sequence of wave random loading, the storm model can be adapted. The storm model leads to slightly conservative estimations compared to those obtained in the as-simulated sequence, under the condition chosen.
- The weather routing affects the fatigue assessment results. The difference in the cumulative fatigue damage between GCR and MTR is at least about 15% under the conditions chosen.
- Additional studies on the development of an advanced wave load sequence model that can consider the effect of whipping/springing vibration are needed.

Acknowledgements The authors would like to acknowledge Dr. Kuniaki Matsuura of Japan Weather Association (JWA) for providing JWA hindcast data.

References

1. Mao, W., Prasetyo, F., Ringberg, J., Osawa, N.: A comparison of two wave models and their influence on fatigue damage in ship structures. In: Proceedings of the 32nd OMAE Conference, vol. 2A, pp. V02AT02A010. ASME, France (2013)
2. Tomita, Y., Matoba, M., Kawabe, H.: Fatigue crack growth behavior under random loading model simulating real encountered wave condition. Mar. Struct **8**, 407–422 (1995)
3. Tomita, Y., Kawabe, H., Fukuoka, T.: Statistical characteristics of long-term wave-induced load for fatigue strength analysis for ships. In: Proceeding of 6th PRADS, vol. 2, pp. 2792–2805 (1992)
4. Tomita, Y., Hashimoto, K., Osawa, N., Terai, K., Wang, Y: Study on Fatigue Design Loads for Ships based on Crack Growth Analysis. ASTM STP 1439 (2002)
5. Kawabe, H., Syuuji, O., Masayoshi, O.: The study of storm loading simulation model for fatigue strength assessment of ship structural member: 1st report new storm loading simulation model which consistent with a wave frequency table. J. Soc. Naval Architects Jpn. **193**, 39–47 (2003)
6. Prasetyo, F., Osawa, N., Kobayashi, T.: Study on preciseness of load history generation based on storm model for fatigue assessment of ship structures members. In: Proceeding of 22nd ISOPE Conference, pp. 709–712 (2012)
7. De Gracia, L., Osawa, N., Mao, W., Ichihashi, D.: Influence of different wave load sequence models on fatigue life prediction of ship structures based on fracture mechanics approach. In: Proceedings of the 35th OMAE Conference, vol. 3, pp. V003T02A030. ASME, Korea (2016)
8. Tamaru, H.: About the optimum route by the weather routing. In: Proceeding of Japanese Society of Naval and Ocean Engineers. JASNAOE (2016)
9. Wan, S., Shinkai, A.: The statistical characteristics of global wave data and appraisal for long-term prediction of ship response. Trans. Soc. Naval Architects Jpn. **90**, 289–296 (1995). (In Japanese)
10. Evans, M., Hastings, N.: Statistical Distributions. Wiley, Inc. Peacock B. New Jersey (2000)
11. Veritas, D.N: Classification Note 30.7—Fatigue Assessment of Ship Structure (2014)

Numerical Model to Analyze a SNCR System to Reduce NO_x

M. Isabel Lamas Galdo, María Jesús Rodríguez Guerreiro, Almudena Filgueira Vizoso, José de Troya Calatayud and Raul Villa Caro

Abstract Taking into account the importance of NO_x (nitrogen oxides) emissions from marine engines and the current increasingly restrictive legislation, this work aims to develop a numerical model to study NO_x reduction. Particularly, a selective non-catalatytic reduction system was designed. A numerical model was developed to analyze several performance parameters. The pressure, velocity, temperature, and NO_x concentration fields were characterized. This numerical model was compared with experimental measurements. The satisfactory results obtained validated the work.

Keywords CFD · SNCR · NO_x · Emissions

1 Introduction

Due to the lean combustion that takes place in diesel engines, these emit low values of carbon monoxide and hydrocarbons. However, their emissions of nitric oxides and particles are considerable [1]. The current environmental situation requires new technologies to control nitrogen oxides, and recently new techniques are being developed. The oxides of nitrogen formed in combustion processes are mainly caused by the reaction of nitrogen present in atmospheric air. For this reason, it is very difficult to avoid their formation. Nitrogen oxides are generated from nitrogen and oxygen at high combustion temperatures. The formation of NO_x increases with the combustion temperature, the residence time of the gas burned at high temperature, and the amount of oxygen present [2, 3].

In the marine field, pollution is controlled by agencies such as the European Protection Agency, European Commission, and the International Maritime Organization, among others. In this regard, the United States Environmental Protection Agency (EPA or USEPA), which belongs to the federal government of

M. I. L. Galdo (✉) · M. J. R. Guerreiro · A. F. Vizoso · J. de Troya Calatayud · R. V. Caro
Universidade da Coruña, A Coruña, Spain
e-mail: isabellamas@udc.es

A. Vega Sáenz et al. (eds.), *Proceedings of the 25th Pan-American Conference of Naval Engineering—COPINAVAL*, https://doi.org/10.1007/978-3-319-89812-4_3

the United States, was created to regulate the environment and its influence on human health. The European Commission also regulates the environment and developed limitations aimed at pollution from ships. The International Maritime Organization (IMO) focuses on marine pollution and other fields such as safety, technical issues, legislation, and so on. According to this legislation, it is extremely important to reduce NO_x in marine engines.

Many NO_x reduction methods have been proposed in the literature [4]. Basically, these can be classified into primary and secondary methods. The difference is that primary methods reduce NO_x while combustion takes place and secondary methods reduce NO_x at the flue gases. The main goal of primary methods is to reduce the combustion temperature due to its importance in the NO_x formation process.

Regarding secondary methods, two procedures are widely employed in the marine field: SCR and SNCR. SCR (selective catalytic reduction) eliminates NO_x contained in the exhaust gas employing catalytic substances to accelerate the chemical reactions that take place. On the other hand, SNCR (selective noncatalytic reduction) does not employ catalytic substances. The main limitation of SNCR is that the temperature must be high if catalytic substances are not employed. According to this, the present work analyzes measures to get a reasonable NO_x reduction. Particularly, ammonia was chosen to reduce NO_x.

2 Kinetic Model

NO is the main species of NO_x [5]. For this reason, the present work focuses on reducing NO. The first research about NO reduction using ammonia was realized in the 1970s and after that several models were proposed in the literature. The most relevant kinetic models are indicated in Table 1.

The present work compares the models of Miller and Bowman [6], Glarborg et al. [7], and Miler and Glarborg [8].

Despite the discrepancies between these models, the main differences can be explained in terms of the branching ratio of the sequence, α. This parameter is defined by the expression:

Table 1 Kinetic models for NO reduction using ammonia

Authors	Number of reactions	Number of species
Miller and Bowman [6]	73	19
Glargorg et al. [7]	104	22
Miller and Glarborg [8]	134	24
Brouwer et al. [9]	2	2
Duo et al. [10]	2	2

$$\alpha = \frac{k_1}{k_1 + k_2} \tag{1}$$

where k_1 and k_2 are the velocity constants of the two main reactions between NO and NO_2, reactions (2) and (3):

$$NH_2 + NO \rightleftarrows NNH + OH \tag{2}$$

$$NH_2 + NO \rightleftarrows N_2 + H_2O \tag{3}$$

The value of the parameter α determines the starting temperature of the reactions. Higher values promote that the first reaction prevails and the reduction process ignites at lower temperatures.

The NO reduction process is mainly initiated by the reaction of NH_3 with OH, through the reaction:

$$NH_3 + OH \rightleftarrows NH_2 + H_2O \tag{4}$$

NH_2 can also be formed by the reaction of NH_3 with O through the reaction:

$$NH_3 + O \rightleftarrows NH_2 + OH \tag{5}$$

NH_2 is highly selective toward NO and causes the overall reduction at an optimum temperature.

Regarding NO, this is eliminated mainly by the reactions:

$$NH_2 + NO \rightleftarrows N_2 + H_2O \tag{6}$$

$$NH_2 + NO \rightleftarrows NNH + OH \tag{7}$$

Figure 1 indicates the experimental results compared to the numerical ones obtained with the models of Miller and Bowman [6], Glarborg et al. [7], and Miler and Glarborg [8]. Regarding the model of Miller and Bowman, the parameter α produces 0.5 at 1100 K. The high value of the branching ratio results in an early ignition of NO/NH_3 reactions. The model of Glarborg et al. [7] sets a lower value of the branching ratio, 0.3 at 1100 K. This lower value leads to ignition temperatures much higher than the experimental values, as indicated in the figure. The model of Glarbord et al. [7] leads to 0.4 at 1100 K and predicts a more coherent ignition temperature than the other models analyzed.

The temperature range at which NO_x reduction is really effective is a narrow range. The reason is that low temperatures promote too slow reactions and the NO_x reduction is low because most of the injected ammonia remains unreacted. On the other hand, high temperatures promote ammonia oxidation instead of NO reduction [11].

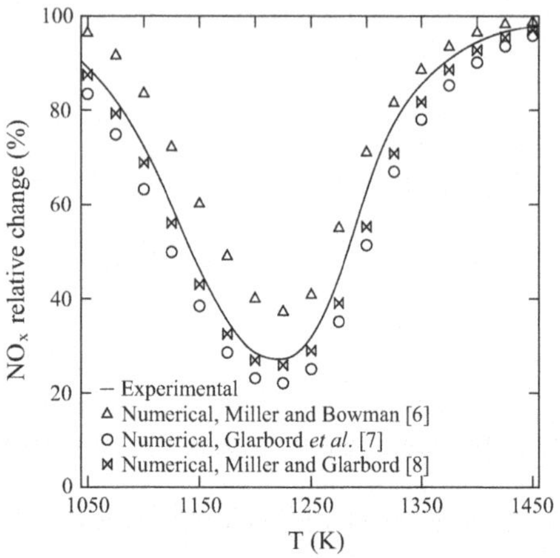

Fig. 1 Comparison between numerical and experimental results

3 Numerical Model

The OpenFOAM software was employed for the simulations developed in the present work. Regarding pressure–velocity coupling, the PISO algorithm was employed. A second-order scheme was used for the discretization of the equations of continuity, momentum, energy, and mass conservation. The time derivatives were discretized through a fully implicit first-order scheme with a constant time step of 0.0005 s. Several convergence tests were realized to confirm that the results were independent of both the time step and grid size. The mesh is shown in Fig. 2.

The governing equations of conservation of mass, momentum, and energy, Eqs. (8)–(10), were solved:

$$\frac{\partial \rho}{\partial t} + \frac{\partial}{\partial x_i}(\rho u_i) = 0 \tag{8}$$

$$\frac{\partial}{\partial t}(\rho u_i) + \frac{\partial}{\partial x_j}(\rho u_i u_j) = -\frac{\partial p}{\partial x_i} + \frac{\partial \tau_{ij}}{\partial x_j} + \frac{\partial}{\partial x_j}(-\rho \overline{u_i' u_j'}) \tag{9}$$

$$\frac{\partial}{\partial t}(\rho H) + \frac{\partial}{\partial x_i}(\rho u_i H) = \frac{\partial}{\partial x_i}\left[\left(\frac{\mu}{\sigma} + \frac{\mu_t}{\sigma_h}\right)\frac{\partial H}{\partial x_i}\right] + S_{rad} \tag{10}$$

In addition, the species concentration equations were solved, Eq. (11):

$$\frac{\partial}{\partial t}(\rho f_k) + \frac{\partial}{\partial x_i}(\rho u_i f_k) = \frac{\partial}{\partial x_i}\left(\frac{\mu_t}{\sigma_\xi}\frac{\partial f_k}{\partial x_i}\right) + \omega_k \tag{11}$$

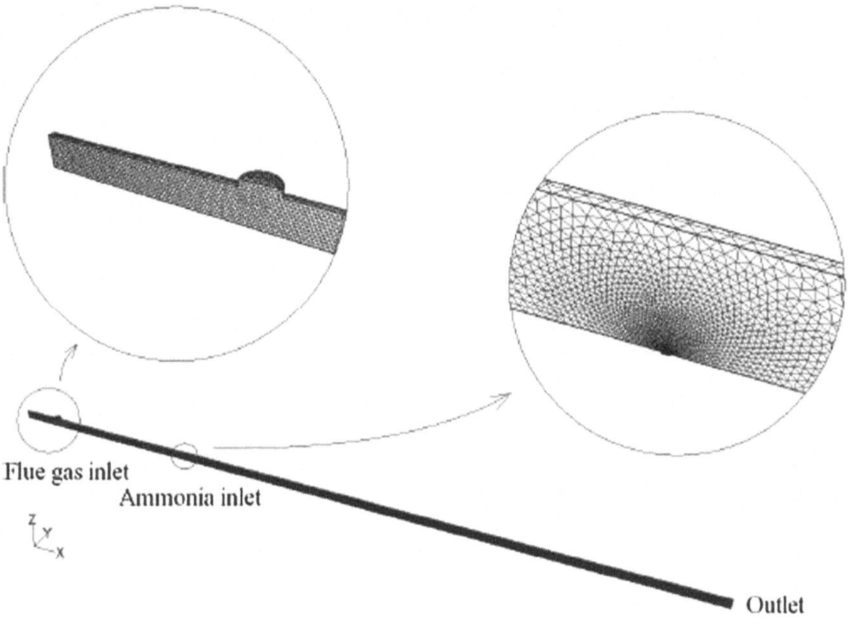

Fig. 2 Computational mesh

4 NO_x Reduction

Once the chemical model has been developed, the influence of various parameters on the NO_x reduction process is as described below.

4.1 *Effect of the Oxygen Content in the Exhaust Gases*

Oxygen is essential for the initial SNCR process. If the oxygen content in the exhaust gases is increased, the performance improves considerably. This effect can be observed in Fig. 3, which represents the NO_x reduction using an exhaust gas at 1000 K.

4.2 *Effect of the Molar Relation NH_3/NO_i*

The NO_x reduction is greater at high NH_3/NO_i relations, as indicated in Fig. 4. Nevertheless, it is important to indicate that high amounts of NH_3 are not suitable due to its toxicity.

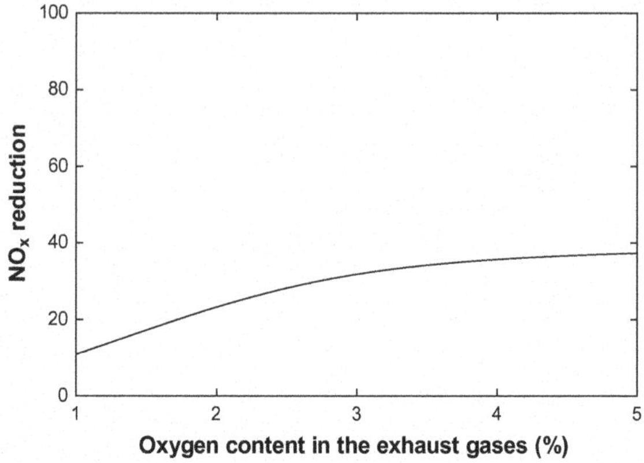

Fig. 3 Effect of the oxygen content in the exhaust gases

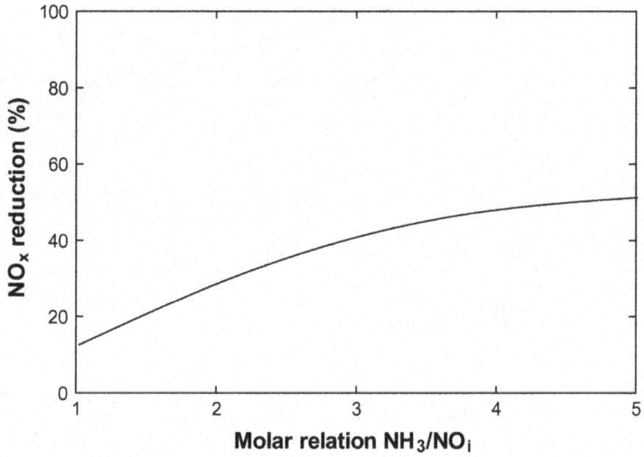

Fig. 4 Effect of the molar relation NH_3/NO_i

5 Conclusions

This document offers an analysis of the SNCR method. The motivation comes from the importance of NO_x reduction in the marine field. It was shown that NO_x reduction improves with the content of O_2 in the exhaust gas and NH_3/NO_i molar relation. Due to the analysis realized in this work, NO_x reduction can be improved but current legislation requires lower values than those obtained in the present work. For this reason, the recommendation in marine ships is to employ SNCR using exhaust gas at high temperatures and thus lower the efficiency of engines or employ SCR. Nevertheless, the main disadvantage of SCR is the cost of catalytic substances.

References

1. Lamas, M.I., Rodríguez, C.G.: Numerical model to analyze NOx reduction by ammonia injection in diesel-hydrogen engines. Int. J. Hydrogen Energy **42**, 26132–26141 (2017)
2. Lamas, M.I., Rodríguez, C.G., Rodríguez, J.D., Telmo, J.: Internal modifications to reduce pollutant emissions from marine engines. A numerical approach. J. Naval Archit. Mar. Eng. **5**, 493–501 (2013)
3. Lamas, M.I., Rodríguez, C.G., Aas, H.P.: Computational fluid dynamics analysis of NOx and other pollutants in the MAN B&W 7S50MC marine engine and effect of EGR and water addition. Int. J. Marit. Eng. **155**, A81–A88 (2013)
4. Lamas, M.I., Rodríguez, C.G.: Emissions from marine engines and NOx reduction methods. J. Marit. Res. **9**, 77–82 (2012)
5. Lamas, M.I., Rodríguez, C.G., Telmo, J., Rodríguez, J.D.: Numerical analysis of emissions from marine engines using alternative fuels. Pol. Marit. Res. **22**, 48–52 (2015)
6. Miller, J.A., Browman, C.T.: Mechanism and modelling of nitro-gen chemistry in combustion. Prog. Energy Combust. Sci. **15**, 287–338 (1989)
7. Glarborg, P., Dam-Johansen, K., Miller, J.A., Kee, R.J., Coltrin, M.E.: Modeling the thermal DeNOx process in flow reactors. Surface effects and nitrous oxide formation. Int. J. Chem. Kinet. **26**, 421–436 (1994)
8. Miller, J.A., Glarborg, P.: Modeling the formation of N_2O and NO_2 in the thermal DeNOx process. Springer Ser. Chem. Phys. **61**, 318–333 (1996)
9. Brouwer, J., Heap, M.P., Pershing, D.W., Smith, P.J.: A model for prediction of selective noncatalytic reduction of nitrogen oxides by ammonia, urea, and cyanuric acid with mixing limitations in the presence of CO. In: 26th Symposium (International) on Combustion, pp. 2117–2124 (1996)
10. Duo, W., Dam-Johansen, K., Ostergaard, K.: Kinetics of the gas-phase reaction between nitric oxide, ammonia and oxygen. Can. J. Chem. Eng. **70**, 1014–1020 (1992)
11. Lamas, M.I., Rodríguez, C.G., Rodríguez, J.D., Telmo, J.: Computational fluid dynamics of NOx reduction by ammonia injection in the MAN B&W 7S50MC marine engine. Int. J. Marit. Eng. **156**, Part A3, A213–A220 (2014)

Hydrodynamic Analysis and Propulsive Arrangement of Two Corvette Hulls with Different Operational Profiles

Rubens C. da Silva, Rodrigo F. Nunes, Thadeu L. C. dos Santos and Kazuo Nishimoto

Abstract This paper presents a few hydrodynamics analyses of two corvette hulls with different operational profiles (corvette 1 with maximum maintained speed of 25 knots and corvette 2 with 35 knots), which designs follow some Brazilian Navy high-level system requirements and are based on modern world military vessels. The total resistance coefficient and wave pattern were determined from computational fluid dynamics simulations using free surface models and ITTC procedures, and the form factors of both ships by Prohaska's method and CFD simulations at low Froude numbers (to neglect the wave resistance). A preliminary numerical analysis of bulbous bow was carried out using CFD, and it was possible to notice that, at high Froude numbers, the ship resistance substantially decreased if compared to the bulbless hull and, with low Froude numbers, the resistance subtly increased. Moreover, the propulsive arrangements of both ships were predimensioned and it was possible to notice the differences due to the operational profiles. Corvette 1 has a combined diesel and diesel propulsion system with two controllable pitch propellers, which optimum pitch was defined using the OpenProp software and the wake field from CFD. Corvette 2, which has a broader range of speed, has a propulsive arrangement of two waterjets and a gas turbine at high speeds, and a CPP with diesel engines when operating at low speeds.

Keywords Corvette · Hydrodynamic · Operational profile · Propulsive arrangement

---

R. C. da Silva (✉) · R. F. Nunes · T. L. C. dos Santos · K. Nishimoto
University of São Paulo (USP), São Paulo, Brazil
e-mail: rubensrcs@gmail.com

K. Nishimoto
Numerical Offshore Tank (TPN), São Paulo, Brazil

A. Vega Sáenz et al. (eds.), *Proceedings of the 25th Pan-American Conference of Naval Engineering—COPINAVAL*, https://doi.org/10.1007/978-3-319-89812-4_4

1 Introduction

The Brazilian coast, which due to its extension of more than 8500 km is called the Blue Amazon, has a great strategic importance for the country. The majority of the national foreign trade (approximately 95%) is circulated by the sea, which makes up almost US$187 billion. The Brazilian Navy intends to renew its fleet with modern vessels capable of carrying out a broad range of naval defense tasks. In this context, corvettes have a strategic importance due to their versatility. Their dimensions are slightly smaller as compared to the traditional frigate combat vessel. However, in their slightness lies the most important feature, serving as a preparatory vessel in crucial wartimes, especially as a stopgap between the larger naval combat vessels.

This paper presents some analyses of preliminary stage design of two corvettes, based on some Brazilian Navy requirements. The study comprised the determination of total resistance coefficient curves and the propulsive arrangement of both vessels.

2 Characterization of the Problem

Corvettes are widely used as one of the main combat ships by many countries. Although these vessels have not been known to survive longer operational durations, technological advancement has made it feasible to further develop them to sustain them for a longer operational lifecycle.

The corvette hulls (Cv-1 and Cv-2) considered in this study had their concept design developed in [1], whose requirements were determined by the Brazilian Navy. The main difference among the vessels is their operational profiles; the Cv-1 has a maximum maintained speed (V_{MM}) of 25 knots and Cv-2, 35 knots. Figure 1 shows the hulls' geometries and some of their main characteristics.

Hydrodynamic analyses are quite important to define some ship features that serve as the base of information at early stage designs. The total resistance coefficient (C_T) curves and the determination of the propulsive arrangement are highly correlated and must be accurately defined to measure the ship design feasibility.

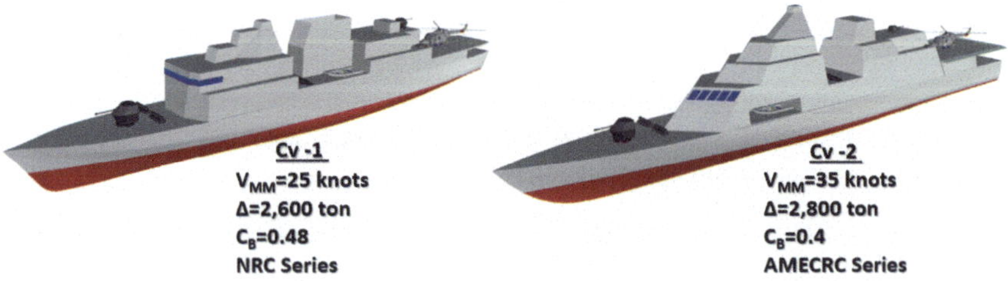

Fig. 1 Cv-1 and Cv-2 hulls [1]

The main results of this paper are the characterization of both corvettes regarding C_T, form factors (k), and propulsive arrangements.

3 Bibliographic Review

3.1 Ship Resistance

To determine the ship resistance a scale model of both vessels was used and computational fluid dynamics (CFD) simulations with free surface settings. Basically, the total resistance of the model (R_{TM}) was calculated and, by the method described in [2] that considers the Froude's comparison, which states that the residual resistance coefficient is the same for model and ship ($C_{RM} = C_{RS}$) at the same values of Froude number (1), and the Hughes' assumption (model and ship have the same k), the ship total resistance coefficient (C_{TS}) is obtained.

According to Todd [2] and after Froude and Hughes considerations, C_t is composed of the frictional (C_f) and the wave-making (C_W) coefficients, given by (2) and (3), respectively.

$$F_r = V(gL)^{-1/2} \tag{1}$$

$$C_F = \frac{0.075}{(\log(R_e) - 2)^2} \tag{2}$$

$$C_W = C_t - (1 + k)C_f \tag{3}$$

where $C_{WM} = C_{WS}$. Thus (1) and (2) are solved from model values, and C_{tS} is obtained by using the C_{WM} and k_M values.

3.2 Form Factor

The form factor (k) of both corvettes was determined by Prohaska's method [3]. In a low-speed region ($F_r < 0.2$) the wave resistance component is assumed to be a function of F_r^n (considering $n = 4$). In a plot of C_t/C_f versus F_r^n/C_F the ordinate ($F_r = 0$) will intersect at $(1 + k)$, enabling the form factor to be determined (4).

$$\frac{C_t}{C_f} = (1 + k) + \frac{F_r^n}{C_F} = (1 + k) + \frac{F_r^n}{C_F} \tag{4}$$

3.3 Bulbous Bow Preliminary Design

The main purpose of using a bulbous bow is the reduction of the wave-making resistance and, consequently, fuel consumption. Some bulb profiles (O, Δ, and ∇) and dimensional parameters are specified byKracht [4], highlighting the length parameter (C_{LPR}= L_{bulb}/L_{ship}) as one of the most important to be considered and the O-type profile as the most suitable for slender ships, also offering space for sonar equipment. Moreover, after briefly analyzing some similar ships (i.e., FFX frigate class), the O-profile bulb with $C_{LPR} \approx 0.046$ was initially chosen for both corvettes.

3.4 Hull–Propeller–Engine Integration

The hull–propeller integration proceeded with the CFD results. As specified by [5], the wake fraction (w) and thrust deduction (t) coefficients were calculated. Such coefficients can be obtained by statistical methods such as the Holtrop method [6], available on NAVCAD [7] software and by using the wake field in the propeller region from the CFD model. The results of the scaled model (w_{TM}) are extrapolated to the full-length ship (w_{TS}) by the ITTC-78 formulation:

$$w_{TS} = (t + 0.04) + (w_{TM} - t - 0.04)\frac{(1+k)C_{FS} + \Delta C_F}{(1+k)C_{FM}} \tag{5}$$

The optimum propeller can be selected from the available systematic series (Gawn and B-series) or using the lifting-line theory to estimate the best hydrodynamic pitch on each point of the blade [8].

4 Methodology

CFD simulations proceeded through the Star-CCM+® software. A fluid domain with a nonstructured hexahedral mesh with 2.1 million elements was developed (Fig. 2) for each vessel. An iso surface was placed at the model draft position to visualize the wave pattern. In addition, both model hulls were 1:14 scaled.

The hull profile curves were obtained from [1], and the bulbous bows were developed on CAD (computer-aided design; Fig. 3). As an early stage design simplification, the cross-section of the bulb was defined as an ellipsis, whose axes' dimensions were arbitrarily stipulated (50 and 25% of the bulb length). Thus other parameters defined by [4] could be calculated, such as the breadth (C_{BPR}= B_{bulb}/B_{ship}) and the section area ($C_{APR} = A_{bulb}/A_{ship}$). Its values are 0.086 and 0.022 for Cv-1 and 0.075 and 0.022 for Cv-2, respectively.

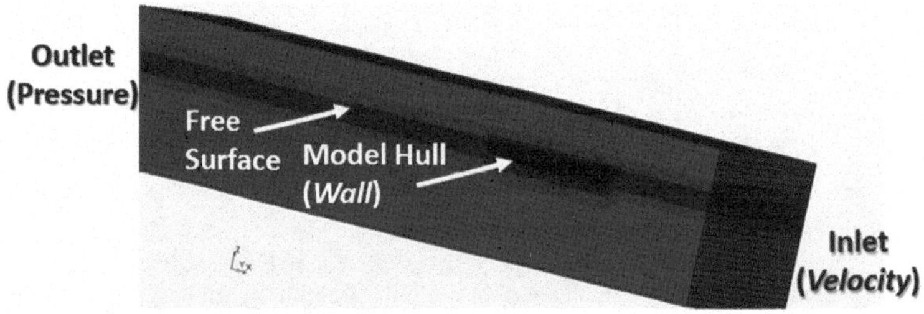

Fig. 2 Hexahedral mesh of the fluid domain

Fig. 3 Bulbous bow and profile curves of each corvette

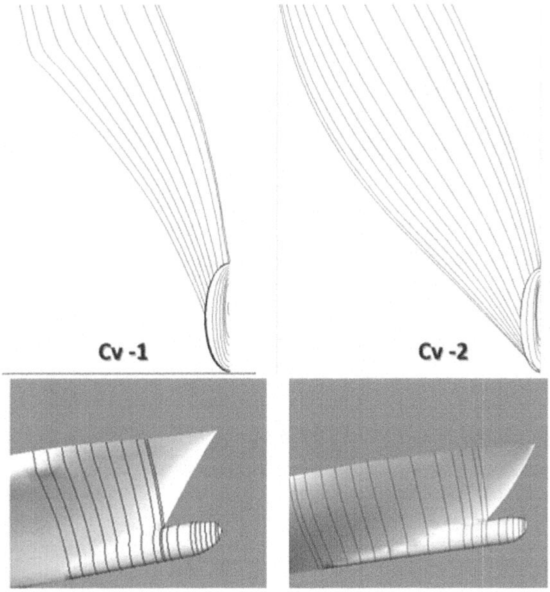

The wake fraction (w) and the thrust deduction (t) coefficients were estimated by the Holtrop method. Thus, by using the NAVCAD software the optimum propeller characteristics were selected. Hereafter, the lifting-line theory was applied using the OpenProp software. From CFD results, the wake field on each point of the propeller plane was obtained (Fig. 4).

The propulsive arrangements were based on the available components on the market that best fit the ships' dimensions, with low specific fuel consumption, and the field proven in modern warships. Particularly for Cv-2, due to its operational profile, at high speeds a normal propeller presents a high level of cavitation. Thus, waterjets were selected for this demand, because they present better efficiencies at high speeds.

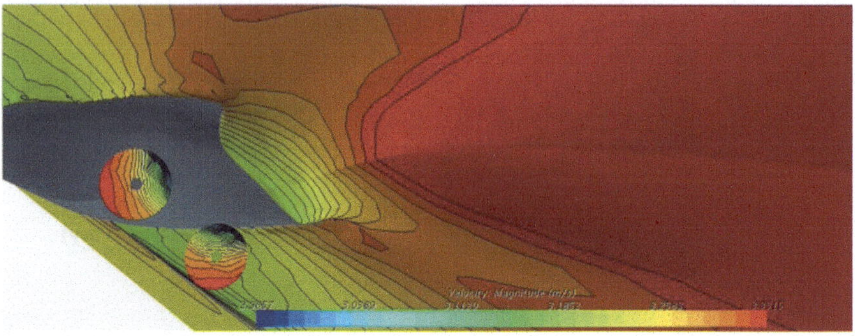

Fig. 4 Wake field from CFD

5 Results

5.1 Resistance

Figures 5 and 6 show the main obtained results from CFD simulations. Figure 5 shows the C_t curve for each vessel and Fig. 6 the wave pattern of the models at the respective scaled V_{MM}.

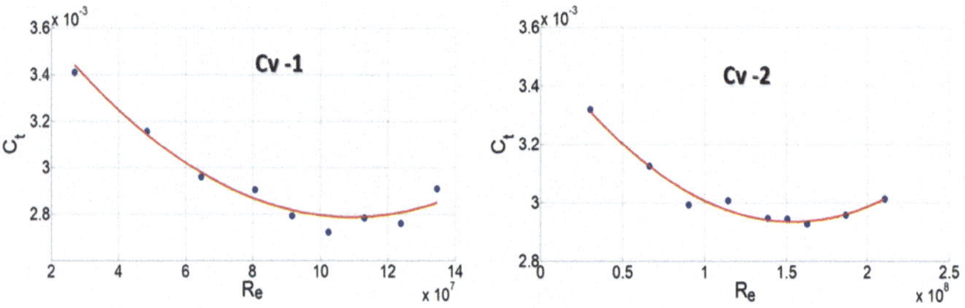

Fig. 5 C_t curves

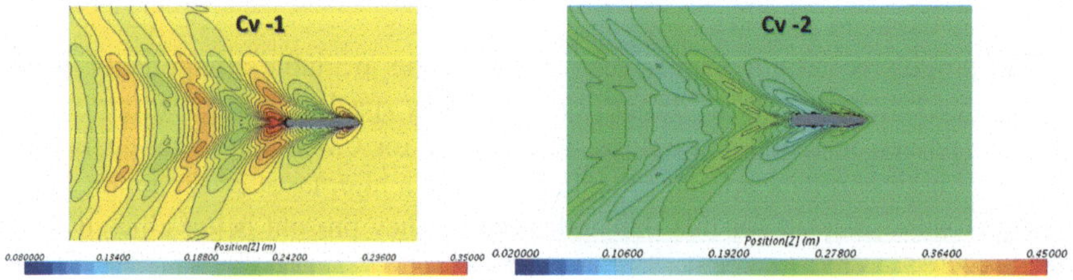

Fig. 6 Wave pattern for each V_{MM}

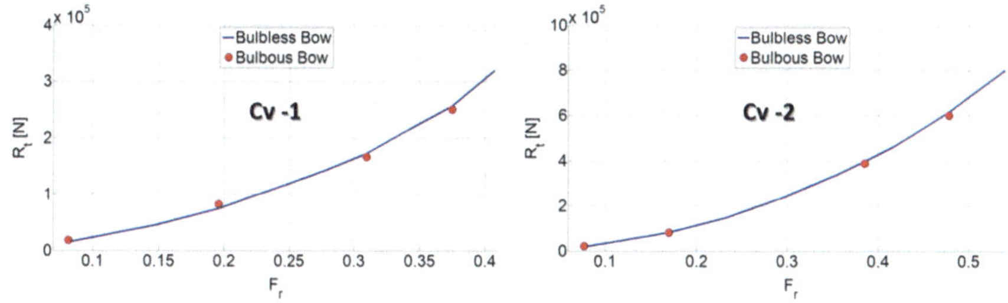

Fig. 7 R_T values for bulbous bows

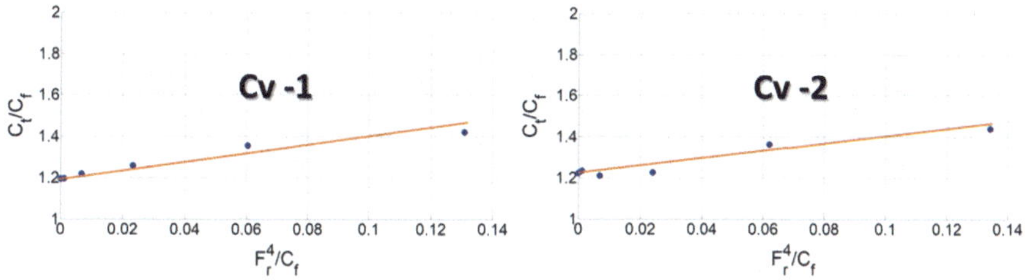

Fig. 8 Form factors

The behavior of R_T with a bulbous bow compared to the R_T curves of the bulbless hulls is shown in Fig. 7. For high F_r the bulbous bow was efficient, substantially decreasing the R_T almost 3.5 and 2.5% for Cv-1 and Cv-2. However, at low $F_{r,}$ the R_T increased 6 and 5% for Cv-1 and Cv-2, respectively.

The form factors were determined analyzing Fig. 8 and based on [3]. At $F_r = 0$, the values of ($1 + k \cong C_t/C_f$) converged to 1.19 and 1.21 for Cv-1 and Cv-2, respectively.

5.2 *Propulsive Arrangement*

Comparing the wake fraction coefficients obtained by the Holtrop method (w_{Holtrop}) and from CFD results (w_{CFD}) to Cv-1, it is noticed that the results were quite similar, with an error of 5% (Table 1).

By using the wake fraction coefficient results, the propeller GAWN AEW with an expanded blade area ratio (A_e/A_0) of 1, 1 was selected as the best model (i.e., the higher efficiency and the lower cavitation level). To a pitch ratio (P/D) of 0.8, the selected propeller presented higher efficiency for various velocities (Fig. 9). In addition, using the same inputs, the OpenProp results were quite similar, defining the pitch ratio for each point of the blade (Fig. 10).

Table 1 Cv-1 wake fraction coefficients

Parameter	Result
w_{tm}	0.0476
t	0.0829
$1 + k$	1.19
w_{CFD}	0.072282
$w_{Holtrop}$	0.076
Error	5%

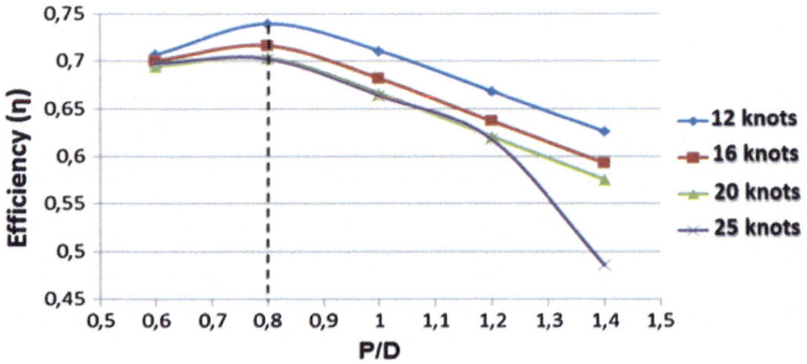

Fig. 9 Efficiency of GAWN AEW propeller

r/R	P/D
0.2	0.6887
0.2628	0.7248
0.4472	0.7708
0.618	0.7832
0.8083	0.7882
1	0.7953

Fig. 10 OpenProp results

Table 2 Brake power results for Cv-1

Speed (knots)	RPM	η	Brake power (kW)
10	156.3	0.7009	644.3
12	195.5	0.6982	1374.5
14	235.7	0.6918	2562.0
16	276.2	0.6851	4309.2
17	290.3	0.6825	5150.4
18	316.5	0.6798	6677.8
20	356	0.6763	9677.8
22	394.2	0.675	13253.8
25	448.2	0.6771	19435

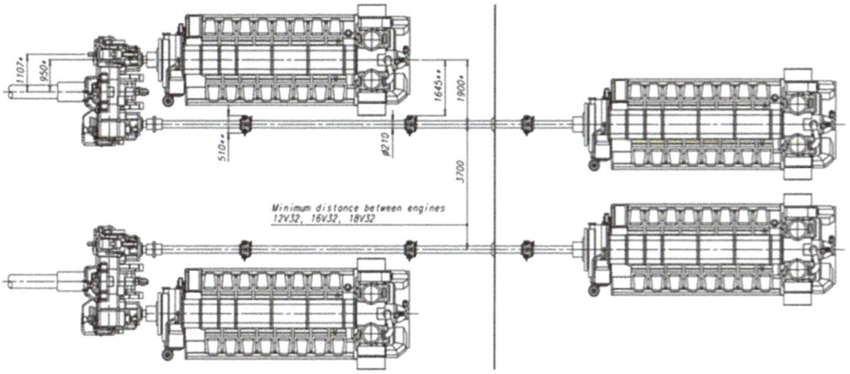

Fig. 11 CODAD arrangement of Cv-1

From the propeller efficiency (η) results, the brake power of the engine was calculated along the velocity range of Cv-1 using NAVCAD software (Table 2).

The CODAD (combined diesel and diesel) arrangement with 4 MTU 16 V 1163 M94 engines [9] was chosen for Cv-1. This arrangement is widely used in warships due to its high reliability, redundancy, reduced fuel consumption, and small required size in the engine room (Fig. 11). The engines have a rated power of 5920 kW at 1200 rpm. In addition, a 4:1 reduction gear was chosen to operate up to 20 knots with only two motors. To reach 25 knots, it is necessary to use four motors and maintain the rotation. So, a controlled pitch system is required to maintain 1200 RPM and attend more power.

On the other hand, a CODOG (combined diesel or gas) arrangement (Fig. 12) was selected for Cv-2. Due to its broader operational profile, at high velocities (up to 21 knots) the use of waterjets is necessary due to propeller cavitation; they are driven by a 30 MW LM2500 + gas turbine [10]. At low velocities a propeller,

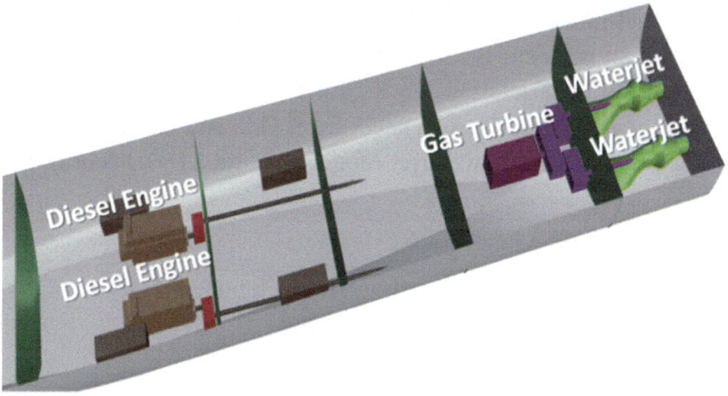

Fig. 12 CODOG arrangement of Cv-2

Table 3 Propulsive arrangement characteristics

Cv-1	Cv-2
Cruise speed	
2 × MTU 16 V 1163 M94	2 × MTU 16 V 1163 M94
Maximum speed	
4 × MTU 16 V 1163 M94	2 × MTU 16 V 1163 M94
	1 × LM 2500 + (gas turbine)
Appendages	
2 × controllable pitch propeller	2 × controllable pitch propeller
	2 × waterjets rolls royce S160

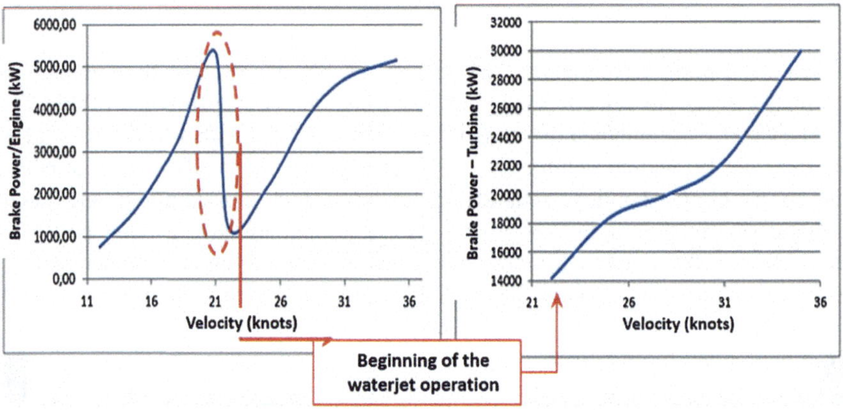

Fig. 13 Power curves of Cv-2 propulsion

driven by MTU 1163 M94 engines, is used. Table 3 summarizes some features of both arrangements.

Analyzing Fig. 13, the moment in which the transition from diesel engines to gas turbine proceeds can be seen. Up to 21 knots the propulsion is made by propellers (at low rpm) and the waterjets.

Specifically, by analyzing the data sheet from [11] and based on the CFD results of total resistance, the selected waterjet model was the S160, according to Fig. 14.

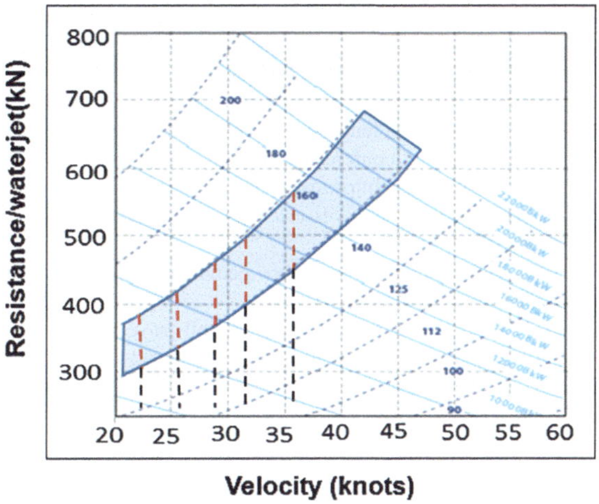

Fig. 14 Selection of S160 waterjet [11]

6 Conclusion

This study focused on the impact of the operational profiles on the design of a military vessel. The main feature that could be highlighted is the peculiarity of the configuration of a propulsive arrangement for high velocities. The use of waterjets increases the design complexity due to the technical requirements of specification, installation, maintenance, and operation.

Further studies must be developed to improve the CFD analyses of waterjets and bulbous bows. A method of optimization and towing tank experiments must be added to the design of a bulbous bow profile that could fit the operational requirements of both considered hulls.

Acknowledgements The authors gratefully thank the Brazilian Navy Naval Engineering Directory (DEN), the Ship Design Center (CPN), the Hydrodynamics Laboratory (LABHIDRO) of the Brazilian Navy Nuclear Development Directorate (DDNM), and the Numerical Offshore Tank of the University of São Paulo (TPN) for their valuable technical support.

References

1. Júnior, J.L.O., Bizarro, R.M., Oliveira, R.C., Nunes, R.F., da Silva, R.C., Santos, T.L.C., Marchione, T.S.L.: Projeto de Concepção de uma Corveta Convencional e de uma Corveta Moderna. Undergraduate Capstone Design. Engineering School of the University of São Paulo, São Paulo (2016)
2. Todd, F.H.: The Model-Ship Correlation Problem. Marine Technology, pp. 152–157 (1966)
3. Prohaska, C.W.: A Simple Method for Evaluation of Form Factor and the Low Speed Wave Resistance. In: Proceedings 11th ITTC, pp. 65–66 (1996)
4. Kracht, A.M.: Design of bulbous bow. SNAME Trans. **86**, 197–217 (1978)
5. Carlton, J.S.: Marine Propellers and Propulsion, 2nd edn. Elsevier, Burlington (2007)

6. Holtrop, J., Mennen, G.G.J.: An approximate power prediction method. Int. Shipbuilding Prog. **29**, 355 (1982)
7. NAVCAD: User's Guide (2009)
8. Epps, B., Ketcham, J., Chryssostomidis, C.: Propeller blade stress estimates using lift line theory. Grand challenges in modelling and simulation. In: Proceedings of the 2010 Summer Simulation Multiconference. Ottawa (2010)
9. MTU Homepage: https://www.mtu-online.com/fileadmin/fm-dam/mtu-global/pdf/applications/military-governmental-vessels/M_Next_Generation_MTU-Series-1163.pdf. Last accessed 10 Aug 2016
10. GE Homepage: https://www.geaviation.com/marine/engines/military/lm2500-plus-engine. Last accessed 10 Aug 2016
11. Rolls Royce Homepage: http://www.rolls-royce.com/~/media/Files/R/Rolls-Royce/documents/customers/marine/waterjets.pdf. Last accessed 16 Apr 2016

Welding and Straightening Simulation of a Deckhouse Structure Using Linear Inherent Strain Method

Ruiz Hector, Naoki Osawa, Murakawa Hidekazu and Rashed Sherif

Abstract Welding is the most widely used assembly method available to industries in the construction of ships and offshore platforms. However, this method always produces a certain amount of distortion that will not only degrade the performance but also increase the building cost of the structure, and it should be straightened. Straightening is performed by mechanical or thermal techniques. The principal mechanical technique is pressing, but it is difficult to apply it to 3D structures such as a ship block. Therefore, mainly thermal techniques are adopted in shipyards. These techniques create irreversible strain (inherent strain) into the component. This is achieved by locally heating the material to a temperature where the heated material with lower yield stress expands against the surrounding cold, higher yield strength material, causing compressive plastic strain in the hot material. When the component is cooled, the heated area shrinks and inherent strain is generated. Spot, line, or wedge-shaped heating techniques are usually applied in thermal straightening. In this study, the modified JWRIAN code by Ruiz is used for performing a sequence of welding and straightening simulations of a deckhouse structure, with thin and opening plates.

Keywords Linear inherent strain · Gauss-Legendre quadrature
Straightening · Finite element method

1 Introduction

Murakawa et al. [1] developed a thermal elastic plastic-based and inherent strain-based welding simulation finite element (FE) code JWRIAN (Joining and Welding Research Institute Analysis). Coarse shell FEs are usually used in the

R. Hector (✉) · N. Osawa
Department of Naval Architecture and Ocean Engineering, Osaka University, 5650871 Suita, Osaka, Japan
e-mail: hector_ruiz@naoe.eng.osaka-u.ac.jp

M. Hidekazu · R. Sherif
Joining and Welding Research Institute, Osaka University, Osaka, Japan

A. Vega Sáenz et al. (eds.), *Proceedings of the 25th Pan-American Conference of Naval Engineering—COPINAVAL*, https://doi.org/10.1007/978-3-319-89812-4_5

Fig. 1 Line heating direction

inherent strain-based JWRIAN analyses. This drastically reduces the manpower needed for modeling and computer resources needed for calculation, and practical application of welding simulation on the production shop floor has been realized. However, it is not easy to perform straightening analysis using JWRIAN because gas heating's inherent strain distributes over a range much smaller than element sizes suitable for welding simulation. Further heating lines on plate fields may be applied at an angle to stiffeners and therefore to element boundaries (see Fig. 1a, b). To avoid having to use much finer meshes, it is needed to develop a numerical technique that can calculate the initial strain force due to straightening inherent strain confined within a narrow region.

In recent studies, Ruiz et al. [2] modified the Osaka University's inherent strain-based welding simulation code JWRIAN so that the inherent strain's equivalent nodal forces are accurately calculated in cases where the inherent strain is confined within a narrow region whose size is smaller/narrower than the element size. The validity of the method and the developed software were examined by comparing a rectangular plate's angular distortion due to gas line heating calculated by three-dimensional thermal–elastic–plastic analysis and that calculated by the developed system. However, a difference was obtained; a part of the difference is due to edge effect and the other part is due to the different natures of solid element models and shell element models Ruiz [3].

2 Elastic FE-Based on Inherent Strain

The distortion and residual stress produced during local heating are caused by the irreversible strain (inherent strain) formed during the plastic deformation process [4–6]. Both distortions and residual stress can be predicted by elastic analysis in which the inherent strain is introduced as initial strain. The total strain ε can be decomposed into the sum of elastic strain-ε^e, plastic-ε^p, thermal strain-ε^T, creep strain-ε^c, and phase transformation strain-ε^t [Eq. (1)]. Noting that deformation and stress are produced by the total and elastic strain, respectively, Eq. (1) can be rearranged into Eq. (2). This equation means that distortion and residual stress are produced by the inherent strain ε^* which consists of plastic, thermal, creep, and that caused by phase transformation. In general, the inherent strain has six components, namely three normal components and three shear components.

$$\varepsilon = \varepsilon^{e} + \varepsilon^{p} + \varepsilon^{T} + \varepsilon^{c} + \varepsilon^{t} \tag{1}$$

$$\varepsilon - \varepsilon^{e} = \varepsilon^{p} + \varepsilon^{T} + \varepsilon^{c} + \varepsilon^{t} = \varepsilon^{*} \tag{2}$$

Normally coarse shell elements are used in the developed code by Ruiz [3]; the initial strain force vector and element stiffness matrix's nonlinear term which includes stress components are integrated using higher-order (e.g., 30 × 30 × 6 for four-node shell elements) Gauss–Legendre quadrature whereas other quantities are evaluated by using ordinary order (2 × 2 × 2) quadrature.

This makes it possible to assess accurately the contribution of narrowly confined inherent strain and inhibit shear locking of the shell element. In JWRIAN, the element stiffness matrix and the equivalent nodal force due to inherent strain are calculated by the following equations.

$$[K_{\mathrm{JL}}] = \int\limits_{V} [B_{\mathrm{JM}}]^{t}[D_{\mathrm{MN}}][B_{\mathrm{ML}}]\mathrm{d}v + \int\limits_{V} \frac{\partial N_J}{\partial x_k}\frac{\partial N_L}{\partial x_j}\sigma_{lk}\mathrm{d}v \tag{3}$$

$$\{F_{\mathrm{inh}}\} = \int\limits_{V} [B_{\mathrm{JM}}]^{t}[D_{\mathrm{MN}}]\{\varepsilon_{\mathrm{inh}}\}\mathrm{d}v \tag{4}$$

where $[K_{\mathrm{JL}}]$ is the element stiffness, $[B]$ the nonlinear displacement-strain matrix of the Mindlin plate for large deflection problems, $[D]$ the stress-strain matrix, N_J and N_L the shape function, x_i the coordinates, σ_{ij} the three in-plane stress components, $\varepsilon_{\mathrm{inh}}$ the inherent strain, and $\{F_{\mathrm{inh}}\}$ the equivalent nodal force due to inherent strain. In the developed code, the second term of Eq. (1)'s RHS and Eq. (2)'s RHS are calculated using higher-order Gauss–Legendre quadrature and other quantities are evaluated by using ordinary order (2 × 2 × 2) quadrature (see Figs. 2 and 3).

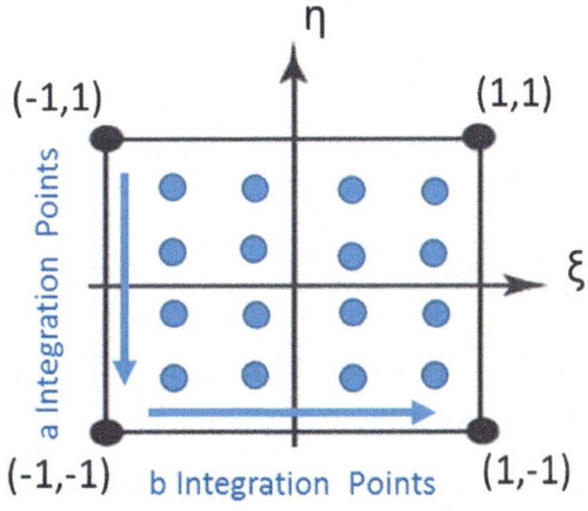

Fig. 2 Shell element (high-order integ. point)

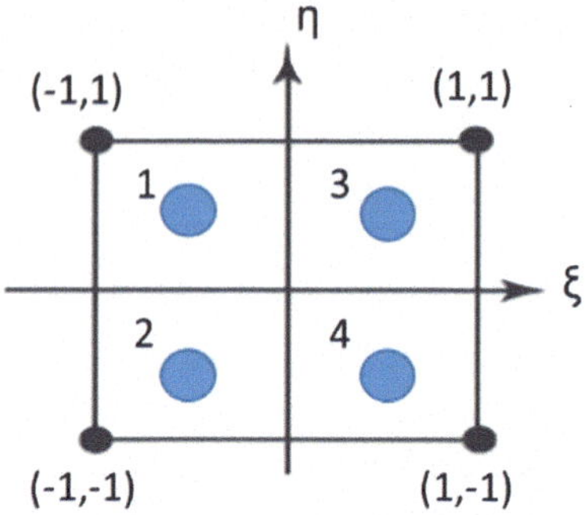

Fig. 3 Shell element (2 × 2 gauss integ. point)

3 Numerical Analysis

Mechanical behavior during welding and local heat straightening are analyzed using the developed code JWRIANSHP [3]. The welding sequence is simulated using inherent deformation values according to the plates' thicknesses, joint processes, and welding condition used in the industry. Inherent strain data used in the local heating process is taken by performing a thermal–elastic–plastic analysis on a flat plate, by using the inhouse code JWRIAN program. These algorithms are already available and well verified against test results [7–9].

3.1 FE Model and Boundary Condition

The first stage of this research consists of computing the distortion of a deckhouse structure caused by a welding sequence using the calculated inherent deformation. Two-dimensional elastic–plastic FE analysis with 4-node shell elements is used. This deckhouse model includes carlings, doors, and windows being considered (see Fig. 4). Different boundary conditions are used; first each member on the plates is welded, so the rigid body motion of each wall is constrained. The following steps supporting below the deck are used (see Fig. 5).

3.2 Welding Analysis

Welding Condition and Sequence. The welding condition is determined by inherent deformation; this is calculated assuming welding conditions normally used in shipyards, plate thicknesses, and joint processes. The welding analysis sequence is divided into four steps, shown in Table 1.

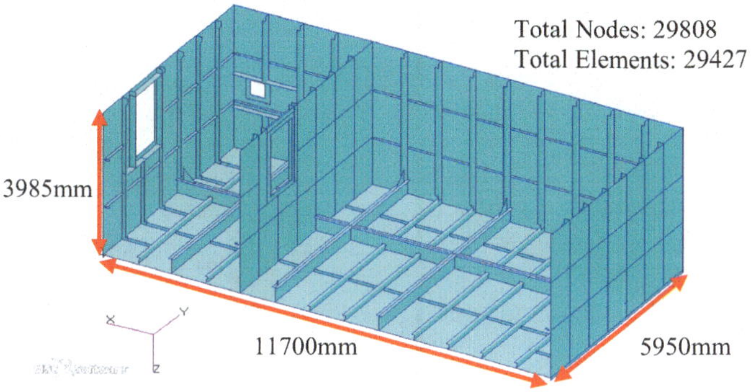

Fig. 4 Deckhouse model

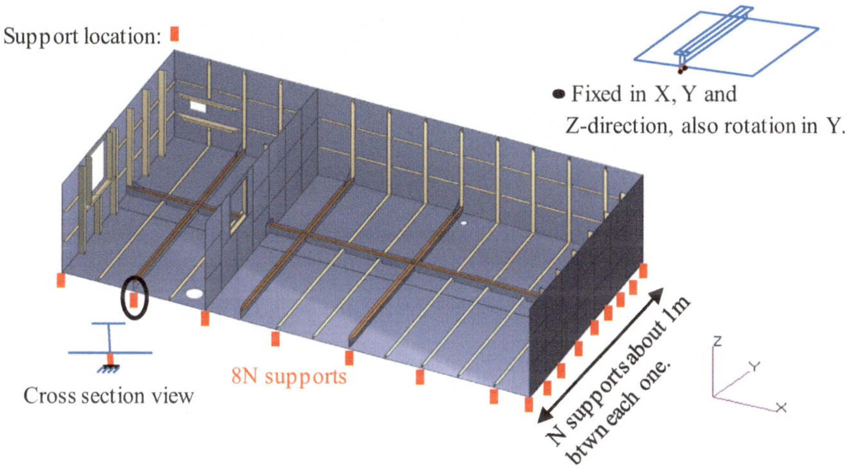

Fig. 5 Boundary condition

Table 1 Welding sequence

Steps	Process
Step 1	(a) L stiffeners on deck are welded to the deck (b) Each member of the wall's plates is welded
Step 2	All panels (walls and deck) are assembled using tack welding
Step 3	All remaining joints are welded, using vertical and horizontal welding as appropriate
Step 4	Turn over, fit, and weld the deckhouse to the lower deck

Welding Results. Using the developed code JWRIANSHP, the welding sequence showed in Table 1 is carried out. The displacement of the entire model is obtained with out-of-plane displacement of the opening wall (see Figs. 6 and 7).

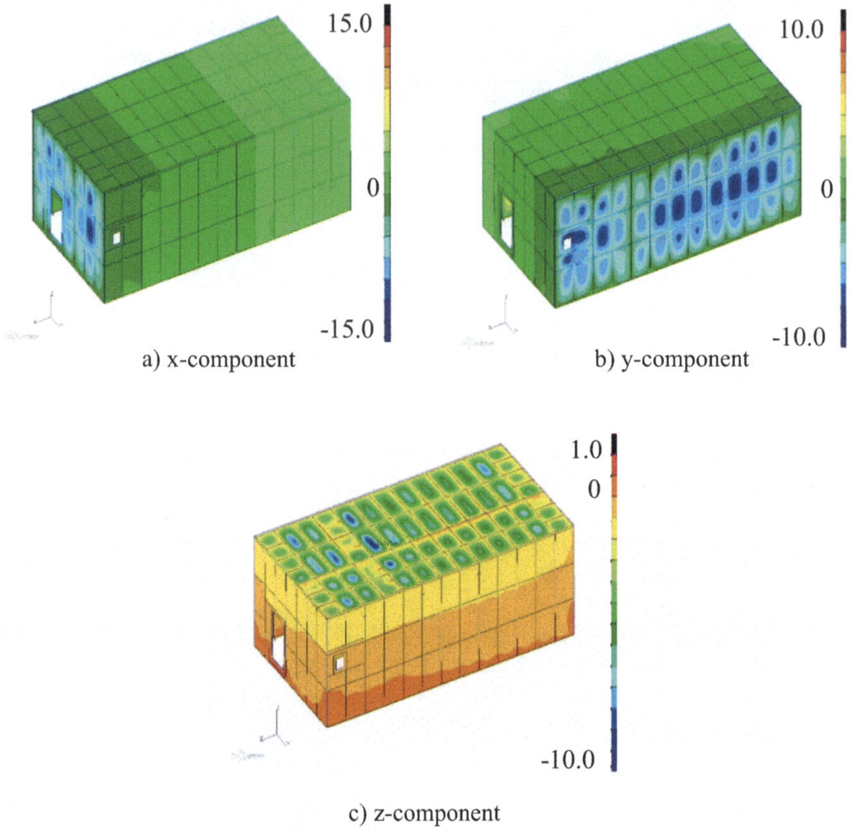

Fig. 6 Full model displacement

3.3 *Post-welding Heat Treatment*

Large out-of-plane displacement is obtained after the welding process (Sect. 3.2), especially around opening areas and between stiffeners and carlings (see Fig. 7). So, for straightening these large deformations, the back heating analysis is carried out.

This back heating process of stiffeners and carlings to the entire model is performed in one step. A study was carried out showing that the sequence is not an important factor, giving a negligible difference between one step and many steps. Each back heating corresponds to the same transverse shrinkage and transverse bending (this last with a negative sign) of each welding used for carlings and stiffeners. The comparison of the full structure between the as-welded condition and back heating is shown, as well as the comparison of the internal wall (see Figs. 8 and 9). Slightly large out-of-plane displacement is obtained after back heating of the internal wall; however, this increment is around the opening area. This large

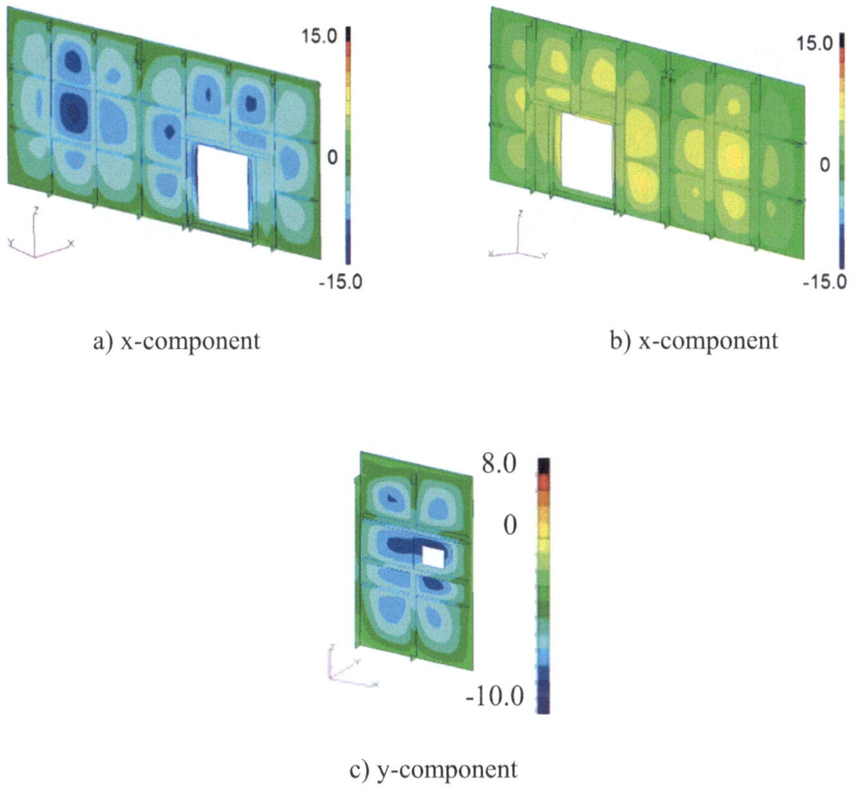

Fig. 7 Out-of-plane displacement on opening plates

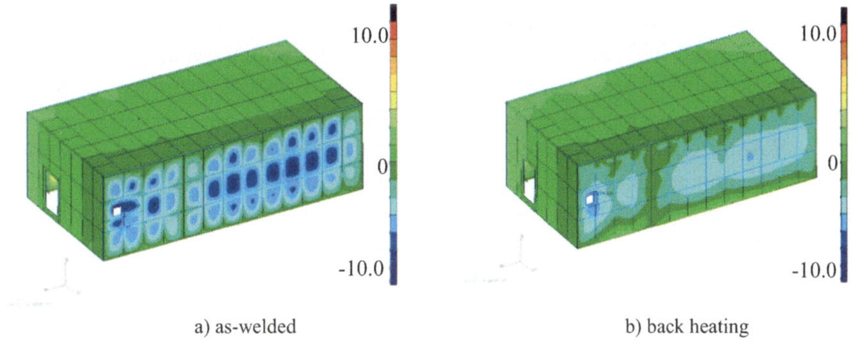

Fig. 8 Comparison of the full model out-of-plane displacement

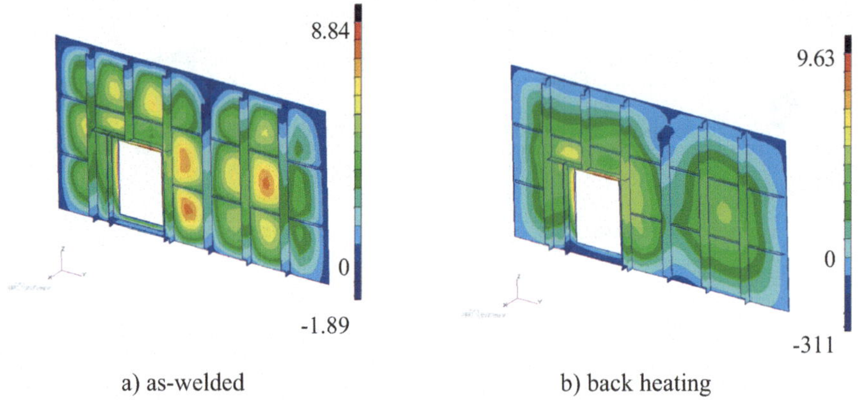

Fig. 9 Comparison of the internal wall out-of-plane displacement

out-of-plane displacement around the opening areas may be improved with local heat treatment.

3.4 Local Heating Using Linearized Inherent Strain Method

A local heating treatment is performed around the opening areas because the large out-of-plane deformation remains in this region. This treatment is performed by using the linear inherent strain method, because the shell FE code used in this study is based on the Kirchhoff–Love plate theory. This theory assumes that straight lines normal to the mid-surface in the undeformed state remain normal to the deformed mid-surface. This requires that the in-plane strain linearly changes with the through-thickness distance from the mid-surface [3].

This linearized inherent strain was obtained by assuming a simple linear equation through the thickness. Then, this calculated linear inherent strain is applied to the region where the local heating wants to be applied.

Local heating pattern. Different patterns normally used in a shipyard were used around opening areas. Two different patterns are shown to straighten the large deformation (see Fig. 10), applying shrinkage or angular displacement according to need.

Comparison between patterns 1 and 2 is shown after the local heating is completed (see Fig. 11). Both patterns show that out-of-plane displacement was reduced substantially. The pattern 1 result shows out-of-plane displacement values near zero, because more shrinkage is given by step 2.

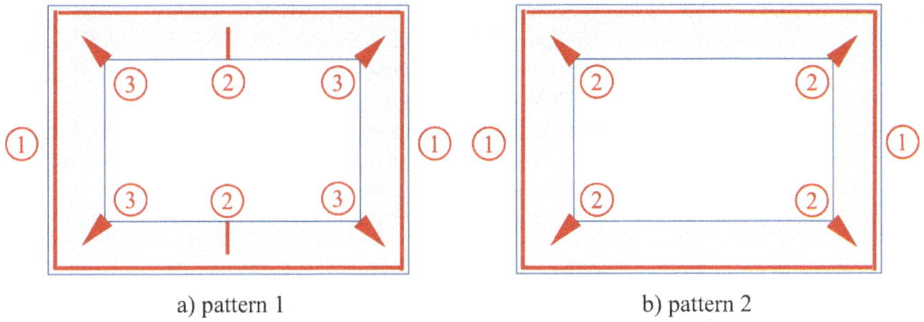

Fig. 10 Local heating patterns

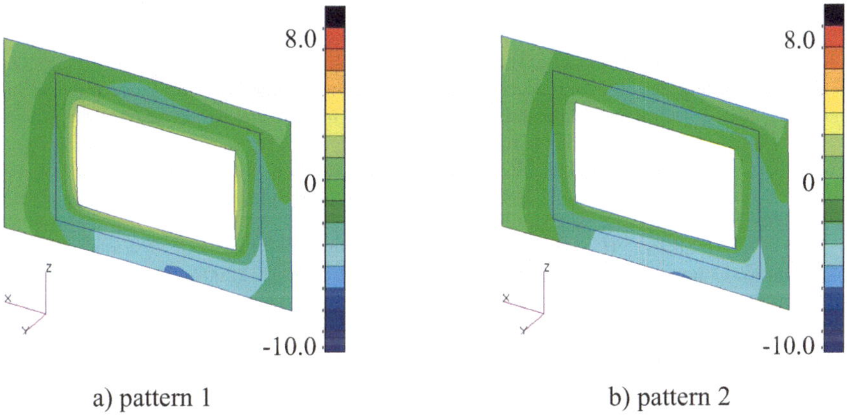

Fig. 11 Out-of-plane displacement after local heating

4 Conclusions

In this study, it has been demonstrated that

- The developed numerical technique, which uses linear inherent strain can be applied to complex structures giving accurate results.
- The working example shows the applicability of the developed method to perform analysis of welding and local heating, giving useful results.
- Different straightening techniques used in shipyards have been reproduced successfully with the developed JWRIANSHP code.
- Performing analysis of buckling and shrinkage caused by local heating is easy with the developed method, once the welding and heating conditions are known by workers.

Acknowledgements This study was performed as a joint research activity of the Global Collaborative Research Center for Computational Welding Science (CCWS) of Osaka University and the research committee on ship construction technology of the Japan Society of Naval Architects and Ocean Engineers (JASNAOE), financially supported by JASNAOE.

The authors would like to acknowledge Mr. Akira Inoue (Mitsubishi Heavy Industry), Mr. Yuji Inoue (Mitsui Shipbuilding and Engineering), Mr. Hiroshi Iwamura (Sumitomo Heavy Industry Marine Engineering), Mr. Eitaro Hara (Kawasaki Heavy Industry), and Mr. Kazuhiro Shimokawa and Mr. Kanaya (Japan Marine United) for their contribution in discussions and field tests.

References

1. Murakawa, H., et al.: Concept of inherent strain, inherent stress, inherent deformation and inherent force for prediction of welding distortion and residual stress. Trans. JWRI. **39**(2), 103–105 (2010)
2. Ruiz, H., Osawa, N., Murakawa, H., Rashed, S.: Prediction of distortion produced in welded structures during straightening process using the inherent strain method. In: Proceedings of the ASME 2016 35th International Conference on Ocean, Offshore and Arctic Engineering, vol. 9, pp. V009T13A007. Busan, South Korea (2016)
3. Ruiz, H., Osawa, N., Murakawa, H., Rashed, S.: Prediction of welded structures distortion due to straightening using linear inherent strain method. In: Proceedings of IIW 2017 International Conference, pp. 86–94. Shanghai (2017)
4. Deng, D., Murakawa, H., Liang, W.: Numerical simulation of welding distortion in large structures. Comput. Methods Appl. Mech. Eng. **196**, 4613–4627 (2007)
5. Murakawa, H., Luo, Y., Ueda, Y.: Inherent strain as an interface between computational welding mechanics and its industrial application.In: Cerjak, H. (ed.). Math. Model. Weld Phenom **4**, 597–619 (1998)
6. Murakawa, H.: Computational welding mechanics and concept of inherent strain for industrial applications. Mater. Sci. Forum **539–543**, 181–186 (2007)
7. Ueda, Y., Murakawa, H., Rashwan, A.M., Neki, I., Kamichika, R., Ishiyama, M., Ogawa, J.: Development of computer aided process planning system for plate bending by line-heating (repot IV): decision making on heating conditions, location and direction (mechanics, strength and structural design). Trans. JWRI **22**(2), 305–313 (1993)
8. Huang, T.D.: Residual Stresses and Distortions in Lightweight Ship Panel Structures, pp. 1–26. Northrop Grumman Technical Review-Journal, Spring/Summer (2003)
9. Blandon, J., Osawa, N., Sano, M., Takaba, S.: Optimization of U-shaped rib bending process using line heating. In: Proceedings of the 27th Asian-Pacific Technical Exchange and Advisory Meeting on Marine Structures TEAM2013, pp. 362–369 (2013)

Optimization of Iron and Steel Products for Shipping

F. J. Correa, V. Fernández, F. Lastra, E. Madariaga and L. M. Vega

Abstract According to the United Nations Conference on Trade and Development (UNCTAD), the sea traffic of steel products in 2016 was just behind the transport of grain for human consumption. According to the World Steel Association, for that year, 473 million tons of steel products were exported, 404 million of them by sea. Of this quantity, steel sections, angles, and shapes represent 25 million tons. We bring up here the solution for a real problem faced by an international steel products company with great importance in the worldwide steel market. Our experience shows us that the steel section transport, usually done by general cargo vessels, implies the charter of oversized vessels. Our solution is to minimize the cargo units' total volume, in order to select the most suitable ship considering volume and cargo hold dimensions, in addition to her deadweight. In this way, we try to reduce the size of the chartered vessels noticeably. Therefore, we have developed an optimization software, described below, and have made a study of the fleet available. Doing this, we have found a new problem: the fleet offered is small in quantity and poorly adapted for the transport of these complex cargoes.

Keywords Shipping · Optimization · Heuristic · Steel cargo · MATLAB®

1 Problem Statement

We want to know what vessel, as small as possible, can carry a steel profiles cargo. And vice versa, to compute if the vessel's holds are big enough to transport this cargo. Thus, we intend to minimize the volume of a set of cargo units, which are made of steel section packages of different sizes, inside the hold of a vessel.

F. J. Correa (✉) · V. Fernández · F. Lastra · L. M. Vega
Cantabria University, 39004 Santander, Spain
e-mail: correafj@unican.es

E. Madariaga
Naval Research Center (CINAV) of Naval Academy
of Marinha Portuguesa, 2810-001 Alfeite, Almada, Portugal

A. Vega Sáenz et al. (eds.), *Proceedings of the 25th Pan-American Conference of Naval Engineering—COPINAVAL*, https://doi.org/10.1007/978-3-319-89812-4_6

The problem can include restrictions on the weight of the cargo per square meter concerning the hold bottom resistance, the discharging ports schedule, or many others.

We must understand the problem as a combinatorial optimization question, due to the high diversification of cargo sizes so that we solve an NP-hard problem. It is impossible to find an optimal solution to this problem due to the high combinatorial degree, as complex in its diversity as its different positions inside the cargo hold. Thus, it is necessary to use metaheuristic advanced algorithms, which can reach suboptimal for the positioning of our cargo units inside the vessel and the operational order, based on the restrictions generated by the cargo.

1.1 Cargo

Buildings or shipbuilding use steel sections extensively, made with different steel qualities and various shapes according to the standardization rules [1]. The steel sections' heights may vary between 40 and 620 mm, and their width can range between 40 and 130 mm. In addition, their longitude can broadly vary between 6 and 24 m. With this mix of dimensions, the weight of steel profiles is between 100 kg and nearly 6 tons [2]. Some steel sections, due to their material (shape and dimensions), may be classified as fragile. It means that these profiles could suffer mechanical deformation during sea transport.

Several steel sections make up packets. The profile or bar numbers per package vary depending on the profile size, and it is very variable, oscillating between 1 and 50. The average weight of these packets is 4.8 tons, although it can vary between 2 and 5.8 tons. The range of heights and widths of the packages is vast.

For sea transport, the stevedores join four or six packages to make up cargo units. These cargo units have two levels separated by stowage wood of 100 mm 100 mm, and we can observe two or three packages per level. For a specific destination port, the cargo units have the same number of packages. However, the sizes and weights of these cargo units can vary broadly.

In this way, we try to stow cargo units made up of steel section packages inside general cargo vessel holds; this is our objective. So we face a complex Tetris.

If we join four packages to form cargo units, the cargo unit will have an average weight close to 20 tons; an average height of 1 m, varying from 0.3 to 1.3 m; and an average width of 1.4 m, ranging from 0.5 to 1.6 m. It is apparent that the relationship between the average height and width of the cargo unit and its longitude, with standard deviations, in both directions, is usually up to 20 cm. The stowage factor is close to 1 m^3/Tm. We must consider a slight longitudinal misalignment of the packages, which can increase the cargo unit total length up to 90 cm. We also consider a small corridor inside the cargo hold, for slinging the packages.

In the case of the cargo units made up with six packages, the average weight will increase to 28 tons, its average width also will grow close to 2 m, and its height will be close to 1.2 m. The stowage factor will change, and we can assume 1.05 m^3/Tm as the new stowage factor taking into account the longitudinal misalignment. The weight per square meter of surface and unit will not exceed 1.3 tons.

On the other hand, this stowage factor is not exact; this is the nature of our problem. For loading a specific steel section cargo in a ship, we must take into account not only the vessel deadweight but also the relation between the hold dimension (length, beam, and depth) and the cargo units. Particular attention must be placed on the relationship between the cargo units' length and the vessel hold length. We must remember two fundamental issues: the cargo must always be stowed longitudinally, and we desire to optimize the cargo hold space.

1.2 Type of Vessel

General cargo vessels carry this type of cargo. Although it can also be loaded in bulk carriers, this will be more complex due to their irregular holds. That is why we only consider solving this problem for general cargo vessels. Thus, for checking if the vessel is suitable for the cargo, we need to know her deadweight, cargo hold capacity, and inner bottom and tweendecks resistant. In addition, we must understand in detail the dimensions and shapes of the cargo hold. Vessels with a single box-shaped hold, perfectly cubic, are the ideal type of vessel for carrying this kind of cargo. In the case of fine hull shape vessels, forward or aft, we consider the fine hull shape area as another cargo hold with a width equal to the smallest beam of this fine area. If the tweendecks are permanent tweendecks, we consider them as cargo holds.

For solving the inverse problem, we have created a general cargo vessel database. We have used the *Fairplay Encyclopaedia* [3] for doing it. We have used some filters for selecting the vessels, as their lifetime, less than 25 years, and their Classification Society which must be inside the IACS (International Association of Classification Societies). In addition, we have selected vessels inside some P&I Clubs. Nevertheless, the *Encyclopaedia* does not show us enough information for our objectives. Thus, we have consulted the Web, vessel by vessel, for their main dimensions. We have obtained the required information for approximately 80% of the ships. We have thrown out vessels whose data were not found.

The *Fairplay Encyclopaedia* contains information for more than 15,000 dry general cargo vessels whose deadweight is more than 75 million tons. Due to our filters, we have reduced this quantity to only 4000 suitable ships. We can say that, after thousands of working hours, we have a database for 1000 general cargo vessels. We have completed a database of 975 ships and 166 fields per vessel until the moment we wrote this study. It contains precise information about all the vessels' cargo holds.

We must say that we have found significant differences between the deadweight and hatch cover dimensions supplied by Fairplay and the information provided by the shipowners' Web pages. We have used the EQUASIS [4] Web for verifying the information in many cases. We have obtained the cargo hold dimensions measured on the plains of the vessels published on the Internet in most cases. We have ignored vessels as potential ships for loading our cargo whose main particulars were not posted on the Internet.

2 Project Background

We have identified more than 150 scientific articles linked to the issue that concerns us after an exhaustive and strict bibliography review. Most of these papers talk about the development of optimization algorithms for container loading, from 1982 to the present. No paper resolves a global stowage problem or deals with the loading of cargo holds.

Two decades ago, the researchers Bischoff and Ratcliff [5] dealt with the problem of cargo in a container using approximations; for solving it they only considered a small number of scenarios within the potential operational situations that might be found in practice.

Nowadays, we understand that we can affirm that we are still in an initial stage of the solution due to the high multitude of different cargo situations that can be found in habitual practice. Within the diversity of approaches given by the literature, more or less similar to our problem, we have discovered optimization problems for maximizing the number of boxes, more or less heterogeneous, to introduce in one or more containers, under different restrictions as to their size and form [6]. Other authors focus on minimizing the number of containers for a group of heterogeneous known boxes [7]. For the stated different problems in the reviewed papers, the authors added some different restrictions related to the container characteristic only in some instances: for example, weight limit and distribution [8], or associated with the cargo, its orientation, stacking, priority, stability, or location [9], among others. In the consulted bibliography, we have stated the experience obtained by the use of the optimization algorithm, even the use of commercial optimization software, in the last case, with poor results due to many possible combinations. This is mainly due to the heterogeneity of the load and the imposed restrictions. Also, high difficulties in the resolution of nonlinear systems because of the stated restrictions also exist [10]. The resolution of this kind of problem with linear programming also implies a great difficulty when we have to deal with a generalization, although if we focus on different particular cases, to a large extent we can reduce the number of problem variables. However, the authors' experience is that solving this type of computational solution takes much time, although the primary task is to verify the quality of the solution obtained by a heuristic algorithm. All the studied authors marked this issue [11].

First, the primary practical objective stated is to solve the optimization problem (minimizing the occupied space) by a large heterogeneous cargo, mainly in two of its three dimensions, inside a cubic-shaped cargo hold. We understand that there may be more cases, but primarily the variations will be more or less heterogeneous relating to the different measures of the cargo units, and the form and number of the cargo hold, including restrictions such as if the hatch covers the complete hold or the existence of spaces out of the square of the hatch.

3 Objective of the Research

The essential objective of this research is the resolution of the problem here proposed through the design and development of a platform with operating software to obtain the optimal or suboptimal solution, depending on the case, of the load and location of highly heterogeneous cargo units in the hold of a target ship.

Our research carries out the development of the platform in two phases. First, we obtain a reliable and safe analytical tool, and then develop a verification tool.

As far as possible, we carry out the second step in parallel to the first, and develop a visual tool for organizing the cargo in the hold.

Based on the physical knowledge of a ship's hold, the software will have the capacity to decide the hold suitability for the transfer of the steel sections and the loading order to reduce the deviations between the theoretical and real positioning, uniformly distributing the load in layers in the hold of the ship.

4 Methodology

The objective is that the cargo fill the minimum space possible. There are two ways for it: to adapt the cargo to the hold of a selected vessel or to choose the smallest possible vessel that can transport it. Our research affirms that the optimal computation entails the formation of cargo units determined by the calculation from the packages that must be carried. This solution, in turn, involves storage logistics in the quay yard, transportation from the factory to the dock, storage in the factory yard, and, probably, a specific order in the production process.

This solution, which would be an ideal one, is not possible in real life. In fact, as we have commented in the background, one of the most significant handicaps in all the articles that we have reviewed is the leap from the theoretical to the real framework. Moreover, in our case, the problem we must solve is a real problem.

Rejecting the global optimum, we must develop a method whose first step is to simulate the formation of cargo units from the packages. A process that has a high degree of randomness, with different subprocesses depends on the human variable: the operator who accommodates the packages in the storage yard in the factory, the one who orders the load in the land transport, and, finally, the one who performs the storage on the dock.

Thus, once we know the processes, from production to the arrival of the profile packages to the pier, and emulate these, we verify that our level of reliability is between 50 and 85%, depending on the greater or lesser heterogeneity of the load.

Consequently, we face the calculation of a solution that will have a certain degree of uncertainty. This calculation will depend on the greater or lesser matching between the dimensions of the cargo units elaborated by our algorithms, using the packing list, and the cargo units developed by the stevedores, who use undoubtedly random methods for doing it.

However, we must highlight the great coincidence, over 95%, between the total volume of the cargo, the volume of freight broken down by the lengths of the steel section packages, and their averages, both overall and separated by lengths, between the actual and emulated cargo units. This empirical verification allows us to search for reliable solutions.

Thus, once we have reproduced the cargo units, and segregated by lengths, with highly reliable values in their totals, the next step is to develop an algorithm which verifies that the cargo can be loaded in the holds of a particular vessel.

There must be two necessary initial filters: first, the deadweight of the vessel. We consider that, in any case, it must always be 1.1 times greater than the total weight of the cargo. We clarify that the ship's size data (DWCC) is not always accessible, so we go to the summer deadweight (DWAT), due to the loading ports.

Naturally, the 10% safety margin is due to the weight of the ship's stores and fuels. In addition, if we want to look for a ship to take the cargo, she must be suitable for it, and her cargo capacity must not be much higher than the total weight of the cargo, for which we must establish an upper limit.

Secondly, we compare the volume of the holds and that of the cargo. As the verification of a determined ship as for the search of a ship, with the two previous steps, we have two filters to delimit the search. Now it is the moment to know if we can embark the cargo and calculate the optimal distribution for each case, which occupies the smallest possible volume.

There are several restrictions: the first is that the axes of the cargo units are aligned with those of the ship; that is, that the length, width, and height of the unit should be embarked in the direction of the length, beam, and depth of the hold.

The second restriction is that there must be transversal corridors that allow stevedores to pass through for the slinging of the units.

The third is that the column weight of cargo units cannot exceed the hold or deck plan strength.

The fourth restriction is that the cargo must be in equilibrium; this means that on a unit of 6 m we cannot load one of 24 m because there will be flights that will cause instability in the stowage.

The fifth restriction is that when there are several discharge ports, we must respect the discharge order, which implies a segregated and ordered vertical and/or horizontal distribution of the load. Apparently, the sum of the lengths of the units forming the base of the cargo, on the plan of the hold, must be less than the length of the hold, or the appropriate combination, if the vessel has several cargo holds.

The stability of the ship and the efforts distribution transcend the scope of this work. However, we understand that our work is not affected by these variables, as long as we speak about single and complete shipments. As a calculation tool we use MATLAB®, version R2017a.

5 Outputs

In the following figures, we represent some of the screenshots of the developed software. This is in response to the methodology enunciated in the previous section in its different steps. In this way, once a loading list in a default Excel format (see Fig. 1), where the separate columns respond to specific fields, is imported, the software will ask us about the number of packages per cargo unit, if the cargo must be considered as fragile cargo, or the order of discharge by ports [12]. With computing times less than 30 s, the program represents a summary of the load units, broken down by lengths. It states the weight and total volume, in addition to the averaged sizes and deviations of the cargo units.

In the next step, the software will ask us if we want to obtain a list of potential vessels (as shown in Fig. 2) to take the cargo or check the suitability of a specific one. Getting the list of ships can take very variable computing times, but always less than one min, depending on the number of ships obtained by applying the first two filters explained above: deadweight and volume.

The program will represent the optimal distribution of the load, with computation times always less than 10 s, after we selected an optimal vessel (see Fig. 3). In Fig. 4, the program generates the representation of the distribution of the cargo load, according to the optimum calculated for the selected vessel.

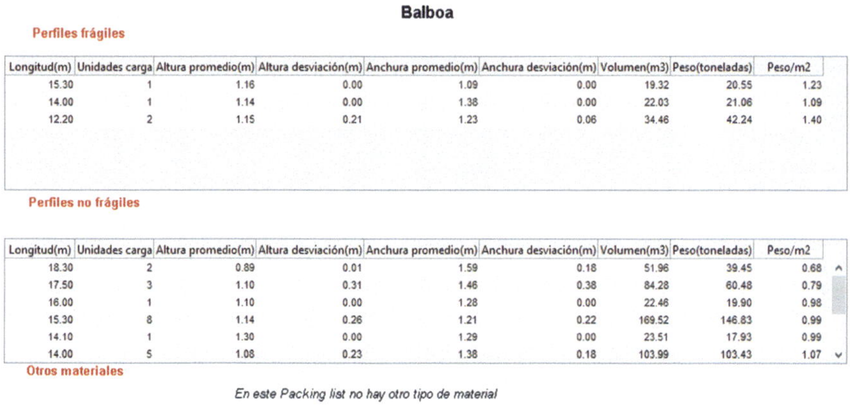

Balboa

Perfiles frágiles

Longitud(m)	Unidades carga	Altura promedio(m)	Altura desviación(m)	Anchura promedio(m)	Anchura desviación(m)	Volumen(m3)	Peso(toneladas)	Peso/m2
15.30	1	1.16	0.00	1.09	0.00	19.32	20.55	1.23
14.00	1	1.14	0.00	1.38	0.00	22.03	21.06	1.09
12.20	2	1.15	0.21	1.23	0.06	34.46	42.24	1.40

Perfiles no frágiles

Longitud(m)	Unidades carga	Altura promedio(m)	Altura desviación(m)	Anchura promedio(m)	Anchura desviación(m)	Volumen(m3)	Peso(toneladas)	Peso/m2
18.30	2	0.89	0.01	1.59	0.18	51.96	39.45	0.68
17.50	3	1.10	0.31	1.46	0.38	84.28	60.48	0.79
16.00	1	1.10	0.00	1.28	0.00	22.46	19.90	0.98
15.30	8	1.14	0.26	1.21	0.22	169.52	146.83	0.99
14.10	1	1.30	0.00	1.29	0.00	23.51	17.93	0.99
14.00	5	1.08	0.23	1.38	0.18	103.99	103.43	1.07

Otros materiales

En este Packing list no hay otro tipo de material

Fig. 1 The output with the summary of the cargo units, segregated by length and by discharge port. The computation time is always lower than 30 s

Nº IMO	Nombre buque	DWT	Eslora bg	Manga promedio bg	Puntal promedio bg	Capacidad	Resistencia bodega
9148180	Nordborg	3714	62.40	11.16	8.62	5669.03	9
9196187	Westborg	3780	62.40	11.16	8.62	5807.50	9
9196204	Sydborg	3780	62.40	11.16	8.62	5807.50	9
9196216	Ostborg	3780	62.40	11.16	8.62	5807.50	9
9137038	Lady Irina	4161	57.20	10.99	10.68	7126.00	10
9671448	Abis Dunkerque	4200	45.20	11.50	10.58	6994.00	15
9671436	Abis Dusavik	4200	45.20	11.50	10.58	6994.00	15
9671486	Abis Esbjerg	4200	45.20	11.50	10.58	6994.00	15
9501708	Abis Albufeira	4200	61.60	11.40	7.94	5576.00	13
9501710	Abis Antwerpen	4200	61.60	11.40	7.94	5576.00	13
9148128	Lady Isabel	4250	57.20	10.99	10.68	7126.00	10
9198628	Peter Ronna	4303	70.35	10.08	6.86	6371.00	15
9289790	Nordersand	4310	70.35	10.08	6.86	6374.00	15
9323651	Ditzum	4310	70.35	10.08	6.86	6374.00	15
9281786	Cimbris	4310	70.35	10.08	6.86	6374.00	15
9195482	Randzel	4310	70.35	10.08	6.86	6374.00	15
9213997	Fehn Polaris	4228	63.04	10.83	8.68	5808.00	15
9279408	Flinterbothnia	3480	60.49	10.30	8.99	5605.00	13

Fig. 2 The output of smaller ships that can potentially take cargo. The computation time can vary between 10 and 60 s

	Datos
LDT	1058
Displacement_Tonnage	4538
FDWT	3505
Dwt	3480
Bale	5.6053e+03
Grain	5605
Decks	1
Tween_Deck	0
Depth	8
Holds	1
Hatches	1
Hatch_Sizes	0
Hatch Size L #1	60.5000
Hatch Size B #1	10.3000
Hatch Size L #2	
Hatch SizeB #2	
Hatch Size L #3	
Hatch SizeB #3	
Hatch Size L #4	
Hatch SizeB #4	
Hatch Size L #5	
Hatch SizeB #5	
hold Size L #1	60.4854
hold Size B #1	10.3000
Hold Size D #1	8.9900
Hold Capacity #1	5.6008e+03

Fig. 3 The output with the data of the selected vessel, more than 150 fields, that can take the cargo

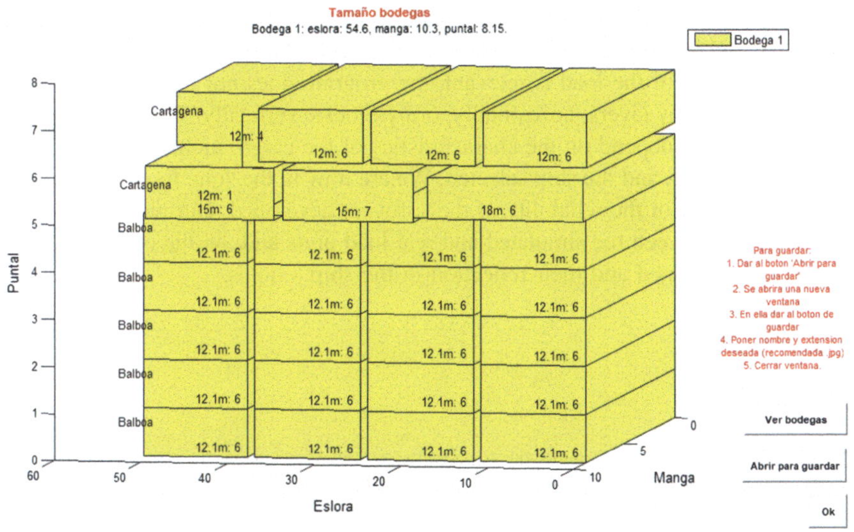

Fig. 4 The output with the representation of the distribution of the cargo load, according to the optimum calculated for the selected vessel

6 Conclusions

The problem of optimization applied to the loading of steel sections in the holds of a ship had never been faced, nor a database of ships prepared, with the particular data of the dimensions of their holds.

The bibliography consulted by the authors of this research has been an excellent guide, but not an example of the resolution of the problem raised here.

For the optimization of iron and steel products for shipping, the need to publicize the plans of the general distribution of general cargo ships is evident. The databases currently marketed have significant differences in their data with those provided directly by shipowners through the Web. It is important to continue expanding the database of ships completed until now. This database is a fundamental tool for solving this type of problem. Being able to obtain a list of potentially suitable vessels of adequate size for the cargo makes it much easier to achieve the best freight, both for the charterer and loader.

It is quite difficult to find general cargo vessels, within a reasonable range of deadweights, close to the weight of the total cargo, to take this type of cargo, with significant heterogeneity in the measurements of the cargo units. This is mainly due to the lack of proportionality between the lengths of the units and the length of the vessel holds.

We underline the usefulness of MATLAB as an adequate language for the resolution of this type of optimization problem of iron and steel products for shipping. The results obtained through this software should serve as a guide for the loading of the vessel, but not as a reference to be strictly followed. We must strengthen and improve the algorithm used inasmuch as we have verified that when the heterogeneity of the load is relevant, the program does not respond. We could obtain a solution by layers, horizontally, or by blocks, vertically. The application of an algorithm will depend on the characteristics of the cargo, including the number of unloading ports, and the ship selected. We are now in the "one foot on the quay" verification phase of the reliability of the software developed. Our research verifies the similarity between the simulated and real load units and, on the other hand, the load models obtained and their reflection in the ship's holds.

References

1. Web Institucional del Colegio de Arquitectos técnicos, e ingenieros de construcción de Barcelona. http://www.apabcn.cat/Documentacio/areatecnica/PDFS_RENART/212677.pdf. último acceso 13 July 2017
2. Web Institucional de la Facultad de Náutica de Barcelona, perteneciente a la UPC. https://www.fnb.upc.edu/mecanica/s11-pw/Prontuario%20-20Perfiles%20Laminados.pdf. último acceso 13 July 2017
3. Fairplay World Shipping Encyclopaedia. Coulsdon Surrey Fairplay. Versión de Julio de 2017
4. Página web de EQUASIS (desarrollada por la Comisión Europea y la Administración Marítima Francesa). http://www.equasis.org/EquasisWeb/public/HomePage. último acceso 15 July 2017
5. Bischoff, E., Ratcliff, M.: Issues in the development of approaches to container loading. Omega **23**, 377–390 (1995)
6. Wäscher, G., Haußner, H., Schumann, H.: An improved typology of cutting and packing prob-lems. Eur. J. Oper. Res. **183**, 1109–1130 (2007)
7. Martello, S., Pisinger, D., Vigo, D.: The three-dimensional bin packing problem. Oper. Res. **48**, 256–267 (2000)
8. Junqueira, L., Morabito, R., Yamashita, D.: Three-dimensional container loading models with cargo stability and load bearing constraints. Comput. Oper. Res. **39**, 74–85 (2012)
9. Ramos, A., Oliveira, J., Lopes, M.: A physical packing sequence algorithm for the container loading problem with static mechanical equilibrium conditions. Int. Trans. Oper. Res. **23**, 215–238 (2016)
10. Wang, N., Lim, A., Zhu, W.: A multi-round partial beam search approach for the single container loading problem with shipment priority. Int. J. Prod. Econ. **145**, 531–540 (2013)
11. Bortfeldt, A., Wäscher, G.: Constraints in container loading—a state-of-the-art review. Eur. J. Oper. Res. **229**, 1–20 (2013)
12. Lützhöft, M.: "The technology is great when it works": Maritime Technology and Human Integration on the Ship's Bridge (2004)

Fire in Confined Spaces: A Pending Task of the International STCW Convention

F. J. Correa, E. Madariaga, A. Trueba, S. García, E. López and V. Fernández

Abstract Work on board vessels or in port facilities creates the need for officers and crews assigned to firefighting teams to have the necessary training to deal with situations of extreme fire behavior. Most of the possible scenarios, in which a fire can be unleashed onboard, are in confined spaces. The current training that the STCW 2010 establishes for any crewmember is the IMO Course 1.20 of Fire Prevention and Fighting and for the officers and controls the IMO 2.03 course of Advanced Firefighting. These courses of sufficiency, at present, do not suppose the acquisition of knowledge or minimum skills to extinguish fires in confined spaces. The training has generalist contents; it tries to cover many topics in a short space of time. This paper analyzes the circumstances of learning and defines a module of specialization on fires in confined spaces to be able to face this emergency.

Keywords Maritime safety · Confined spaces · STCW · Flashover Backdraft · Pyrolysis

1 Introduction

The control of a fire in confined structures requires that the extinguishing personnel have adequate training and preparation with a more profound knowledge of the behavior of fire under these conditions. It is not the same to face a fire in an area outside a ship, where there is visibility, and it is not necessary to use autonomous breathing equipment (the environment dissipates heat, and we have maneuverability

F. J. Correa (✉) · E. Madariaga · A. Trueba · S. García · V. Fernández
Cantabria University, 39004 Santander, Spain
e-mail: correafj@unican.es

E. Madariaga
Naval Research Center (CINAV), Naval Academy of Marinha Portuguesa, 2810-001 Alfeite, Almada, Portugal

E. López
ILUNION, Training Center in Firefighting and Emergencies, 28690 Madrid, Spain

A. Vega Sáenz et al. (eds.), *Proceedings of the 25th Pan-American Conference of Naval Engineering—COPINAVAL*, https://doi.org/10.1007/978-3-319-89812-4_7

possibilities). For the fire to occur in an enclosure of limited dimensions, with little or no visibility, flooded with smoke, with breathing equipment on the back, a mask on the face, and with the heat affecting us in all its intensity [1], the conditions are different. Hence the need to deepen the fire training given to crews about indoor interventions, where fire follows more complex patterns than in outdoor spaces. In addition to technical training, in terms of knowledge about the behavior and development of fire, there is another no less critical formation, which is the psychological one: to face situations as complex as those that occur in confined spaces onboard or in a port facility [2].

When there is a fire emergency in confined spaces (internal parts of ships), there are three phenomena that human beings have always feared ancestrally, which are darkness (and lack of visibility), not able to breathe (nowadays we can overcome this thanks to autonomous breathing equipment), and fire [3]. From these feared parameters are born the psychological problems that can diminish the effectiveness of the intervention teams. The crews can overcome these problems when they arise if they have adequate theoretical and practical training for these emergencies. The theoretical training will lead us to know and to understand what happens and what can happen [4]. The practical training will give us the knowledge of living situations in a realistic way, to experiment and apply the appropriate techniques, which will ensure that crews achieve a high degree of effectiveness and confidence in fire emergency interventions in confined spaces [5]. In practice, the training is essential to follow "the role of fidelity," that is, a degree of realism that generates a fair point of stress, but without generating anxiety and that interferes with the acquisition of skills of the crews. These are why the fire training courses stipulated in the International Convention on Standards of Training, Certification and Watch Keeping for Seafarers (STCW), which emanates from Chapter II-2 of the International Convention for the Safety of Life at Sea (SOLAS), should be adapted and expanded in the study of situations of extreme behavior of fire. Today, it is not included in the official agendas of the model courses of the International Maritime Organization (IMO). The IMO model courses establish a series of generalist concepts, regarding control and extinction of fire in closed spaces, with a frankly sparse teaching load in relation to practical content, 7 hours in the prevention and control of module fires at the basic level (Basic Training in Safety), and 12 in the advanced course (Advanced Fire Fighting Certificate), giving more importance in advanced training to issues of an organizational nature, than to extinction techniques.

1.1 Updates of the STCW2010: Manila Amendments

The diplomatic conference celebrated in Manila, in June 2010, approved a set of amendments updating the 1978 International Convention on Training, Certification and Watch Keeping Standards for Seafarers (STCW) and its associated Code [6]. The approved amendments mark the first significant revision of the instrument since those approved in 1995, which ultimately revised the 1978 STCW Convention.

The shipping companies depend on the competence and proper training of seafarers to ensure the safety of human life at sea, maritime security, and the protection and conservation of the marine environment [7]. The objective of the STCW Convention is to establish the international standards necessary for training centers and educators to develop the skills and competencies required for the profession of seafarer today [6, 8]. The entry into force (on January 1, 2012) of the so-called Manila amendments to the STCW Convention, incorporates essential changes to the STCW Code (Technical Annex to the Convention above). These changes have forced the countries that are part of the IMO to update the contents of the courses to obtain the necessary certificates of sufficiency. This course is essential to develop a professional career in the maritime sector and incorporates the renewal of certificates of prevention after five years (against fires and survival at sea even if you are navigating and practicing as seafarers); seafarers need to update the sufficiency training [9, 10].

1.2 Adaptation of the STCW 2010 to European Higher Education Area: Case of Spain

The Manila Amendments have made a deep change in the training of future officers, chief engineers, and masters of the Spanish merchant marine, as their curricula have been changed and updated to Bologna for internationalization just with the amendments to the International Convention on Training, Certification and Watch Keeping Standards for Seafarers (STCW 2010).

In the case of university training in Spain adapted to the new European Higher Education Area (EHEA), the study plans have implemented this adaptation. Among the seven Spanish centers of university education in Nautical and Maritime Transport (Nautical) and in Marine (Maritime Propulsion), the optimization of resources stands out. The Higher Technical School of Maritime Engineering, belonging to the University of Cantabria, where university bachelors have the proper name of "engineering" encompasses Nautical Engineering and Maritime Transport (Nautical), and Marine Engineering (Maritime Propulsion). Cantabria University has brought the adaptation to STCW 2010 in each of the subjects of the official curriculum. It trains the university bachelors in Nautical Engineering and Maritime Transport (future deck officers, mates, and masters), and in Marine Engineering (future officers, mate and chief engineers), after being approved as a training center by the Spanish Maritime Administration (i.e., the General Directorate of Maritime Affairs). In this way, students who pass these university studies can embark, because they obtain at the same time all professional certificates of sufficiency. One of the most critical aspects of these university careers is cross-training in maritime safety [11, 12]. In these specialized technical subjects, students are trained at the basic level both in marine safety (Fire Prevention and Firefighting according to IMO model course 1.20, Survival at Sea according to

IMO model course 1.19, First Aid according to the model course of the IMO 1.13 and Social Behavior and Responsibilities according to IMO model course 1.21) and in maritime security (Basic Training in Maritime Security, according to IMO model course 3.26). This training is the current basic requirement to be able to board a merchant ship (basic compulsory training for all seafarers). In turn, these courses have an advanced level, which is what is currently needed to embark as a duty officer in any of the specialties (deck, engineer, electro-technical), chief engineer, or master. The advanced maritime prevention courses are: Advanced Firefighting according to IMO model course 2.03, Proficiency in Survival Craft and Rescue Boats other than Fast Rescue Boats, according to IMO model course 1.23 and Handling of Fast Rescue Boats according to the IMO model course 1.24. The onboard officers in charge of maritime safety require constant training and updating [13]. Of this imperative, it depends on successfully achieving decision making in times of crisis in which it is at stake that a mere internal accident could generate a severe external accident, with possible consequences of pollution to the marine environment and substantial economic losses [3, 14, 15].

Fire Training at Two Levels: Basic and Advanced

The STCW Code establishes two levels of training in maritime fire prevention, a basic level and an advanced level. Everything that establishes both the SOLAS Convention and the different conventions that emanate from it, such as the STCW, are levels of "minimum" training, that is, of "sufficiency" for the signatory countries [8, 16]. The formative implementation above the minimum levels defined by maritime codes generates a greater maritime hegemony [17].

This training aims to ensure that seafarers are aware of the risks of working on a ship and that they can respond adequately in the event of an onboard fire emergency. Whereas the STCW 1978 Convention focused almost exclusively on knowledge, the STCW 1995 placed more emphasis on practical skills and competencies supported by theoretical knowledge.

The Manila Amendments to STCW (2010) have aimed to include all the changes agreed upon since 1995, address new technologies, avoid irregularities and interpretations, and eliminate outdated provisions. Maritime authorities have placed particular emphasis on the improvement of the control and communication provisions of the title in Regulation I/7, and compliance with the specific requirements of maritime and coastal navigation.

In this way, in Spain the General Directorate of Maritime Affairs has determined the practices and the time that must be devoted to fire training, as shown in Tables 1 and 2. Students can acquire, via basic training, knowledge about the fire in confined spaces, as can be seen in Table 1, but it is insufficient.

However, officers only reach practical knowledge about managing resources via the advanced training course. For both formations, officers and ratings today do not receive the minimum training necessary to face an emergency by fire in a confined space, such as the interior spaces of a ship.

Table 1 Basic training in safety course (6 h of survival at sea and 10 h of fire)

Competencies	Practical tests
1. Fire Prevention and Firefighting	1.1 Use different types of portable fire extinguishers. 60 min
	1.2 Use the autonomous respiratory system. 60 min
	1.3 Extinguish small fires, for example, electrical equipment, hydrocarbons, and propane fires. 60 min
	1.4 Extinguish significant fires with water using spray and jet lances. 60 min
	1.5 Extinguish fires with foam, dust, or other chemical agent. 60 min
	1.6 Enter a compartment and pass through high-expansion foam equipped with a safety cable but without breathing apparatus. 75 min
	1.7 Fight against fire in closed spaces filled with smoke wearing a self-contained breathing apparatus. 45 min
	1.8 Extinguish fires with mist or other appropriate firefighting agent, in a stateroom or in a simulated machine room where there is a fire and abundant smoke. 90 min
	1.9 Extinguish a hydrocarbon fire with a fogging apparatus and spray nozzles, dry chemical powders, or foam sprays. 60 min
	1.10 Carry out a rescue in a smoke-filled space wearing a breathing apparatus. An approved smoke generation device is worn. 30 min

Fire Does Not Know Levels

Fire is one of the greatest hazards a ship can have. Its effects can be devastating, especially on passenger ships and ships carrying dangerous goods. If the fire is not located in time or is not quickly controlled the risk to the crew and the vessel can become significant. Seafarers have always tried to control maritime emergencies generated onboard or in port by fire. But, in many cases, either due to lack of knowledge (in terms of their behavior), or due to lack of training, or means for their control and extinction, fire in many times has generated a considerable amount of human and material goods losses in the marine environment [1, 18].

One of the characteristics to take into account, before a fire onboard, is that the ship must defend itself only in most cases, because there is no possibility of outside assistance unless the ship is in port or next to another. This need for self-defense means that the availability of means and the training of the crew in fire control techniques and tactics occupy an important part of the training both on land and at sea. The technique is learned in approved training centers in firefighting, but then the tactics must be brought into play, which will be oriented according to the type of vessel and navigation carried out through the obligatory periodic firefighting exercises [5] and stipulated by the regulations (SOLAS exercises about safety onboard).

Table 2 Advanced firefighting course, 8 h practice in resource management

Competencies	Practical tests
1. Control firefighting operations onboard	1.1 Use different types of portable fire extinguishers. 45 min
	1.2 Use of water for extinguishing fires, effects on the stability of the ship, precautions and corrective measures. 15 min
	1.3 Communication and coordination during firefighting operations. 15 min
	1.4 Control of fans, including smoke extractors. 15 min
	1.5 Control of electrical systems and fuel supply system. 15 min
	1.6 Risks of the firefighting process. 60 min
	1.7 Fight against fires involving dangerous goods. 30 min
	1.8 Fire precautions and risks related to the storage and handling of materials. 15 min
	1.9 Treatment and control of the injured. 45 min
	1.10 Coordination procedures with firefighting operations carried out from the ground. 15 min
2. Organize and train firefighting crews (105 min)	2.1 Preparation of contingency plans
	2.2 Composition and assignment of personnel to firefighting crews
	2.3 Strategies and tactics for fighting fires in different parts of the ship
3. Inspect and maintain fire detection and extinguishing systems (90 min)	3.1 Realistic practical exercises in which fire detection and extinguishing equipment and systems used onboard ships are used
	3.2 Practical exercises on the requirements of regulatory and classification surveys
4. Investigate and collect reports on events in which fires occur (15 min)	4.1 Practical exercise of evaluation of the causes that give rise to events in which fires occur by practical exercises

The hazard posed by a fire onboard makes it necessary to maintain specific regulations on their prevention, detection, extinction, and means onboard, given the special conditions of isolation and environment where ships operate. The SOLAS Convention, in its provisions, establishes this regulation, as we have indicated above, thanks to which vessels have become safer. Today we have the knowledge necessary to understand the behavior of fire and generate tactics for its control. Technologically in firefighting, we are at the best moment to date, with very capable means and devices to detect and control a fire, some of them highly innovative and unthinkable in just a 20-year space of time [19]. For this reason, we can treat firefighting at sea, nowadays, thanks to the level of knowledge acquired, as an "art," the art of saving a ship, without the need of its sinking.

The difference between fighting fires at sea and fighting fires on land is fundamental, because at sea, the priority action is to maintain the integrity of the ship and her navigability conditions. On land, a structure affected by a fire can be evacuated, with greater or lesser difficulty and, in this case, we would limit ourselves to try to extinguish it or not to spread to other zones; the structure could even collapse and we would still have control. In a ship, this is much more complicated, because the option is abandonment, which in most cases exposes the staff to an even greater risk. Although this option is the only one onboard in the case of very serious complications, it is a last resort [7, 19].

Ships are like industrial areas; therefore, the sailors live and work inside an industry. Here, hot works can be carried out, heat sources are generated, electricity is produced, flammable products are stored, dangerous goods of different types are transported such as LNG, LPG, oil, chemical products, explosives, and so on. In addition, "she moves"; we work in an industry that suffers an earthquake constantly at the mercy of the state of the sea.

Due to these circumstances, knowledge and fire training by crews in confined spaces is necessary, so that they can develop the necessary skills, so as not to make mistakes that could be fatal [20].

2 Objective

The objective of this study is to define a model course for fire intervention in confined spaces, with problems different in a ship than in a port facility. The priority aim is to increase the training and with it the safety of the personnel who form part of the extinguishing equipment onboard. The objectives that students will have to meet are:

- Knowledge of the nature of fire, fundamental to understanding what it does and how we use it. To apply that knowledge to understand the behavior and development of indoor fires.
- To recognize the signs that the fire sends us during the intervention to recognize possible extreme behaviors of the same and anticipate.
- To carry out a dynamic risk assessment.
- To apply the most appropriate extinguishing techniques, making correct use of water as an extinguishing agent.
- Maximize the safety of intervention teams.

In general, the objective is the safe and efficient control of fires in confined spaces, which have the potential to produce situations of flashover, backdraft, and smoke explosions.

3 Design of the Course: Firefighting in Confined Spaces

IMO precepts for maritime prevention courses are the basis for the design of the fire fighting course in confined spaces. The maximum number of students per course will not exceed 20 (OMI provides for this type of course, a teacher-instructor for every 10 students, in addition to STCW 2010, establishes that to be a teacher-instructor, they shall be officers with an appropriate training in instructional techniques and methods). Course duration will be at least 16 hours, divided into two days. Seven hours of theory in a fire laboratory plus nine hours of practice with real fire in a firefighting center. We have organized the theory and practice hours as follows.

- First day: three hours of theory, one hour of laboratory practices, and four hours of practice in the fire-training field.
- Second day: two hours of theory, one hour of laboratory practice, and five hours of practice in the fire-training field.

In Table 3, we show in detail (in minutes) the load of theory, laboratory practices, and firefighting training field practices necessary to achieve the objectives set in the course of fires in confined spaces.

3.1 *First Day of the Course: Theoretical and Practical Part*

Theory is taught on the first day; it consists of five units, which will be:

Theme 1 Combustion, characteristics of the elements of the triangle of fire. Combustion processes, chain reaction.
Theme 2 Pyrolysis. Fundamentals for the analysis of a fire, indicators, and symptoms.
Theme 3 Characteristics of the behavior of fire indicators.
Theme 4 Flashover. Development and characteristics.
Theme 5 The thermal environment.

At the end of the theoretical topics, the students will have reached knowledge, understanding, and sufficiency about:

How combustion occurs and how its processes develop.

- They will understand the process of pyrolysis and what are the gases resulting from it.
- They will know what are the indicators and symptoms that facilitate the interpretation of how fire can evolve.
- They will know how to differentiate and structure the phases of a fire and to know the characteristics of each of them and in what way an extreme fire behavior can derive.
- They will understand the hazard that thermal environment, in which the firefighting tasks develop, can have for the the human body.

Table 3 Course of firefighting in confined spaces

Competencies	Practical tests
1. Theory in classroom: 300 min	1.1 Combustion, characteristics of the elements of the triangle of fire. Combustion processes, chain reaction. 30 min
	1.2 Pyrolysis: Analysis of a fire, indicators and symptoms. 30 min
	1.3. Characteristics of the behavior of fire indicators. 30 min
	1.4 Flashover development and characteristics. 45 min
	1.5 Thermal environment. 45 min
	1.6 Development of the fire within a compartment. Influencing factors. 30 min
	1.7 Phases of the development of a fire in interior structures. Characteristics and evolution towards extreme behaviors. 30 min
	1.8 The importance of a fire controlled by ventilation. 30 min
	1.9. Backdraft and explosion of smoke, development, characteristics, and differences. 30 min
2. Laboratory practices and laboratory practices abroad: 120 min	2.1 Pyrolysis process of solid fuels, using specific flask, observing the whole process, distillation, products of combustion and inflammation of vapors. 30 min
	2.2 Visualization in simulated structure (Doll House), of the effects of a flashover, depending on the vapor–air mixture, influence of ventilation. 30 min
	2.3 Demonstration through simulated structure (Doll House), designed to the effect, of the appearance and effects of a flashover. 30 min
	2.4 Visualization of a backdraft and an explosion of smoke in an explosion chamber. Effects, consequences, and differences. 30 min
3. Practices in firefighting training field: 540 min	3.1 Demonstration of the effects of fumes and fire gases in model with a simple structure of two decks (plants). Display of heat transmission, and the spread of fire. Effects of ventilation. 80 min
	3.2 Effective use of water in indoor fires. Pulsation method. Offensive attack technique. 80 min
	3.3 Flashover of demonstration, development, indicators, and signs. Effects display. Assessment of the thermal environment. 80 min
	3.4 Flashover of gas control. Effective control of fire gases. 120 min
	3.5 Flashover attack. Application of offensive attack method to fire gases. 120 min
	3.6 Backdraft display in container. 60 min

The practices carried out in the laboratory will be:

Laboratory practice 1	Solid fuel pyrolysis process, using a specific flask, observing the whole process, distillation, combustion products, and vapor inflammation.
Laboratory practice 2	Visualization in simulated structure (Doll House) of the effects of a flashover, depending on the vapor–air mixture, the influence of ventilation.

The practices carried out in a fire-training field and laboratory will be:

Practice in the fire-training field 1	Demonstration of the effects of fumes and fire gases in a model with a simple structure of two decks (plants). Display of heat transmission, and the spread of fire. Effects of ventilation.
Field practice in fire 2	Effective use of water in indoor fires. Pulsation method. Offensive attack technique.
Practice in fire-training field 3	Flashover demonstration, development, indicators, and signs. Effects display. Assessment of the thermal environment.

At the end of the practices of the first day, the students will have acquired the necessary competences and skills:

- To evaluate the process of a fire in confined spaces, recognizing the processes of pyrolysis, heat transmission, and fire propagation.
- In the same way, they will learn to use water in confined spaces, correctly and efficiently, understanding the reason for the pulsation technique.
- To recognize the indicators and signs that inform us of the evolution of fire, having seen from within the process of fire from the start phase to the generation of a flashover, understanding its development and characteristics.
- To understand the importance of personal assurance equipment and its proper use in these circumstances of the thermal environment to which they have been exposed and where the heat to which we are subjected allows us to reach.

3.2 Second Day of the Course: Theoretical and Practical Part

On the second day of this course, we continue with the theory part, integrated by four units, which will be:

Theme 6	Development of the fire within a compartment. Influencing factors.
Theme 7	Phases of the development of a fire in interior structures. Characteristics and evolution towards extreme behaviors.

Theme 8	The importance of a fire controlled by ventilation.
Theme 9	Backdraft and explosion of fumes, development, characteristics, and differences.

At the end of these modules, students will be able to know and assess the stages of fire development in confined spaces as well as the influencing factors. In turn, they will have learned:

- To identify the development of the fire according to the height of the neutral plane, to recognize the symptoms of a flashover
- To understand the differences between fire controlled by fuel or controlled by ventilation
- To recognize the symptoms of a backdraft and assess the situation
- To differentiate between backdraft and smoke explosion.

The practices performed in the laboratory will be:

Laboratory practice 3	Demonstration using a simulated structure (Doll House) designed for the purpose of the appearance and effects of a flashover.
Laboratory practice 4	Visualization of a backdraft and an explosion of smoke in an explosion chamber. Effects, consequences, and differences.

The practices carried out in the fire-training field are:

Practice in fire-training field 4	Flashover of control of gases. Effective control of fire gases.
Practice in fire-training field 5	Flashover of attack. Application of offensive attack method to fire gases.
Firefighting training field practice 6	Backdraft display in container.

At the end of the practices of the second day, students will have acquired competences and skills in:

- The proper technique of applying water to the gas cushion.
- Control the height of the neutral plane and the reason for its oscillations.
- They will know the techniques of control of a flashover if it occurs and they will have skill in the handling of water jets, for the cooling of gases and fumes, knowing their effectiveness.
- They may apply the offensive attack technique inside confined spaces, safely and effectively. They will know how to determine and anticipate the appearance of a backdraft, knowing perfectly their indicators.

Evaluation of Knowledge and Competence

During the development of the course students will have verified their acquisition of knowledge by performing a written test of 20 questions with four possible answers, having to answer 75% of the questions correctly. Passing practical tests is mandatory, and must be completed. The practical evaluation will be done by direct observation of the student's capacity by the instructors evaluating the use and participation in all practice exercises.

4 Conclusions

The development of extreme fire behavior in a fire is a highly hazardous factor for both the ship and its crew, as they involve extreme heat and temperature conditions, which can affect the structural integrity of the vessel and spread with great ease.

Given the structural characteristics of a ship's construction, regarding materials and compartmentalization, the probability of fire emergencies is high.

The appearance of extreme fire behavior in a fire generates environments that are not compatible with life, so it is necessary to know how to recognize fire indicators because the speed with which fire can be produced and its consequences onboard make it a priority to anticipate the situation.

Seafarers, who are part of the firefighting equipment onboard, must receive specific and specialized firefighting training in confined spaces, in order to gain the capacity to react and face these situations safely.

The firefighting training of the seafarers, both officers and ratings, at present, is a formation with generalist contents, in which many topics are covered in a short space of time. The training module that we have designed complements the curricular design of a maritime safety officer, as well as the staff with responsibilities in firefighting on board.

References

1. Hetherington, C., Flin, R., Mearns, K.: Safety in shipping: the human element. J. Saf. Res. **37** (4), 401–411 (2006)
2. Tsui, A.S., Nifadkar, S.S., Ou, A.Y.: Cross-national, cross-cultural organizational behavior research: advances, gaps, and recommendations. J. Manag. **33**(3), 426–478 (2007)
3. O'Neil, W.A.: The human element in shipping. WMU J. Marit. Aff. **2**(2), 95–97 (2003)
4. Lois, P., Wang J., Wall A., Ruxton T.: Formal safety assessment of cruise ships. Tourism Manage. **25**(1), 93–109 (2004)
5. Håvold, J.I., Nesset, E.: From safety culture to safety orientation: validation and simplification of a safety orientation scale using a sample of seafarers working for Norwegian ship owners. Saf. Sci. **47**(3), 305–326 (2009)
6. Mejia, M.Q.: The STCW conference in Manila. WMU J. Marit. Aff. **9**(2), 231–234 (2010)
7. Thebault, L.: Maritime safety culture in Europe. Manag. Law **46**(1), 1–54 (2004)
8. Skjong, R., Soares, C.G.: Safety of maritime transportation. Reliab. Eng. Syst. Saf. **93**(9), 1289–1291 (2008)
9. Bielić, T., Zec, D.: Influence of ship technology and work organization on fatigue. Pomorski zbornik **42**(1), 263–276 (2004)
10. Lützhöft, M.: "The technology is great when it works": maritime technology and human integration on the ship's bridge (2004)

11. Organización Consultiva Marítima Intergubernamental. STCW incluidas las enmiendas de Manila de 2010 (2011)
12. Rodrigo de Larrucea, J.: Las enmiendas de Manila 2010 al Convenio STCW: Un nuevo perfil formativo para la Gente de Mar (2013)
13. Hickethier, A.F., Jia_Shen, H.: Developing cost effective STCW 2010 e-learning courses through international cooperation, a case study. Session A: Keynote Addresses and Status/ Trend of MET:113
14. Schröder-Hinrichs, J., Hollnagel, E., Baldauf, M., Hofmann, S., Kataria, A.: Maritime human factors and IMO policy. Marit. Policy Manag. **40**(3), 243–260 (2013)
15. Håvold, J.I.: Culture in maritime safety. Marit. Policy Manag. **27**(1), 79–88 (2000)
16. Soares, C.G., Teixeira, A.P.: Risk assessment in maritime transportation. Reliab. Eng. Syst. Saf. **74**(3), 299–309 (2001)
17. Grote, G., Kunzler, C.: Safety culture and its reflections in job and organisational design: total safety management. Int. Jo. Environ. Pollut. **6**(4), 618–631 (1996)
18. Rothblum, A.M.: Human error and marine safety. In: National Safety Council Congress and Expo. Orlando, FL (2000)
19. Yang, Z.L., Wang, J., Li, K.X.: Maritime safety analysis in retrospect. Marit. Policy Manag. **40**(3), 261–277 (2013)
20. Cox, S., Flin, R.: Safety culture: philosopher's stone or man of straw? Work & Stress **12**(3), 189–201 (1998)

Superstructure Design: Combination of Fiberglass Panel and Tubular Structure with Naval Steel Hull

Franklin Dominguez Ruiz and Luis Manuel Carral Couce

Abstract Traditionally, a cruising yacht's superstructure has been built using metallic materials such as steel or aluminum. Superstructures built using steel are heavy, need excessive maintenance and have limitations when building smooth shapes. Aluminum superstructures are lighter and need less maintenance, but still have limitations in building smooth shapes. In addition, aluminum superstructures are more expensive than steel ones. This paper proposes the use of fiberglass laminate composite as a material for superstructure construction and exemplifies the methodology used in the design stage and construction of a 36.80 m length cruising yacht with a steel hull. The methodology includes (1) scantling of structural steel elements and fiberglass panels used for composite lamination, (2) laminating procedure for the elements used for reinforcement to avoid deterioration of the steel structure, and (3) fire resistance recommendations for the composite laminate. The advantages of composite laminate are: structural weight reduction, fuel consumption reduction, apparent increase of wind area, superstructure maintenance cost reduction, and inert behavior with the steel structure that avoids galvanic corrosion.

Keywords Composite panels · Hybrid structures · Composite structure

1 Introduction

An active research area in shipbuilding is weight optimization without sacrificing structural safety following the requirements of classification societies. Typically, superstructure weight represents approximately 40% of the total structural weight of the vessel; therefore, the method for reducing structural weight is important.

F. Dominguez Ruiz (✉)
Escuela Superior Politécnica del Litoral, Guayaquil, Ecuador
e-mail: jodoming@espol.edu.ec

F. Dominguez Ruiz · L. M. Carral Couce
Universidade da Coruña, A Coruña, Spain
e-mail: lcarral@udc.es

A. Vega Sáenz et al. (eds.), *Proceedings of the 25th Pan-American Conference of Naval Engineering—COPINAVAL*, https://doi.org/10.1007/978-3-319-89812-4_8

Reinforced fiberglass has mechanical properties that make it appropriate for use in the shipbuilding industry and allow weight reduction.

In general, passenger yacht superstructures need to be large because of the comfort that is needed. Although different materials are used for its construction, steel is the most common. However, two important disadvantages have been reported, namely the material is extremely heavy and maintenance costs are high. Another alternative is aluminum that allows weight reduction, but it is expensive. In the last years, shipbuilders started using reinforced fiberglass panels to build superstructures, reducing structural weight and making construction finishing easier. Also, composite laminates require less maintenance, reduce fuel consumption, increase apparent wind area, and inert behavior with the steel structure. The main drawback of reinforced fiberglass construction is the dissimilar join between the naval grade steel hull and the fiberglass panels because they are different materials. In this study, the yacht superstructure construction using fiberglass panels and tubular steel structure is described.

The project aim was to design a superstructure for a modified passenger yacht whose main hull dimensions were widened and lengthened; see Fig. 1. Originally, the hull was built using steel and its superstructure using aluminum. The new hull dimensions (see Table 1) require a bigger superstructure to be built without increasing superstructure weight, and fiberglass panels with tubular structure combination were chosen considering the advantages described above.

The weight reduction in the superstructure was assessed considering four options where different materials were combined:

1. Fiberglass composite panels and tubular structure, construction made in 2005.
2. Panels and structure using naval grade steel.
3. Fiberglass composite panels and fiberglass structures.
4. Fiberglass composite panels and tubular structures on first deck and fiberglass composite panels and fiberglass structures on upper deck.

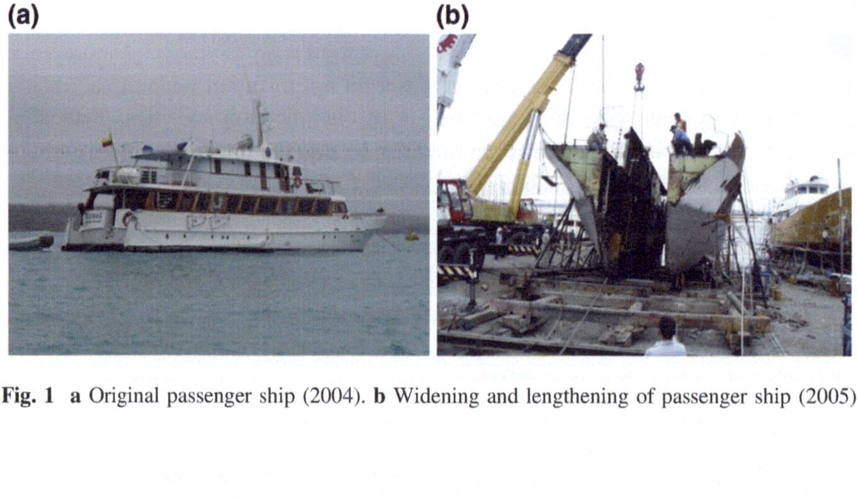

Fig. 1 **a** Original passenger ship (2004). **b** Widening and lengthening of passenger ship (2005)

Table 1 Study case: Main ship passenger yacht data

Ship type	–	Monohull
Service	–	Passenger
Length	m	36.80
Length between perpendiculars	m	30.09
Length to waterline	m	32.27
Depth	m	3.35
Breadth	m	7.70
Draft	m	1.75
Speed	knot	12
Displacement	ton	201.5
Block coefficient	–	0.4709

In each case, scantling calculation was first made using the Special Service Craft–Lloyd's Register (SSC) software; hence the requirements of the Lloyd's Register Classification Society [4] were met. Following, structural analysis was performed using the finite element method (FEM) to verify that the superstructure stress was under the yield point/stress of the given material. Finally, structural weight is compared between options.

2 State of the Art of the Composite Materials Used in the Superstructure

Most of the research on hybrid (steel panels with composite panels) materials focuses on developing panels with high structural strength and low structural weight. For example, Suarez and Herreros [7] developed a hybrid material called MALECON, which consists of metal sheets with reinforced fiberglass lamination. This material manufactured using the infusion method achieved the same strength and properties as steel, but with less density. Shkolnokov [6] proposed a hybrid bonding between steel and composite panels by means of metal sheets. The study focuses on military ship and submarine applications. Ozes and Neser [5] performed experimental tests of hybrid bonding between steel and reinforced fiberglass for different steel roughness. The results reveal that the surface roughness of steel has a significant effect on the bonding performance of steel to FRP combinations and the performance of the resin can be improved by using the primer in an economical way. Boyd et al. [1] performed a local fatigue analysis on a composite steel panel to study the joint between the deck and the superstructure side of a frigate. Concluding that damage caused by plane fatigue loading has little influence on the ultimate failure load of the joint, Klopfer [3] proposed a panel composed of layers of fiberglass and a mesh of metallic wire, which allows welding the metal structure. This join is unreliable for the superstructure and deck. Finally, based on this research, Lloyd's Register offers standards and regulations for the calculation and construction of ships for steel, aluminum, and composite materials [4].

3 Literature Review of the Composite Materials for Ship Superstructure Construction

3.1 Scantling of Fiberglass Panel

Fiberglass panels are dimensionalized using Lloyd's Register standards [4]. In the calculation, the critical layer concept shown in Eq. (1) is used to estimate stresses generated in each lamination layer and to verify that stresses do not exceed the limiting stress factor:

$$\sigma_{\mathrm{ti}} = \frac{E_{\mathrm{ti}} y_i M}{\sum (E_i I_i)} \tag{1}$$

$$\text{Limited Stress Factor} = \frac{\sigma_{\mathrm{ti}}}{\sigma_u}$$

where σ_{ti} is the maximum tensile stress within ply in N/mm^2, σ_u is the ultimate tensile strength of the laminate in N/mm^2, E_{ti} is the tensile modulus of individual ply in N/mm^2, y_i is the distance from the neutral axis to the outer extremity of an individual ply, *M* the bending moment in N*m given by Eq. (2), and *EI* is the total stiffness of the panel.

$$M = \frac{k \cdot p \cdot b}{12} \tag{2}$$

where $k = \frac{\gamma^3 + 1}{\gamma + 1}$, $\gamma = \frac{b_w}{b}$.

The relationship between maximum tensile stresses and ultimate tensile strength should not exceed the values of the limit stress fraction provided by Lloyd's Register [4]. Other parameters assessed in the composite panels are deflection using Eq. (3) and the amount of fiber used in each lamination:

$$\delta = \frac{p \cdot b}{8} \cdot \left(\frac{b \cdot k_{\mathrm{db}}}{48 \cdot D_s} + \frac{k_{\mathrm{ds}}}{G \cdot t_c} \right) \times 10^{-3} \tag{3}$$

where $k_{\mathrm{db}} = 1.5 - \frac{1}{A_R}$, $k_{\mathrm{ds}} = 1.2 - \frac{0.6}{A_R}$

$$D_s = \frac{E_{\mathrm{pi}} \cdot t_i - E_{\mathrm{po}} \cdot t_o}{E_{\mathrm{pi}} \cdot t_i + E_{\mathrm{po}} \cdot t_o} (t_c + t_s)$$

When p is the design pressure in kN/m^2, b is the unsupported panel breadth in mm, D_s is flexural rigidity of the sandwich panel per unit in N-mm, k_{db} is the bending deflection aspect ratio factor, k_{ds} is the shear deflection aspect ratio factor, A_R is the aspect ratio, G is the shear modulus of sandwich core material in N/mm^2, t_c is the core thickness in mm, t_s is the mean skin thickness in mm, t_i and t_o are the thickness of the inner and outer panels in mm, and E_{pi} and E_{po} are the inner and outer skin compressive modulus.

3.2 *Composite Panels with Equivalent Tubular Structure*

We make an adaptation to the equation of equivalent stiffness for the scantlings of tubular steel structures. The critical layer criterion shown in Eq. (1) between laminates is maintained. The objective is to obtain the minimum inertia and rigidity modulus values to meet the required minimum deflection. This deflection is expressed considering the relationship between effective span length and deflection l_e/δ_S, and should not be less than the limit values given by Lloyd's Register [4]. The bending moment has been calculated with Eq. (4):

$$M_s = \frac{s \cdot l_e \cdot p}{12} \tag{4}$$

And deflection is calculated by Eq. (5):

$$\delta_s = \frac{\varphi_s \cdot p \cdot s \cdot l_e^4}{(EI)_s} \times 10^5 \tag{5}$$

When s is the stiffener spacing in mm, l_e is the effective span length of stiffener in m, φ_s is the shear force coefficient, and $(EI)_S$ is the total equivalent stiffness of the panel with a tubular structure.

For the calculation of the equivalent stiffness, estimation of moments is made considering the inertia of the tubular structure (EI_{Tube}), the width of the attached plate (E_{IPL}), the boundary bonding (E_{IB}), and the respective tubular structure fiberglass skin (EI_{LT}). Generally, two layers of Mat 375 are used for lamination of the structure, without considering filler contribution.

$$(EI)_S = EI_{\mathrm{PL}} + EI_{\mathrm{Tube}} + E_{\mathrm{IB}} + EI_{\mathrm{LT}} \tag{6}$$

Due to the criterion of the critical layer applied in Eq. (1), the pipe diameter is modified up to the required structure strength. When greater inertia is required, a double tube with or without separation is used. Figure 2 shows the types of tubular structure used in the present analysis: single tube, double tube, and double tube with separation. The bond between the panel and the tube is made with filler prepared with microspheres, subsequently to perform the tubular lamination to the panel.

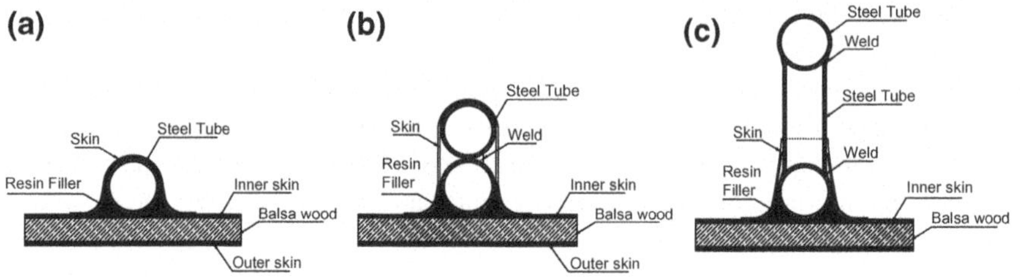

Fig. 2 Structural detail, connection between fiberglass panel and tubular structure: **a** single tube, **b** double tube, **c** double tube with separation

To join the side composite panels to the steel structure, a flat bar is welded to the deck plate and the side panels bolted to this flat bar as shown in Fig. 3. The bolts are made of AISI 316-M9 steel and should be spaced 30 cm apart. For fixing dissimilar bonding between the composite panel and a metal surface, it is recommended to use a welder tube of diameter 3/4 in. at 25 cm from the deck.

To increase the lifetime of composite panels and tubular structures, it is recommended:

- The wooden core is balsa and should be sealed, in order to avoid humidity and termites.
- The fiber layers should be a suitable laminate.

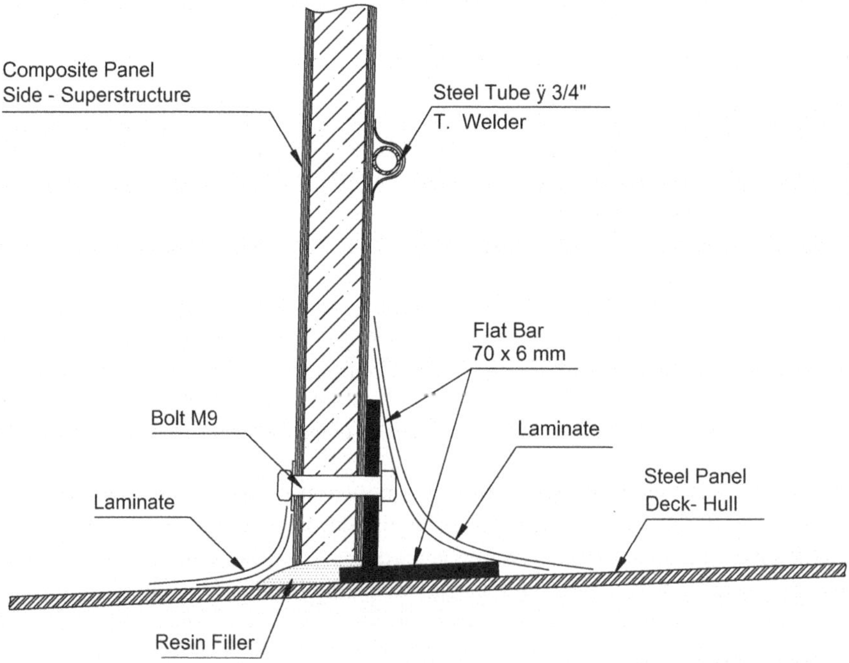

Fig. 3 Structural detail, connection between deck panels (steel) and superstructure composite

- Laminate glass content should be 30% for Matt layers and 50% for woven roving layers.
- The calamine should be removed by SA 2-1/2 grade blasting and should be painted with anticorrosive epoxy polyamine in all steel surfaces where the composite panel will be laminated.
- Steel tubes should be fully welded and painted with anticorrosive epoxy polyamine.
- Steel tubes should be completely sealed to prevent inside corrosion.
- All tubular structure components in the superstructure should be completely laminated to avoid corrosion, using orthophthalic resin with MEK peroxide catalyst.
- Composite panels should be fire retardant. For this, 15% of alumina trihydrate should be added in the gelcoat and in the first laminate layer (Matt) to give a fire retardant property.

4 Proposal for Construction of Superstructure Using Composite Panels

4.1 Proposed Scantling

The proposed lamination plan for Option No 4 is detailed below. The composite superstructure consists of fiberglass reinforced panels with fiberglass type E and balsa core of 144 kg/m^3-H: 7%. The proposed laminate is presented in Table 2.

The scantling of the tubular structure is presented in Table 3. For the sides of level 01 and the floor of deck 02, it consists of ASTM A53 steel tubes, seamless, sch40. For level 02, it consists of a fiberglass structure with omega profiles of a single laminate.

Table 2 Panel lamination schedule proposed for superstructure, option no. 4

	Thickness (mm)	Laminate[a]
Side-01 Dk	33.85	M3M4W8M4-C25-M4W8M4M3
Bulkheads aft-01 Dk	25.70	M3M4W4M4-C18-M4W4M4M3
Deck-02 Dk	25.70	M3M4W4M4-C18-M4W4M4M3
Side-02 Dk	26.28	M3M4W6M4-C18-M4W6M4M3
Bulkheads aft-02 Dk	25.70	M3M4W4M4-C18-M4W4M4M3
Deck-03 Dk	27.16	M3M4M4W6M4-C18-M4W4M4M3
Bulwark-03 Dk	26.28	M3M4W6M4-C18-M4W6M4M3

[a]M3 = Matt 300 g/m^2, M37 = Matt 375 g/m^2, M4 = Matt 450 g/m^2, W8 = WR 800 g/m^2, W4 = WR 400 g/m^2, W6 = WR 600 g/m^2, *C* Core thickness

Table 3 Lamination plan for steel tubular and fiberglass structures: Proposed superstructure, option no. 4

	Structural			Span (m)
	Tubular	Fiberglass	Laminate[a]	
Frames side-01 Dk	2xΦ2″	–	M37M37	1.70
T. welder—side-01 Dk	Φ3/4″	–	M37M37	0.25
Bulkhead stiffener-01 Dk	Φ2″	–	M37M37	1.60
Deck transverse beam-02 Dk	2xΦ2″@ 80 mm	–	M37M37	1.70
Deck longitudinal-02 Dk	Φ2″	–	M37M37	1.10
Frames side-02 Dk	–	H125	M4W6M4W6M4	1.20
Deck transverse beam-03 Dk	–	H150	M4W8M4W8M4W8M4	1.20
Deck longitudinal-03 Dk	–	H125	M4W6M4W6M4	1.10

[a]M3 = Matt 300 g/m^2, M37 = Matt 375 g/m^2, M4 = Matt 450 g/m^2, W8 = WR 800 g/m^2, W4 = WR 400 g/m^2, W6 = WR 600 g/m^2, *C* Core thickness

4.2 Validation Using Finite Element Method

The finite element method is used to validate the proposed scantling. Shell and beam elements are used with quadrilateral and triangular mesh. The beam element has autodivisions of 50 mm length, in such a way that it coincides with the shell mesh. For stress analysis, the hull bottom surface is assumed as supported with spring type supports equivalent to the ship displacement and simply supported supports on the central keel, whereas in forced vibration analysis only spring type supports are used. Vibratory forces are applied in the propeller for the forced vibration analysis; the propeller includes added masses and damping. This calculation is presented in detail in Dominguez [2].

5 Results

Table 4 shows the weight comparison between the different superstructure options. As expected, the largest weight corresponds to the steel superstructure with 16.89 tons, and the composite structure of fiberglass and tubular structure is 9.79 tons.

In Fig. 4, the equivalent stress results obtained with SSC–Lloyd's Register design pressures are shown.

Stress values in MPa for the proposed superstructure options are shown in Table 5. It can be seen that the stress values in the panels are lower than the maximum stress values recommended by Lloyd's.

Table 4 Comparison of structural weight in the different superstructure options

Option	Superstructure	Weight (ton)	% Respect to the Steel Superstructure
1	Original	13.39	79
2	Steel	16.89	100
3	Fiberglass	9.09	54
4	Composite (Proposed)	9.79	57

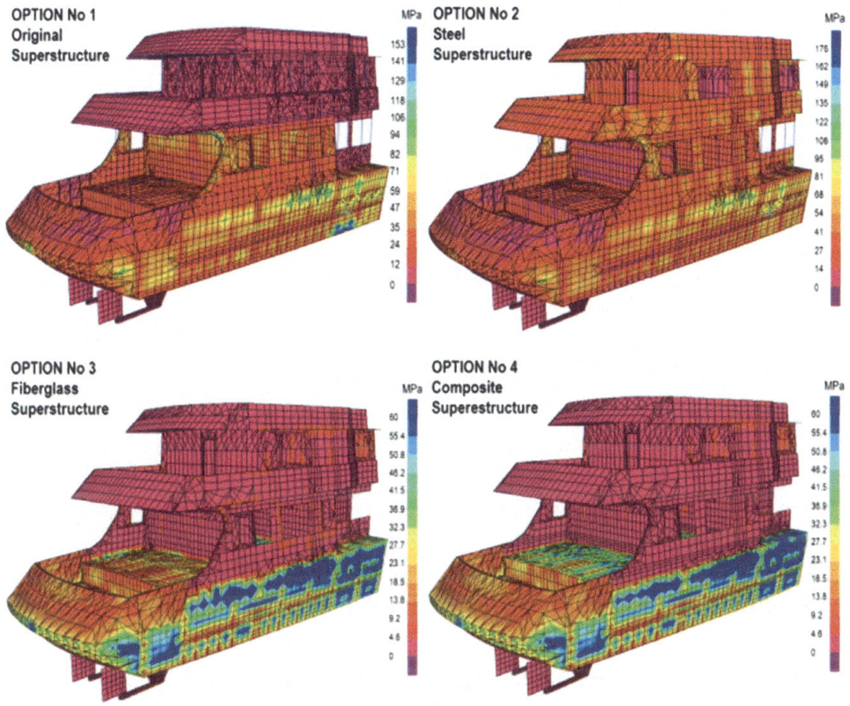

Fig. 4 Results of equivalent stress using finite elements for the options

6 Cost Estimation

The cost estimation for superstructure options is presented in Table 6 and it was estimated considering material and labor costs.

The less expensive construction is the steel superstructure with an approximate cost of $100,127.62 whereas superstructure option No. 4 (composite) has a cost of $155,303.06. The composite superstructure is more expensive because it includes mold cost and needs more time for surface finishing.

Table 5 FEM superstructure stress

Stress (MPa)	Equivalent stress in superstructure							
	Option no. 1		Option no. 2		Option no. 3		Option no. 4	
	σ_{eq}	$\sigma_{Lm\text{-}LR}$	σ_{eq}	$\sigma_{Lm\text{-}LR}$	σ_{eq}	$\sigma_{Lm\text{-}LR}$	σ_{eq}	$\sigma_{Lm\text{-}LR}$
Side-01 Dk	0.83	56.16	43.81	153	8.80	61.88	7.58	61.88
Deck-02 Dk	2.13	65.28	53.40	176	5.00	64.09	5.06	64.09
Side-02 Dk	4.00	56.16	59.37	176	5.56	59.69	4.94	59.69
Bulwark-02 Dk	2.59	56.16	51.54	176	3.90	59.69	3.50	59.69
Bulwark aft-02 Dk	1.02	56.16	31.43	176	2.60	59.69	2.80	59.69
Divisions-02 Dk	2.60	56.16	29.92	176	3.60	58.26	3.60	58.26
Inner hallways-02 Dk	4.46	56.16	3.24	176	6.40	58.26	6.40	58.26
Deck-03 Dk	5.18	61.77	73.90	176	5.68	65.82	5.68	65.82
Bulwark-03 Dk	1.90	59.35	33.85	176	12.15	59.69	12.15	59.69
Bulwark aft-03 Dk	1.30	56.16	17.05	176	3.95	59.69	3.95	59.69

Table 6 Comparison of construction cost of superstructure options

Option		Costs ($)				$/kg
		Panel	Structure		Total	
			Fiberglass	Steel		
1	Original	$ 75,564.14	$ –	$ 44,118.60	$ 119,682.74	8.94
2	Steel	$ 70,765.86	$ –	$ 29,361.77	$ 100,127.62	5.93
3	Fiberglass	$ 125,043.75	$ 41,681.25	$ –	$ 166,725.00	18.34
4	Composite	$ 118,374.75	$ 25,623.84	$ 11,304.47	$ 155,303.06	16.81

7 Comparative Analysis of Options

(1) The superstructure with fiberglass panels (Options 3 and 4) allows more spacing between stiffeners than the steel panel.
(2) The fiberglass superstructure can be molded into any shape and allows adding extra layers in areas of high stress.
(3) The equivalent stress in the fiberglass panels has low values, due to the considerations done in the scantling, to comply with the minimum fiberglass content and criteria of critical layer.
(4) The superstructure proposed option No. 4 (composite) shows 43% weight reduction compared to the steel structure.
(5) The superstructure option No. 3 (total fiberglass) is not recommended despite having a lighter weight, due to the limitation of the dissimilar bond between the steel deck and the fiberglass panels.

(6) The superstructure proposed option No. 4 (composite) requires less maintenance than steel structures, provided all proposed lamination recommendations are accomplished.
(7) The cost of building a composite superstructure is 283% more expensive than steel.

8 Conclusions and Recommendations

Superstructure proposed option 4: The composite superstructures proposed have greater benefits than other materials, but a higher construction cost. This option has been used on yachts for 11 years without reporting structural problems. It is important that classification societies' rules for building with composite materials are met to avoid structural problems. Also, it is strongly recommended to maintain adequate quality control and use proper care in the preparation and handling of the chemical process involved in the build.

References

1. Boyd, S., Blake, J., Shenoi, R., Kapadia, A.: Integrity of hybrid steel to composite joints for marine application. Eng. Marit. Environ. Part M **218**, 235–246 (2014)
2. Dominguez, F., Cali, E., Garcia L.: Forced vibration analysis of the hull girder by propellerexcitation and rudder interation, Progress in the Analysis and Design of Marine Structures, Marstruct, Lisbon, Portugal (2017), Print ISBN: 978-1-138-06907-7; eBook ISBN: 978-1-351-65341-1, https://doi.org/10.1201/9781315157368-11
3. Klopfer, J.: An experimental study of fiberglass composites containing metal wire joints. Thesis, Naval Postgraduate School, California (2009)
4. Lloyds Register: Hull Construction in Composite. Rules and Regulations for the Classification of Special Service Craft, July, 2016
5. Ozes, C., Neser, N.: Experimental Study on steel to FRP bonded lap joints in marine applications. Dokuz Eylul University, Hindawi Publishing Corporation, Turkey (2015)
6. Shkolnikov, V.: Hybrid ship hulls, engineering design rationales. Butterworth-Heinemann, New York, USA (2014)
7. Suárez, J., Herreros, M.: New Fiber-Metal Hybrid Laminated Material (MALECON). Naval Architecture and Shipbuilding Department, Polytechnic de Madrid, España

Methodology for Improvement of the Hydrodynamic Efficiency of an Amazon School Boat Utilizing a CFD Tool

Breno Farias da Silva, Fernando Costa da Cruz, Harlysson Wheiny Silva Maia, Toshi-Ichi Tachibana, Vitor Hugo Macedo Cardoso and Yuri Victor Remígio Guedes

Abstract The Solimões—Amazonas waterway is mainly composed of the Solimões and Amazonas Rivers, along with several tributaries of both, forming a network of almost 7000 km in length. Due to this, the waterway modal stands out in the transportation of passengers and cargo in the Amazon region. In sight of this scenario, the National Fund for Development of Education implemented in the region the "School Boat" project, originating from the Caminhos da Escola (School Path) program, whose objective is to provide a faster, safer, and more efficient transportation of students from the riverside regions to the schools. However, the medium boat model was the subject of complaints from users about the long duration of the journey. In order to analyze the hydrodynamic characteristics of the medium boat, the full-scale sea trials performed showed high residuary resistance values. Hence, it was noted the generation of a large amplitude wave train and a high bow wave height compared with the boat's depth, implied a high fuel consumption for speeds that are incompatible with the design guidelines. These characteristics corroborate the low hydrodynamic efficiency of the hull. Based on this problem, the proposal of this work consists in the analysis of a numeric model for the medium school boat hull, based in CFD, correlated with the full-scale sea

B. F. da Silva · F. C. da Cruz · H. W. S. Maia (✉) · V. H. M. Cardoso
Y. V. R. Guedes
Federal University of Pará, Belém, PA, Brazil
e-mail: harlysson.maia@itec.ufpa.br

F. C. da Cruz
e-mail: fernando.cruz@itec.ufpa.br

V. H. M. Cardoso
e-mail: vitor.cardoso@itec.ufpa.br

T.-I. Tachibana
Polytechnic School of the University of São Paulo, São Paulo, SP, Brazil
e-mail: tatibana@usp.br

A. Vega Sáenz et al. (eds.), *Proceedings of the 25th Pan-American Conference of Naval Engineering—COPINAVAL*, https://doi.org/10.1007/978-3-319-89812-4_9

trials. In this scenario, the hull lines are modified and simulated, in comparison with the original model, in order to reduce the total resistance to advance of the vessel.

Keywords Hydrodynamic · Simulation · CFD

1 Introduction

This work was conceived with the objective of producing a consistent solution to the problem of hydrodynamic efficiency of the hull shape of a riverboat. The studied hull model has specific problems to reach the design speed (11 knots). Based on this, the main objective of this work is to increase the hydrodynamic efficiency of this hull within a range of operating speeds.

In general, the problem is related to the shape of the bow and the keel region. These regions hinder the flow of the incident fluid, the generation of hydrodynamic lift for entering into the planing regime, and still facilitate the generation of bow waves within the displacement regime. In this case, the elaboration of evaluation criteria to identify discontinuities in the hull is necessary, as the design velocity is not reached. Thus, the proposed approach is the application of tools in a computational platform, interconnected to the real-scale tests of the boat.

The speed of service and the wave train profile of the current hull form were determined by sea trials. Based on the geometric characteristics and initial conditions of the sea trial, it was possible to define the computational model of the hull geometry as well as the fluid domain, along with turbulence conditions and the phase model. Under these conditions two numerical hull models were created, one representing the current model and another with proposed geometric variations of the first.

2 School Transportation in the Amazon Region

The proposed hull model has a shape with displacement characteristics, emphasizing that the hull structure is composed entirely of naval steel with sufficient tensile strength to maintain the structural integrity with respect to the forces from the external environment of the vessel, as shown in Figs. 2 and 3.

Judging the inconsistencies with the hull shape criteria as well as the conditions to reach the design speed of the vessel, the need to remodel the hydrodynamic behavior of the hull becomes clear, keeping the design guidelines fixed and the main dimensions unchanged.

For the computational analysis method, it is necessary to validate the shape of the model through computational fluid dynamics (CFD). Table 1 gives the main characteristics of the vessel studied in this work.

Table 1 Main characteristics of the school boat

Characteristic	Dimension
Length Overall (LOA)	7.80 m
Length between Perpendiculars (LPP)	7.52 m
Molded breadth	2.55 m
Molded depth	1.40 m
Design draught	0.67 m
Design speed	20 km/h
Capacity	20 passengers + 1 crewman

Source National Fund for Development of Education (FNDE)

3 Literature Review

With the development of computer-aided design (CAD) and CFD tools, several processes for measuring and analyzing issues regarding the area of naval hydrodynamics can be studied with greater accuracy [2–4]. Theoretical models used as the basis for this work are presented below.

3.1 Experimental Method on Ship Hydrodynamics

According to Tachibana [4], from the assumption that the flow around the hull of a ship is three-dimensional, the process can be analyzed from the individualization of each resistance component. By this means, an adequacy is identified in the area of flows around hull forms, within the area of ship hydrodynamics.

Most of the "pure" resistance components and the existing intersections do not have satisfactory experimental means of verification, being subjected to imprecise numerical measurement and extrapolation processes, with a tradition in experimental hydrodynamics within the area of determination of resistance components.

3.2 Resistance to Forward Motion

The movement of a vessel through a fluid (water), at constant speed, generates two types of forces against the hull, normal and tangential. The force against the movement of the vessel is called resistance to forward motion [1].

The interaction between fluid and structure results in generation and destruction of waves, and this demanded energy composes the wave resistance. A schematic diagram with detailed components of the resistance to forward motion is presented in Fig. 1, based on Tachibana [4].

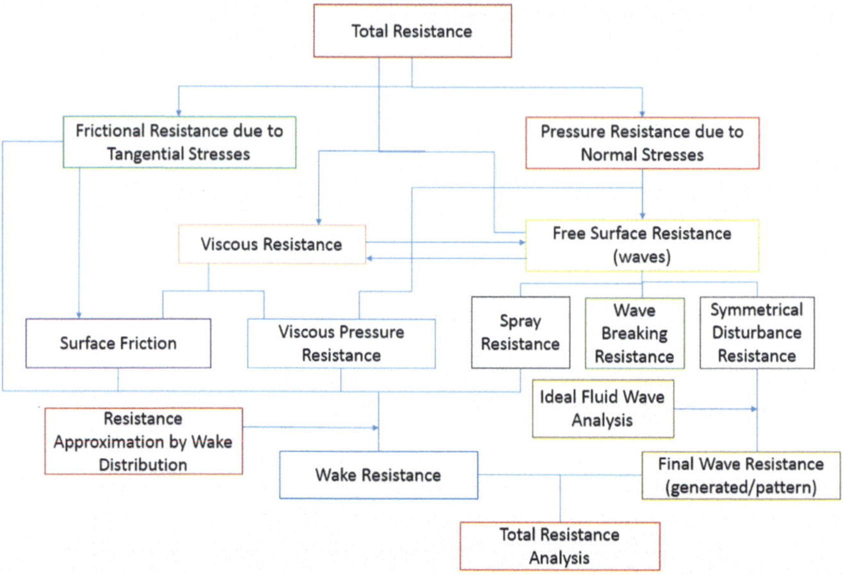

Fig. 1 Schematic diagram of the components of the resistance to forward motion

Due to the no-slip condition of viscous fluids on solid surfaces, when the vessel moves through a fluid at rest, the particles close to the hull tend to adhere to the surface, acquiring the speed of the vessel. The integral of the friction components on the wetted surface of the hull results in the frictional resistance [1].

3.3 CFD Analysis of Vessels

According to Iervolino [5], the computational fluid dynamics is characterized by a numerical simulation of any physical or chemical process where a fluid flow happens. The application of boundary conditions on several finite element models can be recorded on the time and space domain, based on the behavior of several parameters of the flow itself in each of these elements.

Within this procedure, the separation of the model into three main stages is highlighted: preprocessing, solution, and postprocessing. In the preprocessing stage we define the flow model and the boundary conditions of the problem. In the solution step, the resolution procedures are used to determine the desired results. Finally, the postprocessing phase allows the analysis of the results with illustrative graphics.

4 Methodology

4.1 Computational Modeling of the Medium Boat Hull

The full-scale model of the medium boat used on the sea trials is illustrated in Fig. 2. The three-dimensional model was generated from the lines plan provided by the manufacturer.

The profile view of the medium school boat hull and the model with the proposed shape modifications are illustrated in Fig. 3.

The modified hull shape was generated based on the original hull, by removing the side keels at the bow and modifying the flat bottom area to a V-shaped keel, in order to attenuate the flow lines by the hull.

A view of the transverse sections of the bow, displaying the proposed shape modifications is illustrated in Fig. 4.

In Fig. 5, the three-dimensional hull models are illustrated, as exported from the 3D CAD software Rhinoceros 5.0.

Fig. 2 Longitudinal view of the medium school boat

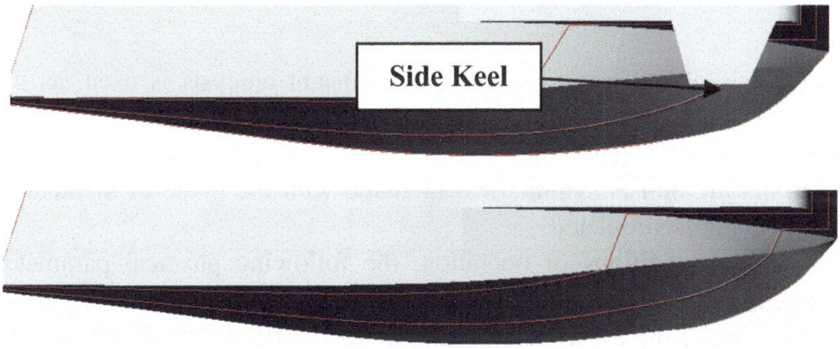

Fig. 3 Profile view of the unchanged hull with side keel (above) and the modified hull (below)

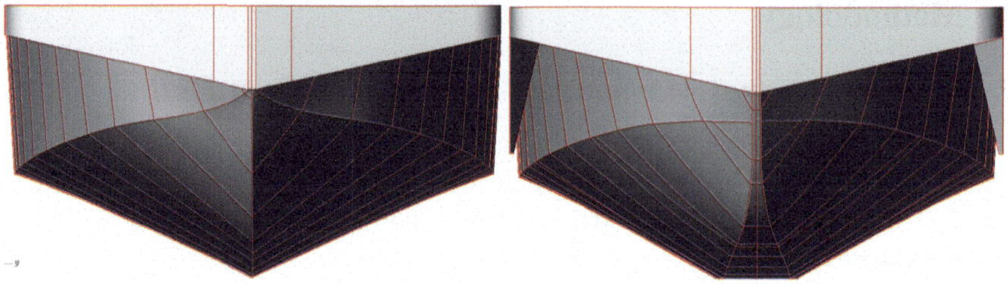

Fig. 4 View of the transverse sections of the bow. Left: the modified hull shape, and right: the unchanged hull of the medium school boat

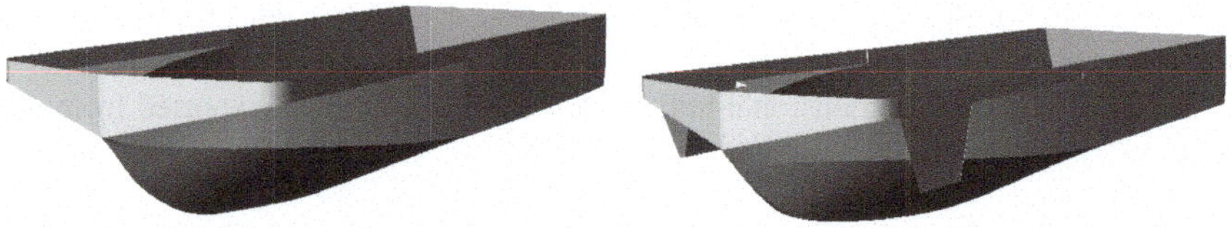

Fig. 5 Three-dimensional hull model with the proposed modifications, with a V-shaped bottom keel on the left, and the unchanged medium school boat hull on the right

With the three-dimensional hull models modeled, the next phase is to define the fluid control volume that the water and air will occupy around the boat in the simulation. The dimensions of the fluid control volume are illustrated in Fig. 6, related to the overall length of the boat, defined as L.

For the criterion analysis, it is important to analyze the wave resistance generated in the case, using a resistance model that simulates the interaction of the hull discontinuities effects on the generation of the wave train, taking into account the vorticity formation in abrupt shape changes in the hull.

4.2 Physical Model Considerations

For the analysis of the wave formation, a transient analysis is used, as the hull geometry is complex and the calculation model used in the permanent analysis generates numerical inconsistencies that must be taken into account. The section that represents the hull is symmetrical in shape with the plane of symmetry, thus only the port side is simulated.

Into the case conditions of operation, the following physical parameters are defined: the gravitational acceleration: $g_z = -9.81\,\text{m/s}^2$, the density of operation: $\rho_{op} = 1.225\,\text{kg/m}^3$, the pressure of operation: $p_{op} = 101,325\,\text{Pa}$, and the reference position of pressure: $z_{ref} = L$, where L is the overall length of the hull. For the multiphase model, the volume of fluid method is used, considering the flow along

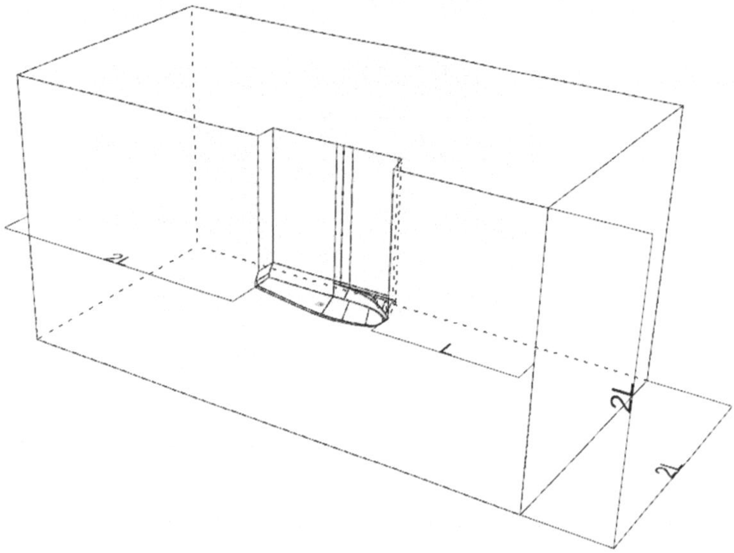

Fig. 6 Dimensions of the fluid control volume surrounding the boat hull, related to the overall length of the boat, L. The control volume is a box with dimensions $4L \times 2L \times 2L$ (port)

the control volume an open channel flow type, and the implicit body force model for force delimitation.

The turbulence model used is the SST K-ω and the Flat method is used for the phase initiation.

5 Results

The definition of the parts of the control volume was adjusted to the model of calculation used by the software Ansys Fluent. The surfaces of the control volume were named according to the regions of analysis. The water inlet represents a cross-section of the control volume, namely the entrance of the fluid flow (air + water) for the analysis. The water outlet corresponds to a cross-section of the control volume where the fluid flow (air + water) exits the channel in the analysis. The symmetry section delimits a symmetry plane for half the control volume. The walls that surround the control volume are defined by a no-slip condition, thus the model has reliability to measure the degree of resistance.

On Fig. 7 it is possible to check the details of the mesh used for the computational simulation in Fluent.

The orthogonal quality of the mesh can be verified, highlighted in the legend. No refinement tool was used around the hull. The mesh was generated with a minimum element size on the order of 5 cm. A total of 452,461 nodes and a total number of 2,577,166 elements were generated.

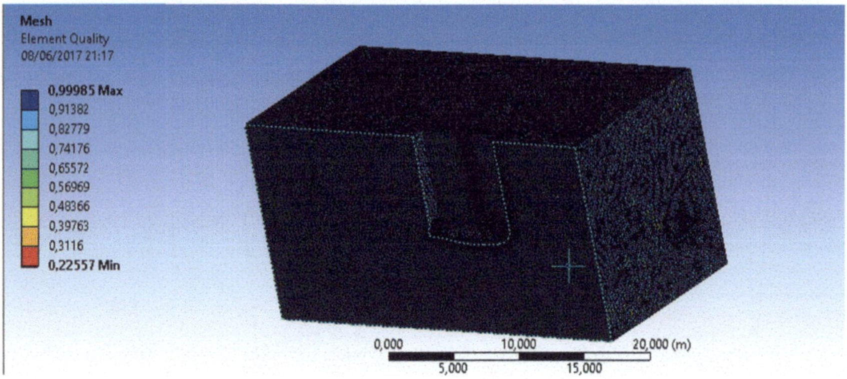

Fig. 7 Indication of the mesh elements present in the fluid control volume around the hull, with the legend indicating the orthogonal quality of the mesh

According to the values indicated in Fig. 7, it can be verified that about 2,096,942 elements presented an orthogonal quality of over 0.75, and 1,510,942 elements presented an orthogonal quality of over 0.85. The quality of the mesh did not reach higher values because the model was based on tetrahedral elements with a random method for filling the control volume.

The angle of curvature used corresponded to 18°, with a minimum gap size of 1.5 cm and a maximum element size of 30 cm. A growth rate of 1.4 was used for the elements of the hull region and for the outermost regions of the control volume.

For the calculation of 1500 iterations of the computational model, a total of 30 iterations per timestep was defined, resulting in 50 timesteps for each velocity. The timestep duration was defined as 0.02 s. The restriction degree used for the residual convergence was on the order of 10^{-5}. Figure 8 illustrates the longitudinal profile of the medium school boat with the free surface wave profile for a velocity of 20 km/h.

The wave elevation in the region of the side keel is illustrated in Fig. 9. The irregularities in the surface of the hull can be noticed due to the mesh quality used in this region.

Based on the results of the three analyses (sea trials and both numerical hull models), the methodology was validated, and based on the proposed hull shape modifications, a maximum reduction of the total resistance of about 41% for a velocity of 25 km/h was reached. Figure 10 illustrates the progression of the total resistance for the range of velocities simulated for both hulls, the medium school boat and the model with the proposed modifications.

Figure 11 illustrates the percentage reduction of resistance for each velocity in the simulated range, varying from 8 to 25 km/h.

The largest reduction rate was identified on the velocity of 25 km/h, as seen in Fig. 11. For this velocity, the corresponding value for the Froude number is equal to

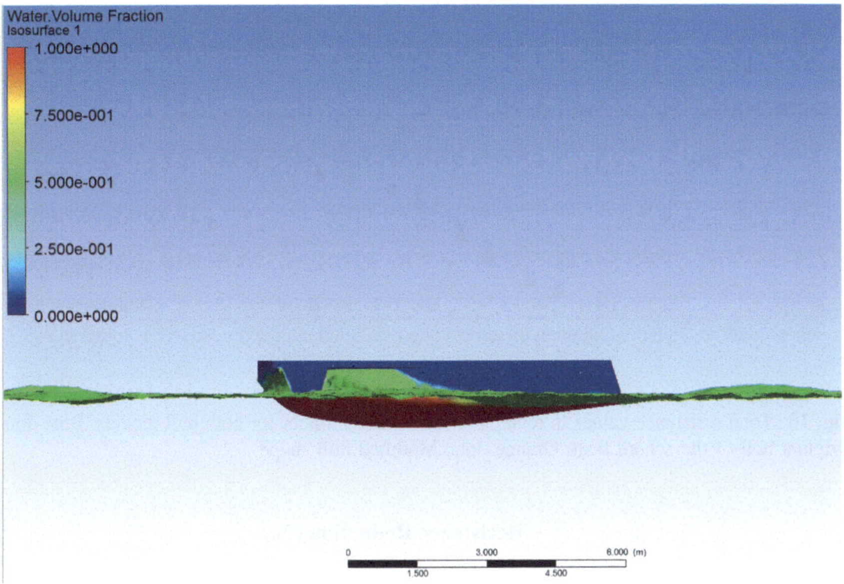

Fig. 8 Free surface wave profile at 20 km/h

Fig. 9 Wave elevation at the bow region

0.794. Relating the Froude number value with the large resistance reduction achieved, a change on the planing mode of the proposed hull is identified. The absolute resistance values are presented in Table 2.

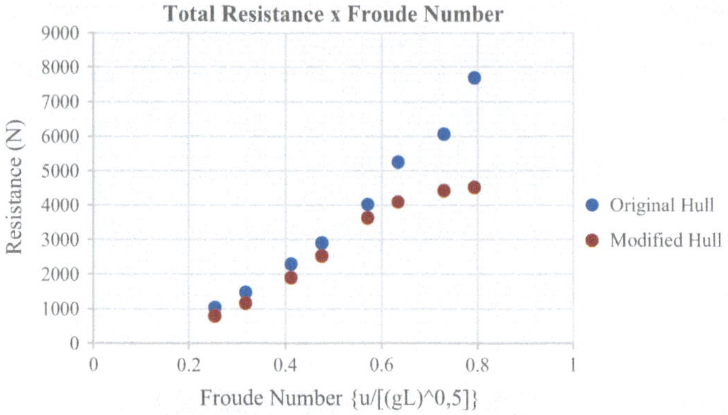

Fig. 10 Total resistance values in relation to the Froude number for both hull models. Blue dots: Original hull of the school boat. Orange dots: Modified hull shape

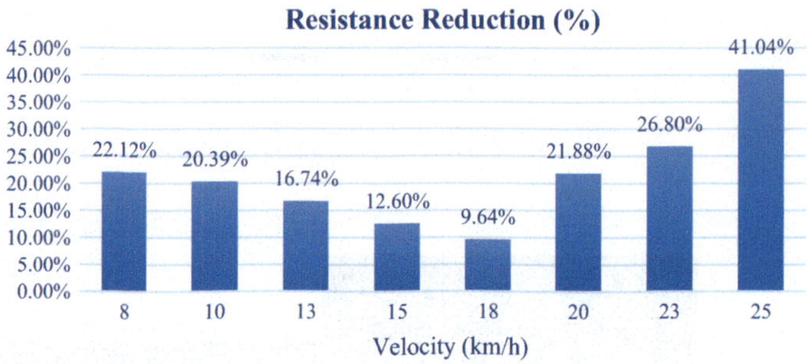

Fig. 11 Percentage of resistance reduction for each velocity

Table 2 Total resistance values obtained in CFD simulation

Velocity (km/h)	Froude # ($V/\sqrt{gL}$)	Total resistance/modified hull	Total resistance/original hull
8	0.254	822.38 N	1055.94 N
10	0.318	1190.79 N	1495.69 N
13	0.413	1926.57 N	2313.86 N
15	0.476	2551.10 N	2919.01 N
18	0.572	3651.21 N	4040.74 N
20	0.635	4117.62 N	5271.12 N
23	0.730	4450.65 N	6079.81 N
25	0.794	4545.24 N	7709.15 N

6 Final Considerations

With the results obtained in this work, a great capacity of variation of the total resistance of the school boat's original hull was identified for subtle changes in the hull shape. The calculated resistance for the model was based on the viscous pressure and wave resistance components of the vessel.

Therefore, the proposed objective of the work was achieved, with a large reduction of the total resistance to forward motion in the range of velocities analyzed. With the present study, a proposal for geometric modification after a hydrodynamic analysis resulted in a maximum reduction of 41.04%. For future studies, a parametric optimization can be done in order to propose further modifications on the bow frames of the boat, related to the wave resistance.

In this way, an even more significant reduction of the resistance can be achieved for the proposed speed range, in order to make the operation of the medium school boat economically feasible. Thus, students residing in more isolated areas of the Amazon basin, where the operation of the large school boat is impracticable, can be benefited with a quality school transport.

We would like also to thank the entities CAPES and FAPESPA for their contribution with research grants to the students, which is of great importance in directing academic work within the area of academic article publication.

References

1. Bertram, V.: Pratical ship hydrodinamics, 1st edn. Butterworth-Heinemann, England (2000)
2. Coelho, D.F.: Evaluation of the hydrodynamic behaviour of a school boat using CFD, pp. 72. Federal University of Pará—UFPA, Belém-PA, Brazil (2016)
3. Graefe, A.V. et al.: Comparison of Aqwa, Gl Rankine, Moses, Octopus, Pdstrip and Wamit with model test results for cargo ships wave-induced motions in shallow water. In: ASME 2015 34th International Conference on Ocean, Offshore and Arctic and Artic Engineering OMAE 2015, Canada (2015)
4. Grinberg, M., Padovezi, C., Tachibana, T.: Utilization of model testing in reduced scale for definition of optimized ship shapes. In: XXII COPINAVAL—IPIN, pp. 28. Buenos Aires, Argentina (2017)
5. Iervolino, L.A.: Study of the resistance to forward motion of a planing vessel of 20 feet: computational approach based on CFD. Undergraduate Thesis, Federal University of Santa Catarina—UFSC, pp. 88. Joinville, SC, Brazil (2015)

SWOT Analysis of Shipyards Located in Santa Catarina

João Luiz Francisco, Ricardo Aurélio Quinhões Pinto and Rui Carlos Botter

Abstract The present paper aims to do the SWOT analysis of a construction shipyard located in the region of Itajaí and surroundings, in the state of Santa Catarina, Brazil. The SWOT analysis has the purpose of identifying the strengths, weaknesses, opportunities, and threats of an organization or industry. In the Internet search and in the Union of Workers of the Metallurgical, Mechanical, Electrical Materials and Naval Construction of Itajaí and Region, five shipyards were identified; four were prepared to respond to the preliminary questionnaires. After the analysis of answers, interviews in loco were performed with directors and managers appointed by the shipyards themselves to represent them. The processing of respondents' data strongly showed the trust relationship with providers, the productive flexibility, and high quality of workmanship. Regarding the weak points, there was a lack of integration between the main players in the naval industry, shipyards, and universities. As threats, the Asian countries and the shipyards of Rio de Janeiro were identified. Among the opportunities, it is worth mentioning the proximity to universities and research centers, which facilitate the development of products, innovations, and technologies.

Keywords Naval industry · SWOT analysis · Brazil

J. L. Francisco (✉) · R. C. Botter
USP—Universidade de São Paulo, Avenida Professor Mello Moraes 2231, Cidade Universitária, São Paulo, SP 05508-970, Brazil
e-mail: jlfrancisco@usp.br

R. C. Botter
e-mail: rcbotter@usp.br

R. A. Q. Pinto
UFSC—Universidade Federal de Santa Catarina, Rua Presidente Prudente de Moraes 406, Santo Antônio Joinville, SC 89218-000, Brazil
e-mail: ricardo.pinto@ufsc.br

A. Vega Sáenz et al. (eds.), *Proceedings of the 25th Pan-American Conference of Naval Engineering—COPINAVAL*, https://doi.org/10.1007/978-3-319-89812-4_10

1 Introduction

Composed of 38 municipalities, the Vale do Itajaí is located in Santa Catarina, in the region formed by the Itajaí-Açu River and its tributaries, which form the Itajaí-Açu River Basin and the Serras do Mirador, Itajaí, and Moema [1]. Along the banks of the Itajaí-Açu River, from the BR101 to the river mouth, in a stretch of approximately 9.5 km in a straight line, is a nexus of shipbuilding. With about 40 companies, the most distant ones are located at an average distance of 500 m from the riverbank and distributed in an area of approximately 17 km^2 with a density of about 2.4 companies per km^2. Considering that the shipyards require large areas for their operation, one can see the great density of companies related to the shipbuilding industry in this region [2]. This location concentrates the construction of tugboats and maritime support vessels and is an important source of jobs [2, 3].

In view of the importance of shipbuilding for the region's economy and the search for the strengthening of this industry, the following questions should be asked: what are the strengths of the construction shipyard in Vale do Itajaí? What are their weaknesses? What opportunities can be leveraged and what external factors can interfere with their business? The SWOT (strengths, weaknesses, opportunities, and threats) analysis, which provides elements for formulating strategies that will generate competitive advantage, answers these questions. Therefore, the objective of this paper was to conduct the SWOT analysis of the construction shipyard located in the region of Itajaí and surroundings, in the state of Santa Catarina, Brazil.

2 Literature Review

The bibliographic review of this paper is divided into two parts. The first presents the emergence of the shipbuilding industry in Santa Catarina, from the second half of the nineteenth century to the present day, when the region became the main Brazilian producer of offshore vessels. Next, the reader will find the definition of the SWOT analysis.

2.1 *The Naval Construction Industry in Santa Catarina*

Itajaí, Laguna, Florianópolis, Tijucas, and São Francisco do Sul, in the second half of the nineteenth century, owned the main ports and market squares of Santa Catarina. The large commercial houses that were established there practiced commerce by means of cabotage navigation and connected the coast of Santa Catarina with the squares of Santos and Rio de Janeiro. In addition to administrative, warehousing, and distribution facilities, many companies had ship repair and production facilities [4].

In the following century, between 1921 and 1930, the production of boats grew in Itajaí [5]. In the late 1960s, a group of technicians and businessmen from Rio Grande do Sul came to Itajaí to meet the South Atlantic Company and, in 1970, founded the EBRASA shipyard. Next, the shipyard CORENA—Metallurgy and Marine Construction S.A., originally from Santos, settled in the region with the objective of building support ships, fishing boats, and research [6].

However, the Brazilian shipbuilding industry, which in 1979 became the seventh largest in the world, went into crisis in the late 1980s. In 1991, the EBRASA shipyard entered into bankruptcy. At the end of the 1990s, with debts of 7 million reais, the company was auctioned repeatedly, until in 2002, it was purchased by the Chilean shipyard Detroit. Also affected by the crisis, the CORENA shipyard was bought in 1994 by the Metalnave Group and became the Itajaí Shipyard (EISA), which quickly saw its business portfolio surpass 250 million dollars.

On October 9, 2006, the US shipyard Navship inaugurated a unit in Navegantes [7, 8]. The enterprise was the most modern in Brazil [7], a prominent position that has been maintained for 10 years as the largest Brazilian manufacturer of PSVs (platform supply vessel) [2, 3]. With an investment of R$220 million, the CBO shipping group began the construction of Oceana Shipyard in 2013. Three years later, the shipyard baptized the CBO Oceana vessel, the first PSV built by the group in Santa Catarina [9–11].

At the moment, from the construction shipyards located in the cities of Itajaí and Navegantes, three are associated with the National Union of the Industry of the Construction and Repair Naval and Offshore (SINAVAL) of the Oceana Shipyard, Detroit Brazil, and Shipyard Navship. Of the 25 contracts for the construction of maritime support vessels and tugboats in Brazil, 17 are owned by shipyards located in Itajaí and Navegantes [3].

2.2 SWOT Analysis

The SWOT analysis is utilized for the development of management strategies [12]. The analysis was developed by Kenneth Andrews and Roland Christensen, researchers at Harvard Business School, and presents the meaning of the four letters that give name to the analysis:

S *Strength*—determines the strengths of an enterprise
W *Weakness*—determines the weakness of an enterprise
O *Opportunity*—determines possible opportunities
T *Threat*—determines external factors that may adversely affect the business [12].

The SWOT analysis consists in identifying the strategies that lead to competitive advantage through the strengths and weaknesses of an organization in order to seize opportunities and neutralize the threats of the environment. In this analysis, two streams of studies have been formed over time, one giving greater emphasis to the environment in choosing the strategy, called structure conduct performance, taking into account the positioning in the face of market threats and opportunities, and another emphasizing the company's internal resources, the resource-based view. Some scholars have focused only on the strengths and weaknesses of organizations, valuing only their internal factors. At the same time, another group seeks in its studies the ideal combination between the analysis of the internal factors of an organization that makes it successful without forgetting the external factors in which it is inserted [12].

3 Methodology

The present study, in relation to its approach, consists of a qualitative research, inasmuch as it involves the obtaining of descriptive data through the contact of the researcher with the situation object of study so that it can be interpreted later. Regarding its objectives, the research has an exploratory character, which aims to increase the understanding of a phenomenon still little known or a research problem not yet perfectly described [13]. Finally, the research method used was the case study, because an "empirical investigation was carried out investigating a contemporary phenomenon in depth and in its real life context" [14].

3.1 Search Process

The research developed in this work was carried out in three stages. The first consisted of a mapping of the maritime industry on the Vale do Itajaí, in which the survey of the shipyards presented in Santa Catarina was carried out through Internet searches. In order to refine the data collected, the shipyards of the Vale do Itajaí were contacted by telephone. Then, the Union of Workers in the Metallurgical, Mechanical, Electrical, and Shipbuilding Industries of Itajaí and Region was visited to obtain more information of construction shipyards of the region.

The second stage of the research process consisted of the application of a questionnaire that was sent to five shipyards indicated in the mapping described above. Of the five identified yards, four were willing to participate in the survey, which can be considered a representative number, and each shipyard indicated the person who would be able to complete the questionnaire.

In the third and final stage of the research, interviews in loco were conducted with representatives of the participating shipyards from the previous stage. It should be noted that during this process, one of the shipyards ended its activities, therefore, three yards were visited. The interviews were conducted based on a questionnaire with 67 questions and in addition to the interviews, documentation and direct observation were used as sources of evidence.

3.2 *Participating Shipyards*

Of the five shipyards identified, four were willing to respond to the questionnaires sent to them and three had their representatives interviewed.

Some characteristics of the shipyards interviewed are presented in Table 1; it is noteworthy that they are treated as “A”, “B”, and “C” yards to maintain their anonymity, as agreed prior to the application of the questionnaires.

4 SWOT Analysis of Shipbuilding Shipyards in Santa Catarina

Tables 2, 3, 4 and 5 show, respectively, the strengths, weaknesses, opportunities, and threats of the shipbuilding shipyards of Santa Catarina identified by the questionnaire. Regarding the aspects pointed out in Table 2, it is worth noting that there is a tendency to outsource services in the yards analyzed. Two representatives from the shipyards stated that there is outsourcing of services and one stated more disagree than agree. In this shipyard that has been reducing the number of outsourced services, two years ago the practice was common, however, with the fall in demand for new vessels, it was decided not to outsource.

Table 1 Characteristics of the shipyards interviewed

Shipyard	Type of vessels	Vessels produced per year	TPB	Number of employees
A	PSV and AHTS	2	4500	910
B	PSV and AHTS	4	4500	1500
C	PSV and tugboats	3	3500	1000

Table 2 Strengths of shipbuilding shipyards in Santa Catarina

Outsourcing of services that were performed by the yard
Industrial equipment and machines are technologically up to date
Flexible production line able to adapt to the requirements of the owner
Strong and reliable relationship with most of the suppliers
High level of integration with strategic suppliers, who participate in planning meetings and know the schedule of construction of the vessels
Contract with suppliers for long periods, thus allowing the schedule to be better planned
Government grants and incentives
Nationalization studies of products/components/parts
Use of the CAD system to carry out the development of vessel designs
Search for new suppliers in the development phase of the vessel project
Knowledge shared through personal interactions
Skilled technical labor in the national market that supplies the shipbuilding industry

Table 3 Weaknesses of Santa Catarina shipbuilding shipyards

Suppliers are not physically installed at the shipyard in the construction phase of a vessel
Few partnerships in the R&D area between suppliers and shipyard
No shipyard has a supplier classification system according to the criticality of its work in production
Suppliers are not responsible for managing the stock of components supplied by them
Difficulty finding qualified workforce to compose the project team
Only one shipyard has a research area in partnership with universities and research centers
No shipyard has the culture to formalize (record) reports and other documents to experience and knowledge of employees
Only one shipyard has masters and doctors on staff
No shipyard invests in universities or technical schools to contribute to the training of the professionals of the sector
No shipyard has a production incentive system
Few partnerships with foreign shipyards to exchange technologies

Table 4 Opportunities for shipyards of Santa Catarina

Partnerships between shipyards, universities, and research centers for the development of products and processes
Opening of technical schools for the generation of skilled labor, as well as new courses at a higher or postgraduate level
Development of research, innovation, and technology for export
Banco Nacional de Desenvolvimento Econômico e Social financing for the modernization of the shipyards
Develop regional suppliers to supply the needs of the shipbuilding industry
Initiatives aimed at strengthening the shipbuilding cluster of Itajaí and region

Table 5 Threats to construction shipyards in Santa Catarina

Asian countries with competitive quality, price, and delivery time
Europe with quality and technology in the production of vessels with high added value, in addition to the development of parts, services, and state-of-the-art technology
Competition of the shipyards of Rio de Janeiro
Dependence on demand by Petrobrás

The three representatives of the shipyards stated that they have a flexible production line, being able to adapt to the requirements of the shipowner. In one of the shipyards visited, due to the decrease in the demand for PSV-type vessels, the production line was beginning to make tugboats. In another shipyard, in addition to PSV vessels, the first AHTS was being built.

The weaknesses of the Santa Catarina shipbuilding shipyards are presented in Table 3. Among them, it should be noted that no yards were found having partnerships with their suppliers in the R&D area. Two respondents said they disagreed with the existence of partnerships and one said more disagreeing than agreeing. In addition, all representatives of the Santa Catarina shipbuilding shipyards have stated that their inventories are not managed by their suppliers.

Table 4 presents the opportunities for shipbuilding shipyards in Santa Catarina. It should be noted that all the interviewees said they disagree more than agree on having suppliers located in the Vale do Itajaí and all said they are interested in increasing the number of suppliers in the region. Having suppliers near the shipyards would contribute to reducing logistics and operational costs, as well as strengthening the shipbuilding cluster of Itajaí and Navegantes, proving to be a great opportunity to increase the competitiveness of shipbuilding shipyards in the region.

Table 5 presents the threats to shipbuilding shipyards in Santa Catarina. In the second stage of this research, conducted in 2014, when questionnaires were sent to the shipyards, all stated that they intended to increase the number of vessels produced in the next 12 months. However, this scenario has undergone a major change, because the shareholders of shipyards involved in *Operação Lava Jato* feel a worsening of their financial conditions and the press reports the difficulties of obtaining financing [3]. The *Lava Jato* operation is the largest investigation of corruption and money laundering ever developed in Brazil, and the amount diverted from Petrobras is on the order of billions of *reais* [15]. A fact that is even more evident, is that during the research for the preparation of the SWOT analysis presented here, a yard closed its activities.

5 Conclusions

The shipbuilding yards of Vale do Itajaí are part of a cluster, which, although in its embryonic stage, makes this industry more competitive and productive. The fact that the shipyards have suppliers located in the Vale do Itajaí influences their operational and logistical costs, and there is interest in increasing the number of suppliers. It is necessary that the prefectures of Itajaí and Navegantes, together with the state and federal government, attract companies to the region that make up the supply chain of the sector. In the analyzed segment, government policies already exist at the municipal, state, and federal levels that foster the growth of the shipbuilding industry, however, they need to be improved and expanded. It should be noted that Japan, China, and South Korea started their shipbuilding industries based on strong incentives and industrial policies focused on technology development, market reserves, and labor qualification.

An opportunity that must be taken advantage of is the proximity of the shipyards to universities and technical schools. Partnerships for internship programs and technical visits by students help to train the workforce. It is worth mentioning that a skilled labor market which meets the needs of the shipbuilding industry is a necessary prerequisite to the productivity and competitiveness of the sector. Research with universities is carried out only with one shipyard, but should be expanded to others, particularly if one considers that the city of Joinville, less than 100 km away from Itajaí and Navegantes, has the naval engineering course of the Federal University of Santa Catarina. There are few partnerships for the development of new technologies between shipyards and universities and research centers, as well as shipyards and their supply chain.

Only one shipyard has a partnership for the exchange of technology with countries considered as a reference. When we remember the development of shipbuilding in the countries that are currently the world's largest producers, China, Korea, and Japan, it is perceived that this was a strategy adopted by them. If shipbuilding yards want to increase their competitiveness and productivity, it might be interesting to follow the examples and strategies of other countries.

Countries that are now recognized worldwide for having their core competencies in the shipbuilding industry have invested heavily in the integration of shipyards, universities, research centers, and supply chains for the development of products, technologies, and production methods that would make them competitive in the international market.

The information management of the shipyards should be developed. No shipyard has formal records regarding the knowledge and experience of its employees. Information exchange and greater employee empowerment are based on personal interactions only. The method may be efficient in the short term, however, it does not guarantee the yard the maintenance of the knowledge developed by its human resources, because the employee can leave the yard.

References

1. Wittmann, A.C.R.: A ferrovia do Vale do Itajaí. Dissertation (Master degree), Universidade Federal de Santa Catarina, Florianópolis (2008)
2. Pinto, R.A.Q.: Proposta de modelo estratégico para consolidação de cluster industrial marítimo. Thesis (Doctorate), Universidade de São Paulo, São Paulo (2016)
3. SINAVAL (Org.). http://sinaval.org.br/wp-content/uploads/Sinaval-Cenário-da-Construção-naval-1-Semestre-2016-27-7-16.pdf. Accessed 24 Aug 2016
4. Moreira, M.R.T.: A construção naval no Brasil: Sua Gênese, Desenvolvimento e o Atual Panorama da Retomada do Setor (1990–2010). Thesis (Doctorate), Universidade Federal de Santa Catarina, Florianópolis (2012)
5. Machado, J.M.P.: O desenvolvimento da construção naval em Itajaí (SC): Uma resposta do mercado local (1900–1950). Dissertation (Master degree), Universidade Federal de Santa Cataria, Florianópolis (1973)
6. Moreira, M.R.T.: A formação de uma vila operária em Itajaí (SC): Uma industrialização interrompida. Dissertation (Master degree), Universidade Federal de Santa Catarina, Florianópolis (2002)
7. INTELOG. http://www.newslog.com.br/site/default.asp?TroncoID=907492&SecaoID=508074&SubsecaoID=818291&Template;=../artigosnoticias/user_exibir.asp&ID=843902&Titulo=EstaleiroNavshipvaiserinauguradonasegunda-feira. Accessed 08 Nov 2016
8. Oliveira, D.L.: O Navegantes que eu conto, 2nd edn. Papa Terra, Navegantes (2012)
9. Secreta De Estado Da Fazenda. http://www.sef.sc.gov.br/clipping/11012013. Accessed 08 Nov 2016

10. Diário Catarinense. http://dc.clicrbs.com.br/sc/noticias/de-ponto-a-ponto/noticia/2016/03/estaleiro-de-itajai-entrega-primeiro-navio-a-petrobras-5110007.html. Accessed 10 Aug 2016
11. Grupo CBO. http://www.grupocbo.com.br/quem-somos/. Accessed Oct 2016
12. Poter, M.E.: Competitive strategy: techniques of analyzing industries and competitors. Free Press, New York (1980)
13. Appolinário, F.: Dicionário de Metodologia Científica, 2nd edn. Atlas, São Paulo (2011)
14. Yin, R.K.: Estudo de Caso: Planejamento e Métodos. Bookman, Porto Alegre (2010)
15. Ministério Público Federal. http://lavajato.mpf.mp.br/entenda-o-caso. Accessed 17 Oct 2016

Expert System for the Execution and Supervision of the Anchorage Maneuver

Luis Manuel Carral Couce, Javier Tarrío Saavedra, Jose Carlos Alvarez-Feal, Laura Castro Santos and Juan Carlos Carral Couce

Abstract Yachting uses the boat, in navigation and anchored, to develop different leisure activities. In this use the vessel remains static a great deal of the time of its total use, which constitutes a notable difference with respect to merchant activity. Maintaining the safe condition of the vessel at anchorage is not easy for the crew due to multiple factors: navigation with reduced crews and/or lack of crew professionalism/experience, tiredness after long navigation, ignorance of the area, or simply lack of attention during this time. These circumstances can be increased when the crew decide to anchor to rest or spend the night and the meteorological conditions vary without adapting the anchorage of the boat to this new situation. The objective of the work is the description of the results of the EAS project in which a system for the automatization and supervision of the anchorage maneuver has been developed. The system is designed to assist the crew in the anchorage by means of partial automation of the maneuver. During the stay at the anchorage area, the system passes to a situation of permanent surveillance that allows the detection of dangerous situations of dragging the anchor, alerting the crew, and recommending an action.

Keywords Ship · Anchorage · Supervision · Control

1 Introduction

The anchorage is a stage, intermediate between the phases of navigation and port, which is presented in the merchant activity in moments of port waiting congestion. The objective is to keep the vessel immobilized for a minimal time presenting, therefore, minor importance compared to the navigation time. A very different

L. M. Carral Couce (✉) · J. T. Saavedra · J. C. Alvarez-Feal · L. C. Santos
Universidade da Coruña, A Coruña, Spain
e-mail: lcarral@udc.es

J. C. C. Couce
Carral Design Engineering Solutions, A Coruña, Spain

A. Vega Sáenz et al. (eds.), *Proceedings of the 25th Pan-American Conference of Naval Engineering—COPINAVAL*, https://doi.org/10.1007/978-3-319-89812-4_11

situation is presented in recreational use, because the boat remains in this anchorage condition an outstanding portion of the total utilization.

During the twentieth century, the concept of sea recreational use was developed, which as one of its manifestations led to the popularization of nautical activity and to the development of a large fleet of yachts. The adaptation of these boats to their program of use, with their consequent differentiation in respect to other types of vessels, constitutes a constant in the technological evolution of ships subjected to the continuous specialization in traffic or activities. In this context, equipment installed in them must be adapted to the peculiarities of their use [1].

The anchor windlass is presented as a main equipment of the ship related to its safety [2]. In nautical terms, small boats with the incentive of coastal navigation, install, each time in lower sizes, windlasses that cover the work of anchorage [1] (Fig. 1). The shipowner obtains with this greater simplicity and safety in the operation of anchorage that could lead to a reduction in the necessary crew [3].

The anchorage maneuver and, in particular, the maintenance of this condition, is a source of concern for the crew of leisure boats. Especially when the maneuver is hampered by circumstances such as operation with reduced crews, lack of professionalism/inexperience of the crew, fatigue after a long navigation, ignorance of the area, and especially when the meteorological conditions vary in time in a dramatic manner [4].

The slide of the anchor on the bottom, known as dragging, is a situation of special risk for the boat. The situation appears when the request of the ship on the anchorage exceeds the gripping capacity of the anchor [5]. It may be due to the increase in demands that wind and current exercise on the ship (worsening of weather conditions), or because of the reduction in the effectiveness of the line of anchorage as a consequence of the increase of the depth with the rise of the tide [6]. To constitute a risk for the vessel, it will be of particular importance to detect this undesirable situation.

The EAS (expert anchoring system) is a complement of the anchoring equipment destined to assist the pattern in the execution and control of this maneuver.

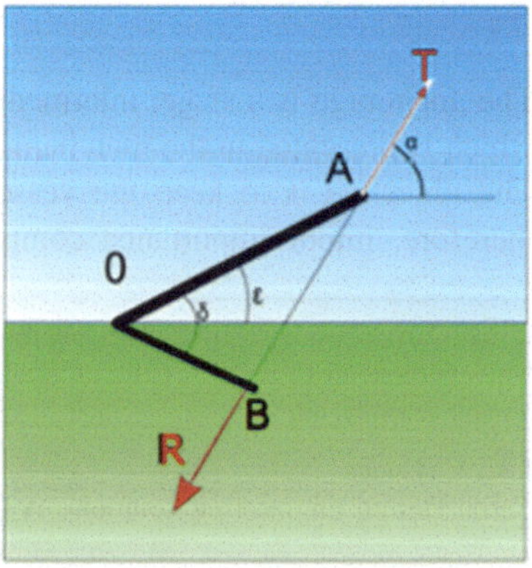

Fig. 1 Burying condition of the anchor and the holding power *R* before values of the angle that forms the shank with respect to the background (angulation). *Source* [8]

Its execution automatically allows the performance of specific operations and helps the crew in decision making by incorporating a surveillance module during the stay in the anchorage intended to detect and alert to dangerous situations.

2 Objective and Content of the EAS Research Project (Expert Anchoring System)

The description of results of the EAS research project, born from a I + D + i activity of an auxiliary shipbuilding company, looked for external collaborations of the University of a Coruña and a shipyard dedicated to the construction of mega-yachts.

The activities carried out during the two years of the research development project are included in Tables 1 and 2.

3 Algorithm for the Preparation and Maintenance of the Anchorage Condition

A vessel maintains its safe condition at the anchorage area immobilized by the action of its anchor line, consisting of a chain length attached to the anchor.

Table 1 Activities carried out within the previous stage of development

Activity	Content
Determining safe anchorage condition	Identification of the necessary parameters: depth, direction, and wind velocity, chain length, speed of the yacht (supplied by the log), chain tension, etc. Positioning of the boat (signal provided by the GPS/Plotter installed)
Conceptual system design	Determination of the instrumentation to incorporate the vessel List of sensors to be included in the deck machinery
Development representative algorithm of system functionality	Ideal conditions of anchorage and safety of the vessel
Design of chain counting system by anchor windlass Redesign of the chain stopper to allow a remote drive	Definition of the sensors to be incorporated in the windlass to measure the length of chain used in the anchorage Modification in the design of standard components of windlasses to incorporate the sensors Replacement of the manual drive locking of the trap by a remote control of the hydraulic or electric type

Source Author

Table 2 Activities carried out in the development stage

Activity	Description
Design and construction of a windlass with remote drive	Incorporation of sensors and long chain count Installation of hydraulic or electric activator on the brake for remote operation
Construction of the chain stopper	
Design and construction of data acquisition units	Identification of commercial equipment adapted to the required calculation and processing Special attention to the buses that support equipment in order to be able to acquire values of the navigation systems Consideration of the price of each unit looking for a final price of the product adapted to the market
Development of data acquisition and interpretation software	Development of user interface software User interface with touchscreen On-screen visualization of the configuration features of the equipment, alarms, and slogans and introduction of the decisions
Essays and tests in workshop	System verification in preinstallation factory onboard.
Tests onboard the ship of the project shipyard	Sea tests aboard the designated vessel seeking to validate the design
Verification of results and validation	Proposals to improve and launch the product to the market

Source Author

Anchorage can be done by using one anchor or two. Normally, and except for extraordinary environmental conditions that suggest another solution [4], it is anchored with a single one in a maneuver called "swing anchored" [2, 6, 7] and can vary, depending on the boat and the area of operation, with the type of anchor [5] and anchorage line (rope or chain) employed. The choice of the type of anchor and the length of chain are the two variants to prevent the feared dragging or trawling of the anchor from happening [6].

In the case of small vessels the chain can be replaced by a plumbed rope [3]. In the same way, the regulation allows anchorings of the mixed type (rope and chain) in boats of small length [3]. With this variation, the anchorage line may be lightened, looking for the double objective of improving the boat's behavior in navigation and to make the maneuver of the anchor's sail easier. In any case, there must be at least one chain length (of equal length to the size of the boat) at the end of the attachment of the anchor to ensure its correct operation in the seabed [8].

The nature of the seabed of the anchorages employed will allow choosing between different types of anchors, all of which are of high holding power; Bruce, Delta, Fortress, Britany, Hall, Danforth, and CQR, among others [9].

Carral et al. [10] indicate that in order to guarantee the optimum functioning of the anchor, and consequently obtain the maximum holding power, two essential aspects must be achieved: to make the anchor shank have a zero angle or close to it, with respect to the horizontal one, and to avoid major variations in the direction of the boat's bow to changing weather conditions (swing at anchor), especially in clay bottoms. The first aspect is of particular importance, because the anchor in a condition of burial and the nature of the given source, significantly decreases its holding power at reduced values of the angle that forms the line in respect to the seabed (Fig. 1) [9]. The way to ensure that the chain's action on the anchor shank is a force exerted horizontally, will cause a chain length to rest on the bottom [6].

Carral et al. [10] enunciate the requirements to be fulfilled by the anchoring system of a vessel that has anchored with a single anchor, to remain in a static balanced position (Table 3). During the stay at the anchorage area, the boat will be subjected to dynamic effects that can affect the condition of the anchorage [8, 10].

In particular, *vertical*-type movements due to the action of waves or tides will not affect the holding power of the anchor whenever the shank rests on the bottom (angle of shank lifting $-\alpha < 5$; Fig. 1), but do affect the request in the vessel's stopper. *Horizontal*-type movements modify the length of the chain in contact with the bottom and therefore could vary the holding power of the anchor. *Rotational* movements around the anchorage (swing at anchor) due to variations in the direction of wind (contrasts) or current (turning angle of the vessel or swing at anchor $-\beta < 45°$), may affect the holding power in the case of soft seabeds, due to the fact that the anchor may not be buried in the new position.

Coll and Quereda [6, 7] assimilate the chain in the situation of the anchoring line like a catenary curve, in particular a half-curved catenary complete, whose vertex corresponds to the point where the chain rests in the bottom (C) and the end to the point of contact of the chain with the ship (hawse pipe or anchor sheave; B, Fig. 2). "Catenary is the curve that forms an ideal thread, of mass distributed evenly and which hangs freely from two points, being the lowest of the two the vertex of the curve."

The maximum holding power of the anchor can be guaranteed whenever we limit the angular values α and β (respectively, lifting the shank and the angle of swing at anchor of the vessel). The first of the limitations will be fulfilled by forcing the lowest point of the catenary (C) to be in contact with the bottom. The limit position for this condition to be fulfilled will occur when point C matches A (Fig. 2). Unfortunately the second condition will not be achieved without the anchor being dug up and losing holding power until a new stable conditioning is reached after a new burial. In all this process the vessel will vary its position,

Table 3 Elements involved in the anchorage and its effect

Element	Effect
Anchor	To counteract the wind and current requests that are exercised on the boat based on its gripping power
Chain stopper	Able to withstand the weight of the catenary that forms the chain suspended to the bottom
Chain	By its length or weight, keep the anchor shank in horizontal position (angular values less than 5°)

Source Author

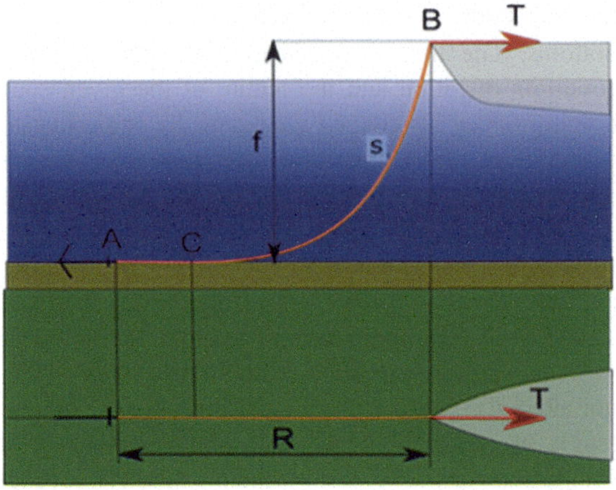

Fig. 2 Permanence of a ship at the anchorage area, indicating the main parameters of its condition. *Source* Author

with the consequent risk. In this situation, the skipper will look for a new anchorage area with greater protection [10].

The triangle of forces represented at the right of the catenary figure (Fig. 3), composed of the forces exerted at the ends of the catenary (T_0 and T) and the weight of the curve section considered T_d (weight per unit length p and S the length of the curve, $T_d = P \cdot S$), will allow us to determine the physical model representative of the anchorage of our boat [7]. The T force corresponds to the effort on the vessel's boat (fixing point of the chain onboard) and it can be broken down into a horizontal force too, which represents the request exerted by the wind and the current on the ship, counteracted by the holding power of the anchor resting on the bottom, and a

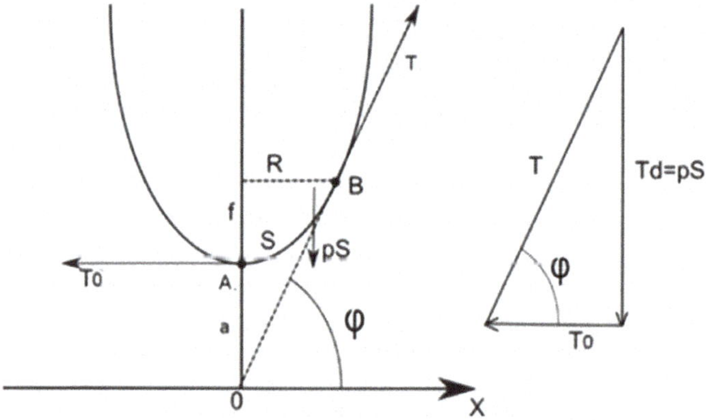

Fig. 3 Half-curved catenary complete, whose vertex corresponds to the point where the chain reclines in the bottom and the end to the point of contact of the chain with the ship (fairlead or anchorage shave), and representation of the triangle of forces that composes the solicitations in the extremes and the weight of the chain. *Source* [7]

vertical component corresponding to the weight of the T_d chain, and which is compensated by the increase in the displacement of the vessel [8, 10].

By means of the composition of forces indicated it will be possible to relate the three parameters that lead to the solution of the physical problem from the value of the solicitation exerted by the vessel and the holding power of our anchor, to determine the length of chain to be extended [6].

The holding power of the anchor is calculated by expression (1), in which k is the holding power coefficient that is determined from a double-input table; the type of bottom and anchor are considered. The coefficient n is a variable depending on the weight of the anchor, which takes the values n = 0.92 for anchors of more than 25 kg weight [11–13] and n = 1.4 [8] for smaller anchors.

$$T_0 = k \cdot P^n_{\text{anchor}} \tag{1}$$

The non ragging condition must always be fulfilled, that is:

$$\text{Anchor Grip Power} = k \cdot P^n_{\text{anchor}} > \text{Calculated value } T_0 \tag{2}$$

Special consideration should be made of those cases in which a certain "lifting" of the anchor shank, the so-called "angulation" may be admissible. A small angular value α (less than 5°), will produce a considerable elevation in the value of the permissible tension at the end of the catenary [8] (Fig. 2):

$$T_\alpha = T_0 \frac{\cos\alpha}{1 - \frac{s}{f}\sin\alpha} \tag{3}$$

Finally, the tension that supports the claws of the chain stopper will be that corresponding to the equation that relates the tension at the end of the catenary (point of attachment to the ship), with the tension at the vertex of the curve, tension to [7], where f = depth, p = unit weight of the chain:

$$T = T_0 + f \cdot p \tag{4}$$

The maximum intensity of wind that the boat can withstand in the anchorage area studied is given to equal the maximum value that can take the T parameter (4), with the expression that relates to the force that the wind exerts on the upperworks of the boat [7] and then clear the value of V. The value corresponding to the length of chain to be extended will be determined by the expression [10]:

$$T_0 = P_a\left(S^2 - f^2\right)/2f \tag{5}$$

$$S = \sqrt[2]{f\frac{2T_0 + P_a \cdot f}{P_a}} \tag{6}$$

where S = total length of chain, f depth in meters, and F = depth in meters (considering the height of the bow), P_a = apparent weight of each meter of chain, and T = force that acts on the boat. The previous expression can be written as a function of the ratio N, (S/F), which represents the relationship between the long chain length and the existing depth [10].

$$N = \sqrt[2]{1 + 2\frac{T_0}{P_a \cdot f}} \tag{7}$$

Equation (7) clearly indicates that the value N depends proportionately on the force that acts on the boat (T_0), and inversely on the depth of the place. In no case can a present value be generalized (the value of $N = 3$ is usually handled in a worse manner) [10].

The maximum depth at which the vessel may be anchored for a given length of the anchorage line (S) and a given value of the force is given by the positive root of the equation in f, of Eq. (8):

$$f_{\max} = \sqrt[2]{S^2 + \left({}^{T_0}/{}_{P_a}\right)} - \left({}^{T_0}/{}_{P_a}\right) \tag{8}$$

The wind action for motor boats is obtained from the expressions collected in [10]. k_1 = Constant that depends on the protection of the anchorage, Vv speed of the wind in m/sg, B = beam of the boat, H = effective height from the waterline to the roof of the higher cabin with beam over $B/4$.

$$T_v = \frac{3}{64} k_1 \cdot V_v^2 \cdot B \cdot H \tag{9}$$

Fraysse [8] advises the use of the formula published by the American Boat and Yacht Council, due to its good estimation of the wind effect on sailing boats, with the consideration of the effect of the rigging and the mast, with L = total length of the boat, V_v = wind speed in knots:

$$T_v = 0.0089 \cdot L^{1.66} \cdot V_v^2 \tag{10}$$

In the cases in which the marine stream is to be determined, it will be determined according to the wet surface of the vessel [14], following expression (11), when C_f = coefficient of friction resistance acting on the underwater, ρ = fluid density, V_c = stream speed, and A_m = wet hull surface.

$$T_c = C_f \cdot \frac{1}{2} \rho \cdot V_c^2 \cdot A_m \tag{11}$$

Estimating the value of C_f through the ITTC–57 and operating we would obtain:

$$T_c = 258.33 \times 10^{-3} \cdot (d \cdot \delta + b) \cdot (B + 2T) \cdot V_c \cdot \left(V_c \cdot L_{pp}\right)^{4/5} \quad (12)$$

where d, b are coefficients that depend on the type of prow (cylindrical or bulbous), δ block coefficient, and L_{pp} length between perpendicular, B beam, and T draught. In the case of joint action of wind and current, the expression of the total force action will depend on the angular values α_1 and α_2 which form the wind and the current incidents with the central line of the ship in the position of final balance [7].

$$T_0 = T_v \cdot \cos\alpha_1 + T_c \cdot \cos\alpha_2 \quad (13)$$

4 Description of the System and the Process to Be Controlled

The decking equipment of ships can be automated in their main functions and remotely controlled in various ways [15, 16]. In this line the EAS system groups together a set of devices constituting a system of intelligent supervision and control of anchorage maneuver. In general, the EAS allows the preparation of the maneuver to carry out the sequence of maneuvers necessary for the execution of the anchorage and then determines the actions to be followed to maintain the situation of anchorage in safe conditions. Finally, the system is a continuous monitoring of the parameters determining the condition of the anchored ship.

4.1 EAS Components

The system consists of the functional units described (Fig. 4; Table 4). **Conventional navigation instruments**: Depth sounder, speed sensor, and anemometer. The depth sounder provides depth information. The speed sensor gives us information on the value of the current acting on the vessel. Finally, the anemometer shows the speed and direction of the wind. These are used to deduce whether the ship is on the wind and from the speed readings to estimate the approximate force exerted by the wind on the dead work of the ship.

Anchor windlass, additional instrumentation, and remote maneuvering elements: When handling the chain, the windlass will perform a long chain length measurement and tension. The length measurement will be achieved by means of the installation of a sensor in the warping end of the windlass, which generates a fixed number of pulses for each turn, totalling the full length of the distance (Fig. 5). As for the tension in the chain it can be obtained by means of two procedures: measuring the torque in the motor (either hydraulic or electric) [15]

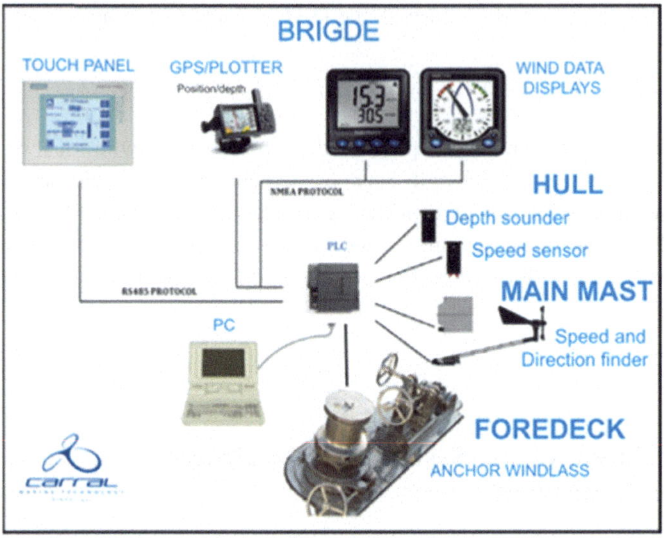

Fig. 4 System block diagram including instrumentation, the CPU, the interface screen, and anchor windlass connection. *Source* Author

Table 4 EAS system components and functions

Components	Functions
Conventional navigation instruments	Preparing the maneuver
Anchor windlass, additional instrumentation, and remote maneuvering elements	Running the sequence of operations
Signal capture and processing unit	Condition maintenance
User interface screen	Continuous monitoring of anchoring parameters

Source Author

or installing a strain gauge on the tool. The remote maneuver elements are formed by electrical or hydraulic activator, which by acting on the tool allow the chain to be immobilized. This prevents the windlass from being in tension for long periods of time, with the inconvenience of being released when a maneuver is required.

Signal capture and processing unit: This element will consist of a programmable automaton, in charge of the acquisition of the signals and of its processing for the making of the correct decisions at any time.

User interface screen: There is a touchscreen where the skipper of the boat can enter slogans, acknowledge the information that the system manages, and monitor the alarms.

In the maintenance condition of the anchorage a mimic will show the main parameters of the condition. This screen is connected to the processing CPU.

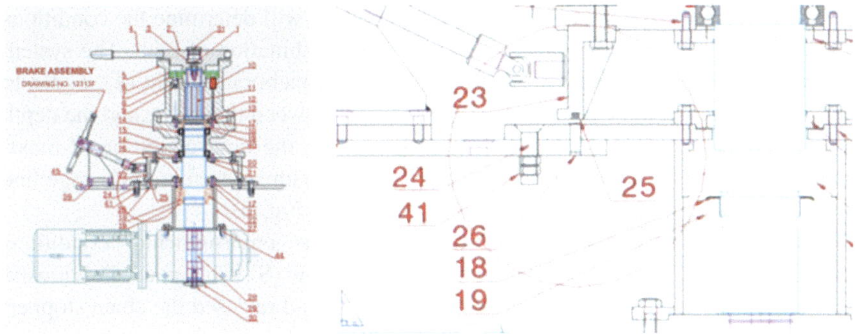

Fig. 5 Overview of a section of one of the vertical anchor windlasses in detail. Magnet (25) and magnetic sensor (26) view. *Source* Carral Engineering Solutions

4.2 EAS Functions and Screens

For each of the functions that the system develops (Table 4), the corresponding screens are displayed on the user interface screen (Fig. 6).

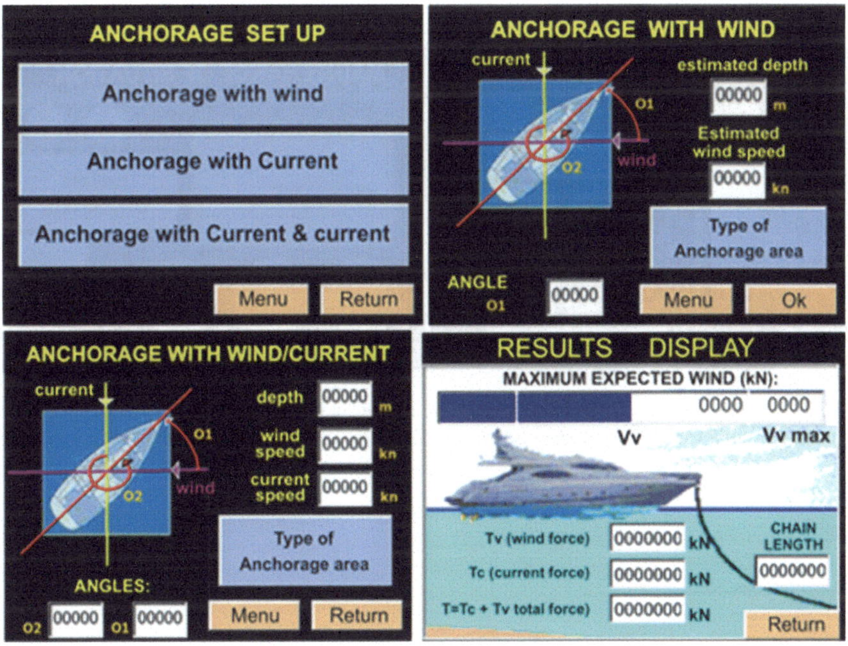

Fig. 6 Screens corresponding to the preparation of the anchorage maneuver. *Source* Carral Engineering Solutions

Preparation of the Anchorage: For a given place will determine the conditions of anchorage in the wind situations, current, or a combination of both. The system will calculate the force exerted by the vessel on the anchorage chain (T_v, T_c, T_0) to the amount of chain to be extended (s) and relation between this value and the depth of the anchorage (s/f). In the same way it determines the limit condition of maximum wind speed (V_{max}) that the boat can withstand with its current anchorage line in the anchorage selected (condition of no dragging; Fig. 6).

Execution of the Anchorage: Considering the environmental conditions existing in the chosen place, the amount of chain to row (S) is obtained. Previously, the windlass started, engaged, debraked the wildcat, and released the chain stopper; the system allows the chain to be extended until the length corresponding to the calculated value "S" is reached and the maneuver is terminated. As the last process, and as in the previous screen, the system will calculate the maximum wind intensity that the ship can withstand in that anchorage, informing the skipper (V_{max}; Fig. 6).

Maintenance of the situation: From the information received from the anemometer (Vv) and the sounder (f) the variables of the anchorage (s-amount of long chain) are recalculated, proposing the performance on the windlass. In the same way there is an alarm triggered when the wind condition (V_v) exceeds the safe situation calculated previously (V_{max}; Figs. 7 and 8).

Continuous monitoring of the anchoring parameters: see Fig. 8.

Fig. 7 Screens corresponding to the execution of the anchorage and the maintenance of the situation. *Source* Carral Engineering Solutions

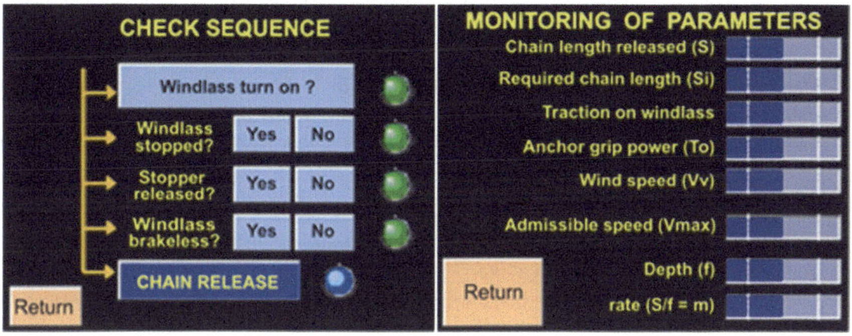

Fig. 8 System screens, continuous monitoring of anchoring parameters. *Source* Carral Engineering Solutions

5 Conclusions

Executing the anchorage maneuver and the control of the situation of permanence at the anchorage constitutes an important source of concern for the crew of leisure boats associated with the equipment that helps the crew in these tasks. Therefore, the development of a system associated with this equipment can be of commercial interest.

Determining the anchorage rode (anchor and anchorage line) is given by knowledge of the forces involved in the operation: those due to the action of the wind and sea on the boat, and the interaction of the anchor with the seabed. With the anchorage rod considered, determination of vessel safe condition will mean combining these factors with the nature of the seabed and the depth at the anchorage.

The designed system brings together a set of devices constituting a monitoring and control of the intelligent type of anchorage maneuver. In general, the EAS will allow the preparation of the maneuver, carry out the sequence of maneuvers necessary for the execution of the anchorage, and then determine the actions to be followed to maintain the anchorage situation in safe conditions. Finally, the system performs a continuous monitoring of the parameters determined by the condition of the anchored ship.

Acknowledgements The work carried out is part of the EAS research project: An expert in the manoeuvre of anchorage of recreational boats. PGIDIT06DPI009E research project of the XUNTA de GALICIA, within the Program of Technologies of Design and Industrial Production.

References

1. Carral Couce, L.: Particularidades en el diseño de molinetes de anclas destinados a embarcaciones de recreo—Revista de Ingeniería Naval, nº 809 (Nov 2003)
2. Comas Turnes, E.: Equipo y Servicios. Servicio de publicaciones de la Escuela Técnica Superior de Ingenieros Navales, vol. 1 (1985)
3. Carral Couce, L., Martinez López, A., Diaz Casas, V.: Diseño de yates con líneas de fondeo mixtas. Madrid Diseño de Yates Mayo 2010 (2010)
4. Poiraud, A., Ginsberg-Klemmt, A., Ginsberg-Klemmt, E.: The Complete Anchoring Handbook. Mc Graw Hill, USA (2008)
5. Sagarra, R.M.: Maniobra de los buques. Ediciones UPC Barcelona, Barcelona (1994)
6. Coll Alas, M.: Fondeo, teoría matemática—Revista de Ingeniería Naval, marzo 1980, (1980)
7. Quereda Laviña, R.: Análisis de esfuerzos en la cadena y el ancla de buques fondeados a la gira. Revista de Ingeniería Naval **68**(769), 93–102 (2000)
8. Fraysse, A. 2017.: Sail Rode Forces. Link at http://alain.fraysse.free.fr/sail/rode/forces/forces.htm, last accessed June 2017
9. Puech, A.: The Use of Anchors in Offshore Petroleum Operations. Technip, Paris (1984)
10. Carral Couce, L., Fraguela Formoso, J.A., González Filgueira, G., Rodríguez Guerreiro, M.J., Carral Couce, J.C.: Algoritmo para el mantenimiento de la condición de fondeo en Yates—Congreso Copinaval, Montevideo Octubre 2009 (2009)
11. Bruce Anchor Group.: Disponible en: www.bruceanchor.co.uk/ last accessed 13 June 2017
12. Vryhof.: Anchor Manual 2015—The Guide to Anchoring. In http://www.vryhof.com/flipbooks/am2015/#1/z, last accessed February 2017
13. Hodges, T. Springer, B.: Pick the Right Hook-Product Test—Anchors. Motor Boats, pp. 51–58 (May 2007)
14. Saunders.: ITC—International Towing Conference, Madrid-España (1957)
15. Carral Couce, L., Carral Couce, J.C., Fraguela Formoso, J.: Operation and handling of escort tugboats with the aid of automatic towing winch systems. J. Navig. **68**, 1–71 (2010)
16. Carral, L., Lara Rey, J., Alvarez-Feal, C., Carral Couce, J.: Winch control gear for CTD sampling with a system to compensate vertical motion heave when manoeuvring in rough seas. Ocean Eng. **135**(2017), 246–257 (2017)

On the Sinkage of Ships

Luis Pérez-Rojas, Adriana Oliva-Remola and Misael Goicoechea

Abstract All ships acquire a dynamic trim and sinkage when they are in motion. These dynamic parameters depend on the hull shape and the velocity. The ship at rest already has a specific trim, the difference between the draft forward and the draft aft, due to its loading condition. Once in motion, and because of the pressure distribution along the hull, the ship reaches a new equilibrium condition, which is stationary if the ship's velocity is kept constant and no sea state is considered. Although there are many publications devoted to optimizing the trim to obtain the minimum drag, the same cannot be said regarding the sinkage that appears to have been forgotten from any consideration. This work is dedicated to analyzing the sinkage phenomenon that the ship suffers when it is in motion. The aim is to quantify the prejudicial effect that the sinkage has over the drag resistance, both from the numerical and experimental points of view. Finally, the last purpose is to evaluate the effect of the hull shape on this phenomenon, whose assessment allows for the reaching of a hull shape optimization.

Keywords Drag resistance · Sinkage and trim · Towing tank trials CFD

1 Introduction

Circa 200 BC Archimedes of Syracuse in his principle explained why ships floated. The demonstration is simple and can be understood by applying the basic principal of the hydrostatic [1]. The equilibrium position of a ship is that which equals the integration of the fluid pressure over the hull and the weight of the ship. When the

L. Pérez-Rojas (✉) · A. Oliva-Remola · M. Goicoechea
Universidad Politécnica de Madrid, ETSI Navales, Avenida de la Memoria, 4, 28040 Madrid, Spain
e-mail: luis.perezrojas@upm.es

A. Oliva-Remola
e-mail: adriana.oliva@upm.es

A. Vega Sáenz et al. (eds.), *Proceedings of the 25th Pan-American Conference of Naval Engineering—COPINAVAL*, https://doi.org/10.1007/978-3-319-89812-4_12

ship moves, the pressure distribution changes, and if the hull velocity is kept constant a stationary state is reached, leading to a new equilibrium between the weight and the buoyancy.

As indicated in [2], this change in the pressure distribution depends on the velocity, which will cause the ship either to sink or to emerge, and change its trim. At low velocities, it is foreseeable that the ship sinks and submerges the bow. When the velocity increases, the bow motion changes, and around 0.30 Froude number (Fr) the bow emerges significantly, and thus the stern submerges. The stern sinking is much greater than the bow emersion; therefore, the ship displays an important stern trim.

The *ITTC Dictionary of Hydromechanics* [3] defines sinkage as "the steady state lowering of a ship's position of floatation in the water; to be distinguished from heaving, which is an oscillatory motion." Trim is defined as "the steady state of a ship's angular position in the centerplane; to be distinguished from pitching, which is an oscillatory motion."

Even though dynamic trim has always been related to drag resistance optimization, this has not happened with sinkage. Taylor [4] noted in 1943 that for a common merchant hull form the additional stern trim generates an increase of the resistance at low velocities, and a reduction when increasing the velocity. At low velocities, the increment of the aft draft makes the ship fuller, with the correspondent increment of the resistance and the separation of the boundary layer. However, at high velocities, this increase of resistance is highly compensated by the reduction of wave resistance, because of the finest entrance of water in the bow.

Saunders [5 (Fig. 29D, p. 417)] shows, as an indicative value, that sinkage exists up to 0.6 Fr. Over this range, the hydrodynamic forces have already developed, and the ship enters the planing regime. He also highlights that more sinkage is presented as the midship volume increases; that is, sinkage increases when the prismatic coefficient reduces.

The assertions of D. W. Tailor about trim are inquisitive [4]. In 1922, he indicated that, according to the towing tank trials performed in the facilities currently called the "David Taylor Model Basin," the trim changes did not affect the effective power. Furthermore, if any improvement was obtained it was not because of a possible reduction of the resistance, but rather due to an improvement in the propulsion efficiency. In 1943 he nuanced his disclosure, noting that if any improvement due to trim was demonstrated it was so small that it lacked any practical relevance.

Nevertheless, recent research exists regarding the optimization and influence of trim. Lv et al. [6] use potential codes to calculate the trim that minimizes the wave resistance of a ship model; Iakovatos et al. [7] study the influence of trim on the resistance of six ship models performing towing tank tests; Sherbaz and Duan [8] analyze the resistance of a container ship changing its trim and obtaining its optimum value; and Sun et al. [9] optimize the trim of a container vessel using the optimization method known as the response surface method (RSM), introduced initially by Box and Wilson [10]. In all these works the result is a resistance optimization of around 5%. Reichel et al. [11] advocate reductions of 2–3%, but are able to reach reductions of 15%.

The extensive literature about trim hull optimization is reduced considerably when dealing with sinkage. Yang et al. [12] show results of the Serie 60 and Wigley ship including sinkage. Saha et al. [13] study the sinkage phenomenon in shallow water using computational fluid dynamics (CFD) simulations. It may be interpreted that when dealing with trim, the combined effect of trim and sinkage is considered. Or maybe it is commonly assumed that sinkage does not have an influence on effective power, and therefore the study of this phenomenon is not relevant at all.

Concerning the difficulty of the sinkage determination using numerical simulations, the ITTC Ref. [14] is of particular interest. According to this publication, the validity of CFD to predict the drag resistance has reached a high level of accuracy, especially when using the results in the design stage. However, the deviations of the simulations for trim and sinkage are much larger.

The present work aims to deepen knowledge about sinkage, quantifying its value and the increment of resistance that it entails. The relationship between hull shape and sinkage value is also studied.

2 Research Background

Our renewed interest in sinkage comes from a commercial project performed at Canal de Ensayos Hidrodinámicos de Navales (CEHINAV), relating to a ferry vessel whose hull shape was exceptionally flat. Towing tank trials were performed, and from the analysis sinkage values of up to 40 cm were obtained, that is, a sinkage equivalent to 10% of the draft of the model. Initially, these surprising results were linked to the flat hull shape. However, due to the lack of research studies that justify it, it was deemed necessary to realize a comparative analysis with other ship types.

In this context, four ships tested at CEHINAV were selected: a trawler-fishing vessel, a tuna-fishing vessel, an oceanographic vessel, and a RoPax. The analysis was based on the comparison for the different ship types of the sinkage, as a percentage of the draft (Fig. 1), and the relationship of the displacement of the ship in motion and that of the ship at rest, as a percentage of the displacement of the ship at rest (Fig. 2).

From the analysis, it can be seen that the ferry does not present such an exceptional sinkage as was initially thought. Moreover, the ferry is in the lower part of Fig. 1. At low Froude numbers the sinkage phenomenon is similar for all ship types. However, as velocity increases, the values start to diverge. This leads to clear differences between the models, whose differences can come to 20% of the sinkage.

Additionally, sinkage can be equated to an excessive load because the increase in draft produces a variation in the hydrostatic pressure. Figure 2 illustrates that, for the selected range of Froude numbers, the vessels that presented higher sinkage were also the ones that experienced larger displacement variations, except for the ferry.

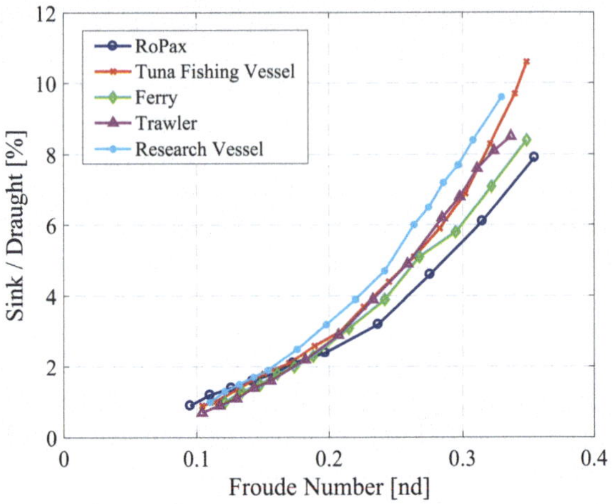

Fig. 1 Sinkage as a draft percentage versus Froude number for the different ship types analyzed

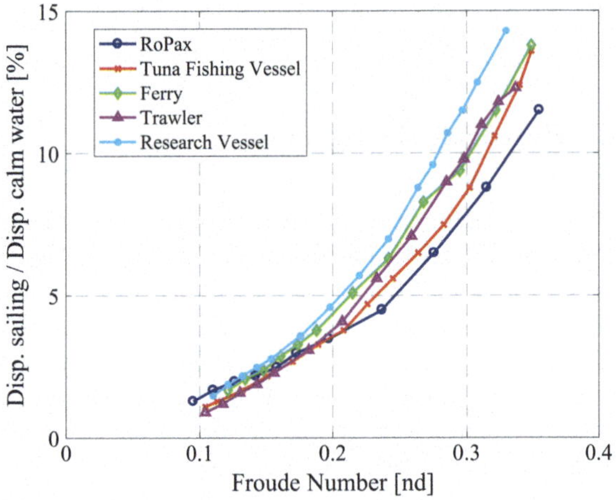

Fig. 2 Displacement relation between the ship in motion and the ship at rest versus Froude number, for the different ship types analyzed

Nevertheless, not all the ships under analysis sail at the velocities comprised in the Froude number range studied. If the results outside this range are taken into account, extraordinary results are obtained. For instance, for the ferry ship the increase of displacement is 22%, and at the maximum velocity tested the largest sinkage is obtained, which corresponds to 13%.

One may ask what quantitative influence the sinkage has on the resistance. Table 1 provides the increments of the drag resistance due to the previous sinkage values. As may be seen, sinkage values of the same order of magnitude present significant drag resistance variations, from 2.5% for the tuna-fishing vessel, and up to 17.5% for the trawler-fishing vessel. Regardless, the resistance increments in the majority of the cases are relevant.

Table 1 Increment of drag resistance due to sinkage for the different ship types analyzed

Vessel type	Fr	Displacement [%]	Resistance [%]
RoPax	0.35	7.7	4.4
Tuna-Fishing vessel	0.35	10.6	2.5
Ferry	0.35	8.4	8.3
Trawler	0.34	8.4	17.5
Research vessel	0.33	9.6	8.3

Given the difficulties of comparing ships of such different hull shapes and different ranges of navigational speeds, in the present work simpler hull shapes are studied. These hull shapes present similarities intended to find relationships between the hull shapes and the sinkage and drag resistance values obtained.

3 Analysis of the Simple Hull Shapes

As indicated previously, the aim is to investigate the relationship between sinkage and hull shapes. This is addressed using three simple hull shapes, which were studied both experimentally and numerically.

3.1 The Selected Hull Shapes

The selected models possess the same length (150 cm), the same breadth (30 cm), and the same height (24 cm). All of them possess, along almost the entire length, a constant transversal section, perpendicular from the aft up to 112.5 cm. In other words, they consist of a bow body with an extension of 25% of the length, and a cylindrical body with the remaining 75%.

In the cylindrical body, the first and second models (Model 1 and Model 2) have pentagonal and semicylindrical sections, respectively. The section of the third model (Model 3) corresponds to a square.

One of the main variations of the hull shapes that we study is the bottom shape, in particular the deadrise. In Model 1 the deadrise is greatest; to the contrary, Model 3 presents zero deadrise.

All models were analyzed in the range of Froude numbers in which Saunders [5] predicts that sinkage occurs. The upper limit was established as 0.5 Fr, whose value is slightly larger than the velocities at which maximum sinkage should occur. The lower limit was 0.1 Fr, which is the minimum value tested in the ship types selected for the previous analysis.

To select the displacement for each of the three models, different criteria may be considered. Each criterion must take into account a parameter fixed for all the models. From one side, it may be established the displacement is such that all the models present the same draft. This criterion ensures that the models at rest have the same length at the waterline, as the main dimensions and the bow profile angles are equal. Another criterion may be to consider the same wetted surface area for all the models. A third criterion may be used in order to maintain the same displacement, which consists of equalizing the submerged volume in each of the three models. This last procedure was used by Valle [15] to perform stability studies, as the restoring moment is the restoring arm multiplied by the displacement.

Despite the fact that previous studies do not consider any correlation between the sinkage curves and the other parameters specified previously (length waterline, wetted surface area, or submerged volume) [16], we hypothesized that the length at the waterline might have the greatest influence on the sinkage phenomenon. For instance, consider how the boundary layer's thickness is intimately related to the length of a ship. Moreover, as a consequence of all the models having the same length at the waterline, the nondimensional coefficients are equal as well; therefore, the hydrodynamic scenario for all the models is approximately the same.

3.2 *Model Characteristics*

As indicated previously, all three models have similar geometric particularities. This allows us to isolate the effect that a specific hull shape has on the sinkage. The similarity between the models is clearly shown in Table 2, where the main dimensions of the models are illustrated. Notice the equality of all the dimensions, except the parameters that define each model: the deadrise and the bottom radii. Figure 3 represents the three models graphically.

Table 2 Main particularities for each simple model

Parameters	Units	Model 1	Model 2	Model 3
Length overall (L)	[m]	1.50	1.50	1.50
Length at the waterline (Lwl)	[m]	1.36	1.36	1.36
Maximum breadth (B)	[m]	0.30	0.30	0.30
Height (H)	[m]	0.24	0.24	0.24
Draft (D)	[m]	0.15	0.15	0.15
Submerged volume (V)	[dm^3]	40.84	44.48	53.06
Block coefficient (Cb)	[nd]	0.67	0.73	0.87
Prismatic coefficient (Cp)	[nd]	0.89	0.89	0.87
Waterline area coefficient (Cwl)	[nd]	0.94	0.94	0.94
Midship section coefficient (Cm)	[nd]	0.75	0.82	1.00
Bow profile angle	[deg]	36.62	36.62	36.62
Deadrise	[deg]	26.57	–	0.00
Bottom radii	[cm]	–	16.25	–

3.3 CFD Simulations

The analysis of the models using numerical simulations may be considered a simpler and cheaper way to obtain results than conventional tests. To compare, an estimation of the hydrodynamic performance for one of these models using CFD can be obtained in less than a day if enough computational resources are available (in our case: 7 CPU Intel i5 of 2.6 GHz). Meanwhile, performing towing tank tests implicates a long process that begins with the construction of a model.

The CFD software used is Star-CCM+, which has an intuitive interface and plenty of information available for the user [17].

To validate numerical simulations the experimental and numerical results were compared. In Figs. 4 and 5, the resistance and sinkage obtained by the two methodologies are illustrated for Model 1 and Model 2, respectively. From the

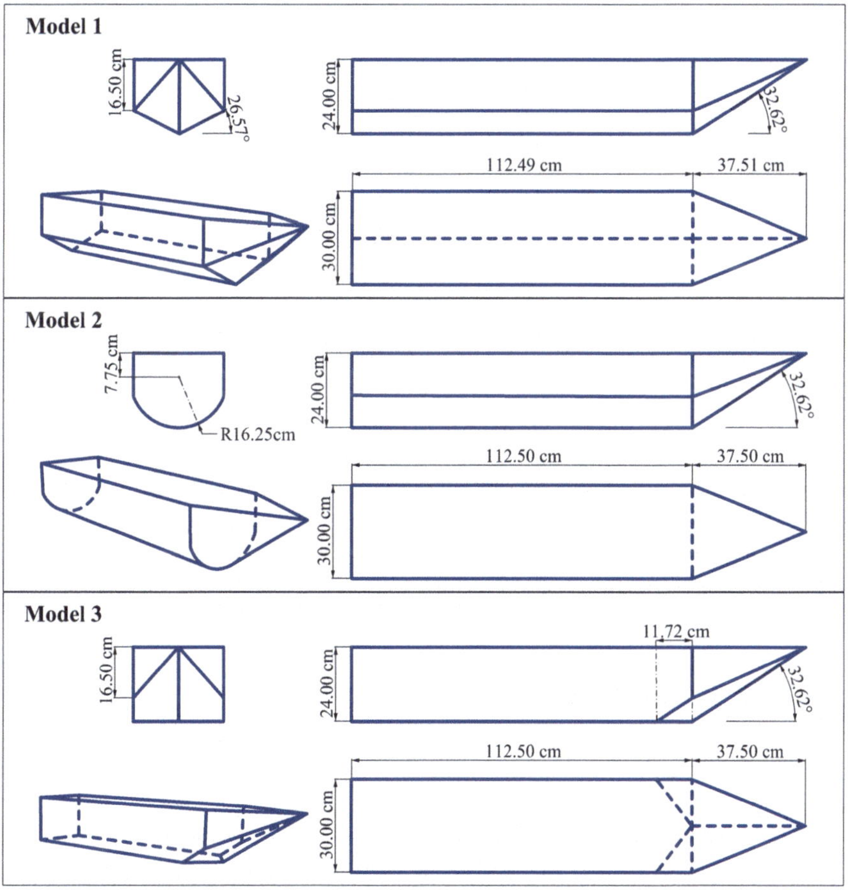

Fig. 3 Model projections and main dimensions

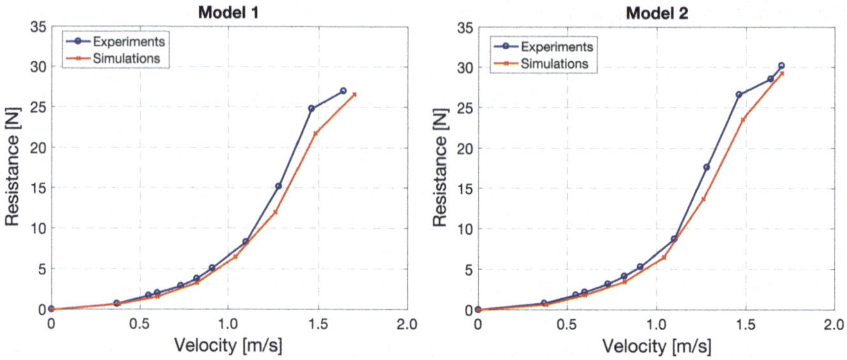

Fig. 4 CFD drag resistance validation (Model 1, left; Model 2, right)

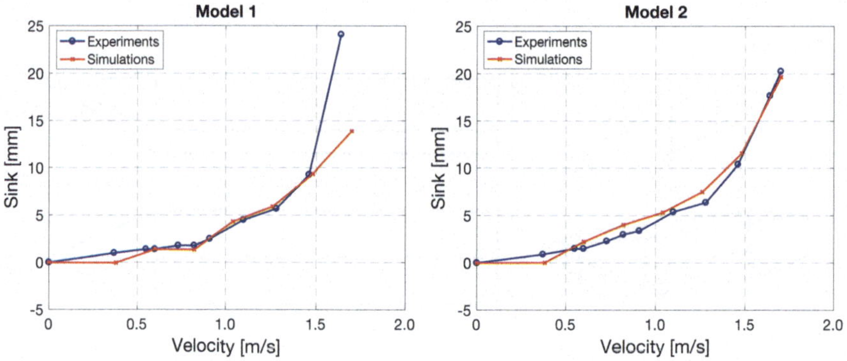

Fig. 5 CFD sinkage validation (Model 1, left; Model 2, right)

comparison, an uncertainty analysis can be assessed. For the drag resistance the uncertainty is 10%, and for sinkage 12%. This is well within the range of Froude numbers common for experimental and numerical tests, that is, between 0.10 and 0.35. In light of the last ITTC report [14], caution must be exercised when dealing with a quantitative analysis of the results, focusing preferably on qualitative ones.

3.4 *Analysis of the Results*

After having confirmed that the numerical results coincide with the experimental ones, at least from a qualitative point of view, the former results may be used to accomplish the aim of establishing a relationship between sinkage and hull shapes.

Figure 6 shows the total drag coefficient and the sinkage as a function of Froude number. In this figure it may be observed that Model 1 presents the lowest sinkage

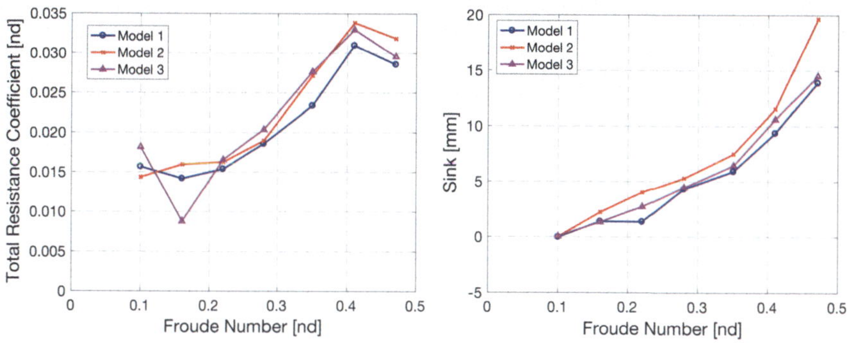

Fig. 6 Total drag coefficient and sinkage of the models as a function of Froude number

and resistance, whereas Model 2 presents the highest resistance, with Model 3 presenting the greatest sinkage.

It may be thought that the flat bottom area of Model 3 is the root cause of having the greatest sinkage. However, the similarities between the sinkage curves of Models 1 and 3 suggest that the sinkage may be penalized because of the pressure distribution created by the bow shape. We have reached this deduction because the only shared characteristic between Models 1 and 3 is that they share the same bow body.

4 Conclusions

From the results shown in this work, the following conclusions may be stated.

- The sinkage of ships is an undervalued factor when compared to its "brother," the trim.
- The sinkage values are relevant as they may increment the displacement of a certain ship by up to 14%.
- The sinkage generates an increase in the drag resistance that can be up to 17.5% in some cases.
- Based on the information obtained from the study, it appears that the bow hull shapes have more influence on the sinkage than the bottom area.

Lastly, the present study invites readers to conduct further research to continue deepening knowledge into sinkage.

References

1. Biran, A.B., López-Pulido, R.: Ship Hydrostatics and Stability, 2nd edn. Butterworth Heinemann, Oxford (2014)
2. Lewis, E.V.: Principles of Naval Architecture, Volume II: Resistance, Propulsion and Vibration, 1st edn. Society of Naval Architects and Marine Engineers (SNAME), Jersey City (1988)
3. Ferrando, M., Benedetti, L., Derradji, A., Johnson, B., Kobayashi, K., Morabito, M.G., Pérez-Rojas, L., Sena Sales, L., van Rijsbergen, M., Woodward, M., Park, J.T.: Dictionary of hydromechanics. In: 27th International Towing Tank Conference (ITTC), pp. 1–113. ITTC, Copenhagen (2014)
4. Taylor, D.W.: The Speed and Power of Ships: A Manual of Marine Propulsion, 1st edn. Wiley, New York (1943)
5. Saunders, H.E.: Hydrodynamics in Ship Design, vol. I, 1st edn. Society of Naval Architects and Marine Engineers (SNAME), New York (1957)
6. Lv, X., Wu, X., Sun, J., Tu, H.: Trim Optimization of Ship by a Potential-Based Panel Method. Adv. Mech. Eng. **5**, 1–7 (2013)
7. Iakovatos, M.N., Liarokapis, D.E., Tzabiras, G.D.: Experimental investigation of the trim influence on the resistance characteristics of five ship models. In: Soares, C.G., López-Peña, F. (eds.) Developments in Marine Transportation and Exploitation of Sea Resources: IMAM 2013, vol. 1, pp. 23–32. Taylor & Francis Group, Abingdon, UK (2014)
8. Sherbaz, S., Duan, W.: Ship trim optimization: assessment of influence of trim on resistance of MOERI container ship. Sci. World J. **2014**, 1–6 (2014)
9. Sun, J., Tu, H., Chen, Y., Xie, D., Zhou, J.: A study on trim optimization for a container ship based on effects due to resistance. J. Ship Res. **60**(1), 30–47 (2016)
10. Box, G.E.P., Wilson, K.B.: On the experimental attainment of optimum conditions. J. Roy. Stat. Soc.: Ser. B (Methodol.) **13**(1), 1–45 (1951)
11. Reichel, M., Minchev, A., Larsen, N.L.: Trim optimisation—theory and practice. Int. J. Mar. Navig. Saf. Sea Transp. (TransNav) **8**(3), 387–392 (2014)
12. Yang, C., Löhner, R., Noblesse, F., Huang, T.T.: Calculations of ship sinkage and trim using a finite element and unstructured grids. Int. J. Comput. Fluid Dyn. **16**(3), 217–277 (2002)
13. Saha, G.K., Suzuki, K., Kai, H.: Hydrodynamic optimization of ship hull forms in shallow water. J. Mar. Sci. Technol. **9**(2), 51–62 (2004)
14. Hino, T., Carrica, P., Broglia, R., Bull, P., Kim, S.-E., Li, D.-Q., Wan, D., Rhee, S.-H., Saisto, I., Viola, I. M.: Specialist committee on CFD and marine hydrodynamics: Final report and recommendations to the 27th ITTC. In: 27th International Towing Tank Conference (ITTC), pp. 522–567. ITTC, Copenhagen (2014)
15. Valle-Cabezas, J.: Estudio Teórico Experimental de las No Linealidades del Amortiguamiento en el Movimiento de Balance de Buques (Theoric and Experimental Study of the Roll Damping Non-Linearities). Ph.D. Thesis. ETSI Navales—Universidad Politécnica de Madrid, Madrid (1998)
16. Toby, S.A.: The mystery of sinkage and trim for high performance craft. In: The 5th Chesapeake Power Boat Symposium, pp. 1–18. Chesapeake Power Boat Symposium, Annapolis, Maryland (2016)
17. Star-CCM + User Guide, https://thesteveportal.plm.automation.siemens.com/, last accessed 12 December 2017

Nontraditional Towing Tank Tests

M. Salas, C. Cifuentes, G. Tampier and C. Troncoso

Abstract Traditionally, towing tanks had been used primarily to assess the resistance of hulls in calm water, by means of towing geometrically similar scaled models at equal Froude number, usually without appendages such as rudder, propeller, and so on. When wave-making capability is available, towing tanks can also be used to determine seakeeping occurrences, such as green water effects and slamming, for a range of wave frequencies and amplitudes. This is not different for the University Austral of Chile Towing Tank, which has been involved in resistance and seakeeping tests of a variety of hulls, from fishing vessels to ferries, barges, and passenger ships, over the past 40 years. The development of marine industries other than shipbuilding, such as aquaculture, maritime connectivity of isolated geographical regions, and the harvesting of marine energy from tides and waves, have demanded a new set of tests with adequate models to properly replicate the physics of the full-scale case in the towing tank.

Keywords Towing tank · Nontraditional tests · Hydrodynamics

1 Towing Tank Description and Main Characteristics

The University Austral of Chile Towing Tank (Fig. 1) is a unique laboratory of its kind in Chile; it has played a key role in the development of the naval industry for more than 40 years. The tank main dimensions are 45 m in length, 3 m wide, and 2 m deep. It has an aluminum carriage running on lineal rails mounted with high precision, in order to guarantee accurate measurements. The towing carriage is equipped with a load cell, accelerometers, and ultrasonic sensors; if required surface and underwater cameras can also be installed. The carriage reaches a speed of up to 5 m/s and is driven by a servo motor of 0.75 kW.

M. Salas (✉) · C. Cifuentes · G. Tampier · C. Troncoso
Instituto de Ciencias Navales y Marítimas, Universidad Austral de Chile, Valdivia, Chile
e-mail: msalas@uach.cl

A. Vega Sáenz et al. (eds.), *Proceedings of the 25th Pan-American Conference of Naval Engineering—COPINAVAL*, https://doi.org/10.1007/978-3-319-89812-4_13

Fig. 1 University Austral towing tank

One end of the towing tank is provided with a regular wave generator, allowing it to generate waves up to 0.2 m in height. The tank walls and bottom are constructed entirely of steel and the structural frames are freely sustained by a system of flexible supports in order to withstand seismic movements.

The laboratory has a workshop for the manufacture of scale models to be tested. This workshop is equipped with a three-axis CNC milling machine (Fig. 2) capable of producing models up to 2.5 m in length, in materials such as wood, plastics, and aluminum.

The tests carried out in the towing tank allow the determination of advance resistance of hulls in calm water and in waves, visualization of flow lines, and dynamic response of ships and floating devices under regular waves, among others. The information obtained allows the selection, for instance, of the engine, propeller, and rudder of a ship, optimization of forms, and the behavior estimation of ships and buoyant artifacts in waves.

Fig. 2 CNC model milling machine

Recently the laboratory has been awarded a grant worth US$ 200,000 from the Chilean Government Office for the Promotion of Science. The grant will be used to provide new scientific equipment, namely a new wave generator that will be able to generate regular and irregular waves.

2 Wave Energy Converter Tests

Chilean energy demand has been growing over the past few decades and it is expected to remain so as the country develops. Traditional renewable energy sources as hydroelectric power are either fully exploited or face fierce opposition from environmental groups. In the present situation more than half the energy the country requires is produced by thermal plants burning imported and expensive oil, coal, or gas. The Chilean government is fully aware of this vulnerability and is making efforts to increase the contribution of nonconventional renewable energy sources (NCRE). It is intended that by 2025, 20% of the energy produced in the country will come from NCRE.

Recently important funding was awarded to create a Marine Energy Research and Innovation Centre (MERIC), a consortium of universities and research centers with the objective of promoting research in the field of nonconventional marine renewable energy (NCMRE). The University Austral de Chile through its Institute of Naval and Maritime Sciences (ICNM) is a partner of MERIC dedicated to investigate devices to capture wave energy, the so-called WEC (wave energy converter) mechanisms [1] and tidal energy harvesting.

Wave energy potential in Chile is among the largest in the world, with an energy potential that has been estimated between 164 and 240 GW [2]. Although the development of WECs has seen important advances in the last years, these are still considered to be in an early stage of development when compared to other NCRE technologies such as wind, solar, or hydro power.

Application of WEC technologies in Chile requires several aspects to be taken into account, considering local resource characteristics, potential environmental impacts, and/or available facilities for investigation, construction, installation, and maintenance. For these reasons, local capabilities for the development and adaptation of technologies are needed, with experimental scale model tests playing an important role. This section shows the evaluation of a scale model of a WEC; the response of different wave conditions was measured, as shown in Fig. 3. The experimental configuration for the converter consisted of two ultrasonic meters to record the WEC and incoming wave's movement. An adjustable damper was connected to a load cell and data acquisition software (LabView) to measure the force. Experimental results were compared with numerical results obtained by WAMIT, in the form of transfer function and output power [3]. The comparative data show good agreement for the variation of a set of parameters of wave

Fig. 3 WEC tests

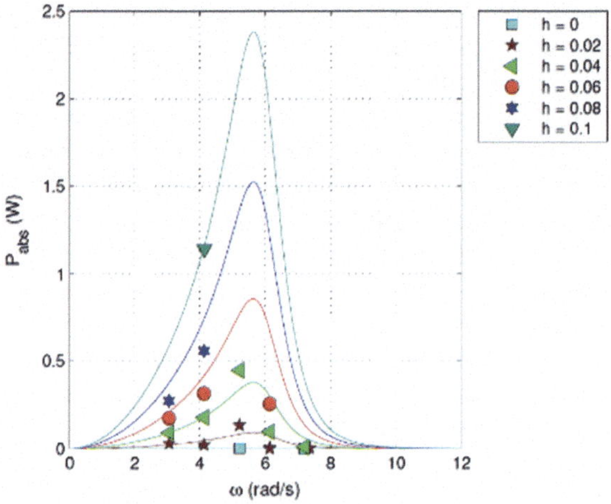

Fig. 4 Numerical and experimental results

frequency, wave height, and damping factors, as shown in Fig. 4. These developments are the first steps towards the specialization of this tank in experimental hydrodynamics of marine renewable energy technologies.

3 Floating Ports Modeling

Southern Chile is a region of extreme access difficulty; the existence of thousands of islands and canals makes the construction of roads extremely difficult. In many cases maritime transport is the only route of supply, therefore, appropriate ports and conditions must be provided to ensure the connectivity of those regions. The government of Chile has developed a connectivity program for remote regions that, among other solutions, has led to the construction of floating ports.

Conditions of southern Chile inland seas make the installation of floating ports advisable. According to Tsinker [4], these conditions are:

1. There are protected waters, where waves generated by the wind or the movement of ships do not exceed 1 to 1.5 m in height.
2. Where soil quality does not allow fixed structure foundations.
3. The change in sea levels is a high product of great tidal difference.
4. In seismic zones, because floating docks are less affected by earth movements during an earthquake.

A floating port is mainly a collection of pontoons with appropriate land access and a suitable mooring system to secure it in the operating location. Depending on the port's design requirements, it can be designed using various pontoon configurations, including one or more modules attached by a continuous cover or a pivoting mechanism. Although it is a simple hydrostatic problem, careful attention should be paid to the displacement of rolling loads on deck, so as to avoid displacements or rotations of the pontoons that may prevent proper operation; see Figs. 5 and 6.

In addition to the static analysis of the pontoons' loading, dynamic effects should also be considered, and environmental loads such as waves or the effect of wind should be taken into account. Dynamic analysis of vessel motion is well documented, however, for floating structures the problem has additional variables. Due to the complex interaction between floating structures, anchoring, and the fluid, it is advisable, in addition to analytical studies and numerical simulations, to carry out experimental tests with scale models. Tests should consider characteristic waves

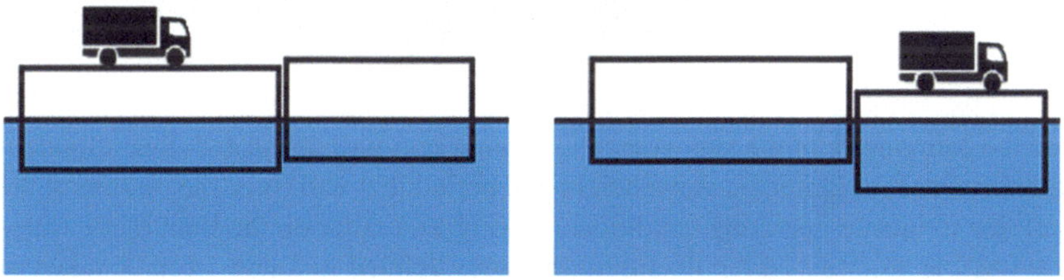

Fig. 5 Effect of moving load across pontoons with different buoyancy

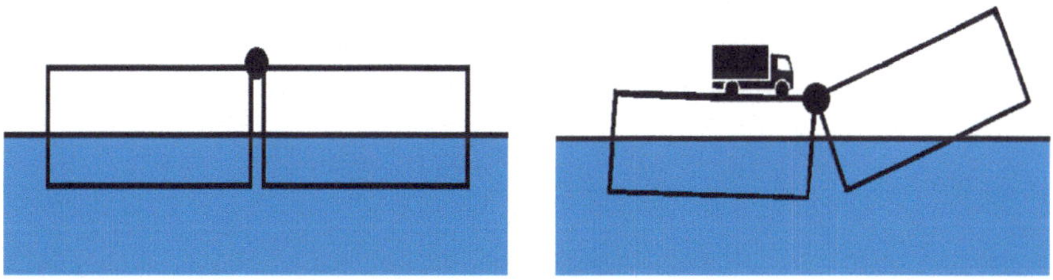

Fig. 6 Effect of moving load on deck formed by hinged pontoons

Table 1 Criteria for acceleration and roll (NORDSFORK, 1987)

Description	RMS vertical acceleration (g)	RMS lateral acceleration (g)	RMS roll motion (°)
Light manual work	<0.20	<0.10	<6.0
Heavy manual work	<0.15	<0.05	<4.0
Intellectual work	<0.10	<0.05	<3.0
Transit passengers	<0.05	<0.04	<2.5
Cruise liner	<0.02	<0.03	<2.0

Fig. 7 Typical fetch wave. $H_{1/3} = 0.52$ m, $T = 2.99$ s

and the bathymetry of the site where the floating port will be located. It is important to determine the significant wave height and period that will force the closure of the port due to unsafe operating conditions. Taking as a criterion the limit that ensures comfortable working on the floating port, limits defined in Table 1 can be taken as an example.

In addition to the adoption of criteria to close port operations, the floating structure must be designed to survive extreme conditions that will ensure its life, for example, 100 years survivability. Figure 7 shows tank wave tests for a floating port model for Southern Chile, under typical conditions imposed by the zone's fetch. Figure 8 shows tests with waves with a 100-year return period.

Fig. 8 100-year wave. $H_{1/3} = 1.38$ m, $T = 4.15$ s

4 Aquaculture Net Modeling

The development of large-scale aquaculture in countries such as Norway and Chile has raised the need to develop studies, both numerical and experimental, to determine the forces and corresponding deformation of flexible structures such as nets used in floating cages [5]. Aquaculture centers, mainly breeding salmon, operate with square or circular cages (Fig. 9), installed in protected areas such as bays and fjords, to avoid the effect of waves and currents present in the open sea. The feeding of fish is carried out by automated systems installed on a pontoon, as shown in Fig. 10. The basic components of the cages are: a buoyant part that provides a flotation reserve to support the weight of the net and shapes the net that encloses the water volume with the fish. Hanging from the net are some weights that help to keep the net shape undeformed, or at least less deformed by environmental loads, mainly currents.

Fig. 9 Circular cage fish farming center. *Source* www.sermar.cl

Fig. 10 Control and feeding pontoon

The study of the hydrodynamic behavior of fish farming cages is complex due to the dependence between the drag force on the nets and the deformation of them. Similarly, for circular HDPE (high-density polyethylene) cages, the rings forming the buoyancy volume are deformed by the wave loads, directly influencing the tensions in the mooring lines.

In addition, the current velocity is reduced when passing through the nets, a phenomenon known as the shadow effect; this adds complexity to numerical and experimental modeling [6]. This effect works directly on the available volume inside the cages. Various studies, including the behavior of the fish inside the cages, biofouling on the nets, and flow into the breeding centers have been developed by various specialists to improve the production and increase the safety of these structures.

There are several numerical methods to estimate the hydrodynamic response of cages. Nets have been modeled using bar and panel elements in a finite element model; other finite element models use bars and buoys to adjust the submerged weight of the grid to create a reduced model of elements that simulates mechanical properties and net inertial forces in the physical model. An example of results obtained by the bars and buoys method can be seen in Fig. 11, which also includes the study of element density in direct comparison with the resulting deformation from experimental tests.

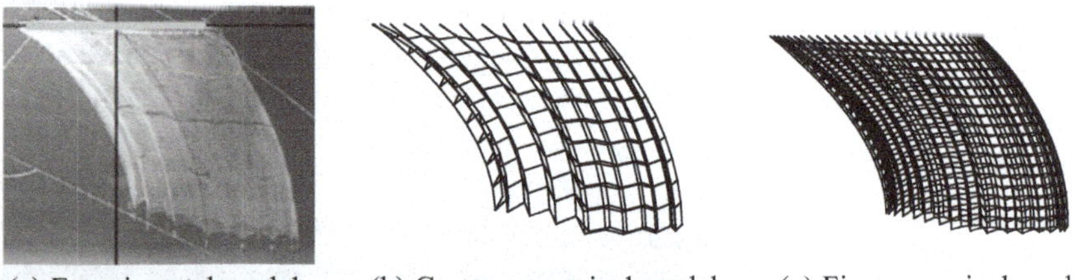

(a) Experimental model (b) Coarse numerical model (c) Fine numerical model

Fig. 11 Net shape deformation under current loads

As for the response of the cages to waves, it has been determined that the forces on the mooring lines are directly related to the wave height. When wave and current effects are simultaneous, depending on the combination of height and wave period and current velocity, the effect of the wave load can be of the same importance as the current load, which implies a careful modeling of both effects.

In exposed areas, fish farms should operate autonomously with minimal human intervention and the cage anchoring system will be shared with the platform containing equipment and fish meal. Results obtained for systems exposed to extreme conditions of waves and currents, both in numerical models and field tests, demonstrate the feasibility of the installation and operation of these structures and open the door for the future development of the fish farming industry in exposed zones. Open sea conditions are optimal for fish due to the constant exchange of nutrients in the water column. The almost unlimited availability of space would allow the production of a greater volume of fish, in addition to the opening to new species of greater commercial value.

5 Conclusions

Towing tanks are well-known experimental facilities, mainly dedicated to hull resistance and seakeeping tests, however, they can be adapted and implemented with instrumentation to perform a variety of experiments proved useful to many marine fields, such as aquaculture, marine energy harvesting, and floating ports among others.

Whenever possible, experimental measurements should be calibrated using benchmark tests, in order to avoid uncertainties in the predicted behavior of full-scale developments.

References

1. Tampier, G., Grüter, L.: Development of an experimental test bed for wave energy converters. In: WTE'14 First Workshop on Wave and Tidal Energy, Valdivia, Chile. 29–31 October 2014
2. Monárdez, P. et al.: Evaluation of the potential of wave energy in Chile OMAE2008. In: International Conference on Offshore Mechanics and Arctic Engineering (2008)
3. Tampier, G., Salas, M., Troncoso, C., Grüter, L.: Wave energy converter and hydrokinetic rotor developments in Chile. In: 10th Annual Energy Harvesting Workshop. Blacksburg, USA, September 13–16, 2015
4. Tsinker, G.P.: Floating Ports (1986)
5. Salas, M., di Girolamo G.: Engineering applications in Chilean aquaculture. J. Ocean Technol. **5**(2) (2010)
6. Cifuentes, C., Kim, M.H.: Numerical simulation of wake effect in nets under steady current. In: Proceedings of the ASME 2015 34th International Conference on Ocean, Offshore and Arctic Engineering OMAE2015, 2015, pp. 1–10

Using Computational Fluid Dynamics to Improve Hydraulic Design of an Internal Element in a Gunbarrel Tank

Mayra Agustina Pantoja-Castro, Jose Marcio Vasconcellos, Benjamín Portales-Martínez, Ángel Gómez-González, José Manuel Domínguez-Esquivel and Francisco López-Villarreal

Abstract Once oil has been extracted from land or marine deposits, it is processed by three recovery techniques (primary, secondary, and tertiary) to eliminate or separate all the desired components from those that are not, such as water and salts. In fact approximately 15% of the well potential is recovered from the primary technique associated with the process of extracting oil via the natural rise of hydrocarbons to the surface of the earth, or using pump jacks mainly. The second includes the injection of gas or water to the well to displace the oil and move it from its resting rock, and the tertiary technique, also called enhanced oil recovery, is used to extract the remaining oil from reservoirs where primary and secondary techniques are no longer cost effective. EOR includes use of thermal, gas chemical hydrodynamics, or combined EOR. This technique includes use of the Gunbarrel tank, which can obtain high efficiency and identify the variables of operation; it is important to model this process considering a distributor as a horizontal tube located at the bottom of this tank where the oil flows through small holes. The improvement of the hydraulic design of the distributor in a separator tank using computational fluid dynamics, reporting six cases, was carried out in this work.

M. A. Pantoja-Castro
Universidad Juárez Autónoma de Tabasco, 86040 Villahermosa, TAB, Mexico

J. M. Vasconcellos
Universidad Federal do Rio de Janeiro, Rio de Janeiro 21941, Brazil

B. Portales-Martínez · Á. Gómez-González · J. M. Domínguez-Esquivel
F. López-Villarreal (✉)
Instituto Mexicano del Petróleo, 07730 Ciudad de México, CDMX, Mexico
e-mail: iq.pacolopez@hotmail.com

B. Portales-Martínez · F. López-Villarreal
Consejo Nacional de Ciencia y Tecnología, 03940 Ciudad de México, CDMX, Mexico

A. Vega Sáenz et al. (eds.), *Proceedings of the 25th Pan-American Conference of Naval Engineering—COPINAVAL*, https://doi.org/10.1007/978-3-319-89812-4_14

The results show a preliminary model design of this internal distributor that improves when used with 10 holes starting with a diameter of 6×10^{-4} m for the first hole increasing each one by 6×10^{-5} m.

Keywords Desalted · Dehydration · Emulsions · Simulation

1 Introduction

Oil is a mixture of different types of hydrocarbons composed mainly of carbon and hydrogen. Since this discovery society has benefited from associated products as fuel and raw materials that serve as the basis of pharmaceuticals, cosmetics, and footwear, among others.

Some researchers are focused on the development of processes or techniques for maximum extraction of this nonrenewable natural resource [1, 2].

The oil production process begins with the extraction of hydrocarbons from the deposit using natural flow and artificial lift. The first uses the pressure of the deposit itself to achieve the extraction and subsequent transport of the hydrocarbon from the well to the surface, whereas with artificial lift it uses the help of mechanical, electric, or submersible pumps or pumping stations, among others [3, 4].

One pursuit in each process or technique (primary, secondary, and tertiary recovery), is to recover the maximum possible amount of oil in the reservoir from natural and artificial uplift.

The primary recovery starts with natural lifting and use of pumps; there is no addition of mass or energy inside the deposit; however, at a certain point of production this technique is inefficient and oil extraction remains at around 15% or less. When this process is expensive, the addition of components to the well is necessary, resulting in the secondary recovery that is used when the reservoir energy is not enough and pumping is not convenient, making it necessary to take some measures such as the injection of a less expensive fluid than oil.

Because a large amount of mainly heavy oil (approximately 60%) is trapped in the pores of the reservoir rocks, it is necessary to use some treatments also known as recovery oil methods that involve the production of heavy oil, which is injected with compounds typical of the deposit and also different from this one with the target of modifying the physical–chemical conditions of the crude oil and fluids inside the well [4, 5].

An increase in the production of oil reserves from light to heavy compounds is the main objective of each oil industry, and over the years they have adapted to the new extraction technologies [6].

In Mexico and other countries, the largest amount of oil recovered is heavy crude oil, and this variety makes it difficult and expensive to produce and transport. Due to its low mobility and flow at room temperature, this sometimes leads to search and model different ways to study the behavior of recovery processes [6].

Our goal in this work was to obtain a preliminary design model of an internal in a Gunbarrel tank (tertiary recovery) using computational fluid dynamics (CFD).

2 Methods

This work was studied using CFD where the geometry of a cylindrical Gunbarrel tank was built, considering specifications such as: 0.28 and 0.30 m in height and diameter, respectively, and a nominal capacity of 1.6×10^{-4} m^3/s. This tank was fed with an emulsion with a water content of less than 15% wt in crude oil. The flow velocity of the emulsion was 1.8×10^{-4} m^3/s, and the speed wash water flow used was 1.27×10^{-5} m^3/s.

The height of the washing water was measured from the bottom of the tank considering a value of 0.80 m, oil spill nozzles were approximately 3 m high and brine collectors were at the bottom of the tank.

The simulated model was developed using the parameter of the fluids involved in the process: water content (w), dehydrated oil (o), emulsion (e), fluid densities, and height of wash water, emulsion, and dehydrated crude.

On the other hand, it is specified that this tank has three nozzles, one inlet and two outlets all of 0.02 m in diameter of Schedule 40 steel. At the lower part of the tank there is a tube that functions as a distributor and contains small holes, which were analyzed using different conditions of diameters starting with 1.6×10^{-3} m and length of 0.010 m. In this study distribution and quantity of holes are considered and the results of six cases are presented.

3 Results

In this work some simulations were carried out and only six case studies presented considering a tank with a nominal capacity of 14.30 m^3, height of 3.05 m, and an outer diameter of approximately 2.74 m; data are based on the specifications given by the API 650 standard for oil storage tanks.

This tank has a radial distributor with 8 distribution branches of 0.60 m, each with 10 holes of 1.27×10^{-3} m diameter.

For all the studies analyzed in this paper, a transient analysis was made to observe interactions of the interfaces of the fluids over time.

Results of Cases I and II showed the water and oil interfaces and velocity profiles, whereas from Cases III to VI speed profiles were analyzed in detail within the distributor and their speed profile analyzed; these studies were done to obtain the optimal conditions and to be able to model the process in 2D.

3.1 Case I

Initially, to reduce the computation time required for the simulation of this process, the geometry of the longitudinal slice of the tank was used in 2D and to evaluate the necessary computation time a first test was performed with a less dense mesh. This generates fewer elements to analyze; calculations are facilitated but sometimes the resolution is not very good.

Figure 1 shows a mesh used for study Case I, which consists of 6640 triangular elements. The computation time was 259 s, using 1.28 GB of physical memory.

3.2 Case II

In order to check the water–oil interface and identify velocity profiles inside the tank, it is necessary to increase the density of the mesh. This act also increases the size of the problem and the computation time of the simulation. In Case II an extremely fine meshing consisting of 130,920 triangular elements was used (Fig. 2). Meanwhile a computation time of 51,570 s, and a physical memory of 6.87 GB was required. The snapshots of the water–oil interface at different times and speed profiles inside the tank at different times were analyzed. This study allowed us to notice that the exit of the emulsion for each of the holes in the distribution branches, where flow occurs mainly at the center of the distributor, caused the flow to be uneven throughout the branches, which is not a desired process. The velocity profiles corroborate this information showing that fluids move mainly to the center of the tank.

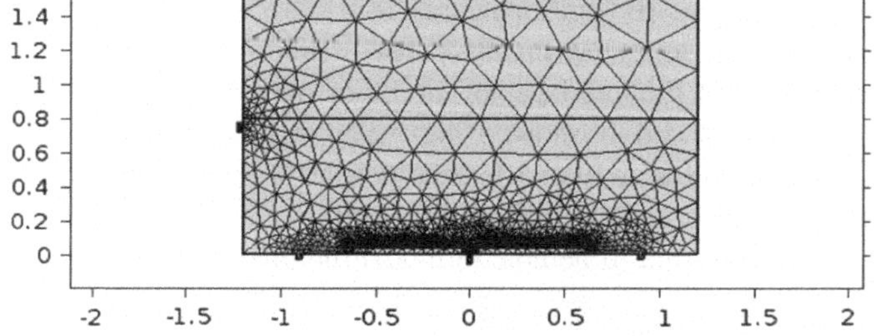

Fig. 1 Mesh representation to analyze Case I using 6640 triangular elements

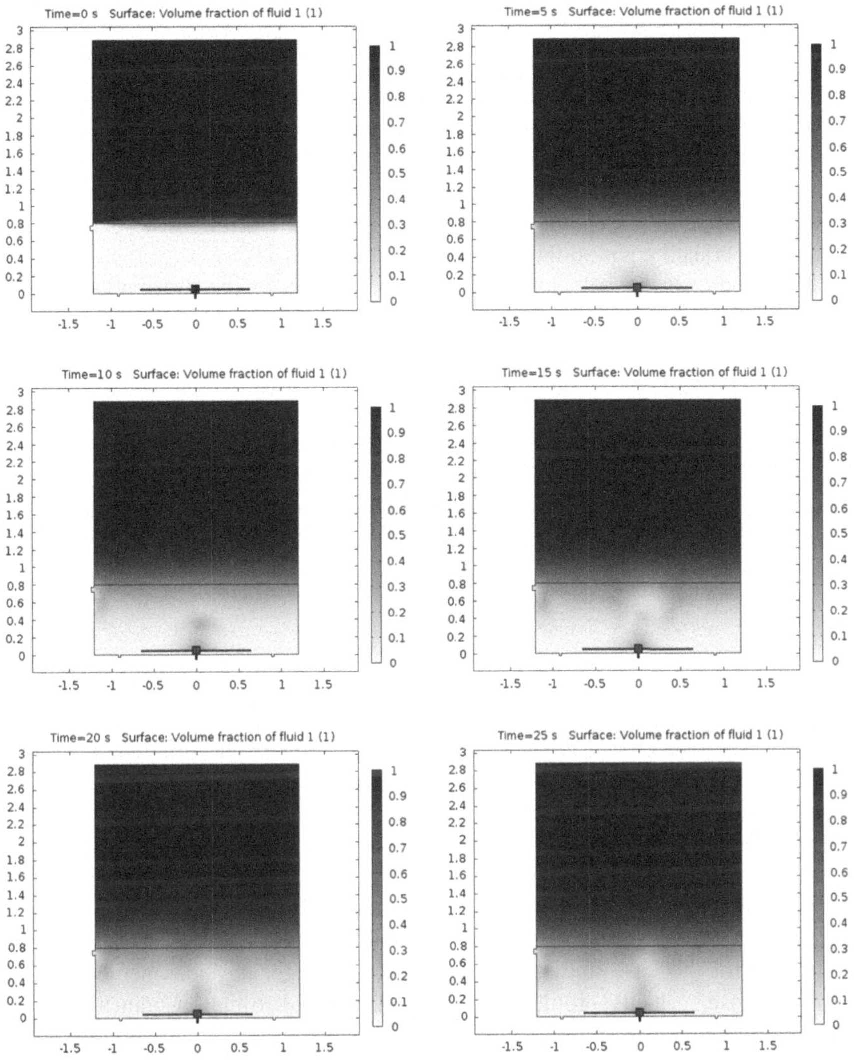

Fig. 2 Snapshots of one case for oil–water interface at: 0, 5, 10, 15, 20, and 25 s

On the other hand, it can also be observed that the washed oil confers stability to the flow inside the tank. The washed oil has high residence times (from 43,200 to 86,400 s) allowing the oil–water emulsion to separate. Some simulations were carried out changing the meshing of the designed geometry; in this work only the most representative results are shown. This design process desired to make the emulsion evenly distributed in the wash water bed.

3.3 Case III

In this case steady-state analysis of the emulsion flow was made through one of the distribution branches. Geometry of these simulations was developed in 3D and a profile of flow velocities was analyzed along the distribution branches.

Velocity profiles in the distribution branch were studied using an isometric and lateral perspective, respectively, considering the diameter of the orifices of the distribution branch equal to 0.012 m.

On the other hand, velocity profiles along the axis of the distribution branch were simulated, and their results showed that flow velocity along the axis of the distribution branch decreases in each hole.

In general, the results of these analyses have indicated a more pronounced decrease in flow velocity where the holes of the distribution branch are located; in the same way it can be observed that after the fourth orifice, the flow velocity of the emulsion has decreased by half with respect to the initial speed at the entrance of the branch. For this reason, it is proposed that the holes closest to the center of the distributor have a smaller orifice in comparison with those holes located in the farthest radius and this gives rise to the next case of study.

3.4 Case IV

For the study, this case was proposed with a diameter of the orifices of the distribution branch varying linearly with each hole from 0.00012 to 0.0012 m, respectively, with increments of 0.00012 m. Again, the profile of flow velocities along the axis of the distribution branch is analyzed.

Due holes on the left are smaller, the meshing in these areas must be denser, and therefore the number of tetrahedral elements to be analyzed was increased. A meshing consisting of 3,671,360 tetrahedral elements was used and a computation time for this case study was 1204 s using a bigger physical memory.

In this case, velocity profiles were studied in the distribution branch from an isometric and lateral perspective, respectively. A study of the velocity profiles along the axis of the distribution branch was made, where it can be seen that the flow velocity along the axis of the distribution branch decreases by half after 8 holes.

3.5 Case V

With the conditions and results obtained in the previous Case (Case IV), the tank was simulated again, incorporating the modifications made to the distribution branch. This simulation was done in a transient state for a time interval of 0 to 50 s. The diameter of the distribution orifices was varied from 0.00012 to 0.0012 m, respectively, with increments of 0.00012 m using a meshing consisting of 186,406 triangular elements.

According to the simulations obtained it can be seen that unlike Case II, flow occurs mainly at the ends of the distribution branches, the emulsion does not flow through the orifices near the center of the distributor, and distribution of emulsion droplets inside the tank is not uniform. This causes preferential flows and a decrease in the efficiency of the mass transfer speed between the two phases.

3.6 Case VI

To carry out this analysis, it was proposed to modify the distribution of each hole diameter in the distribution branches. Each branch has 10 holes; the diameter of the first hole is 0.0006 m, and the increments are 0.00006 m. In this way the measurements of the distributor orifices are in the range of 0.0006 to 0.00114 m, respectively.

A mesh for this simulation consists of 154,913 triangular elements.

Figure 3 shows the mesh for this case was made an approach in the center of the distributor and you can see that the meshing is denser in areas where the flow is higher.

Figure 4 shows the snapshots that describe the tracing of the water–oil interface through the time horizon. Likewise, analysis of velocity profiles developed inside the tank was carried out over time for this case. It can be seen that the flow of emulsion through the orifices of the distributor is uniform, and the flow rates are better distributed compared to the distributor configurations shown in the previous study cases. An improvement in the distribution of flows is related to the improvement in the efficiency of mass exchange between the emulsion and the washing water. A distributor configuration shown in this case corresponds to the improved hydraulic design of the distributor in the Gunbarrel dehydrated desalination tank.

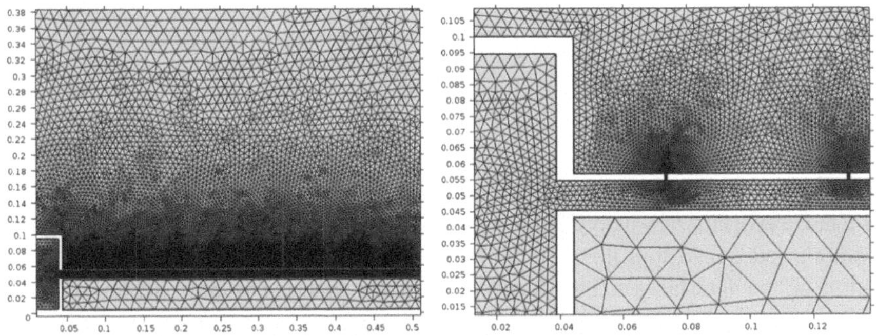

Fig. 3 Mesh details used in Case VI

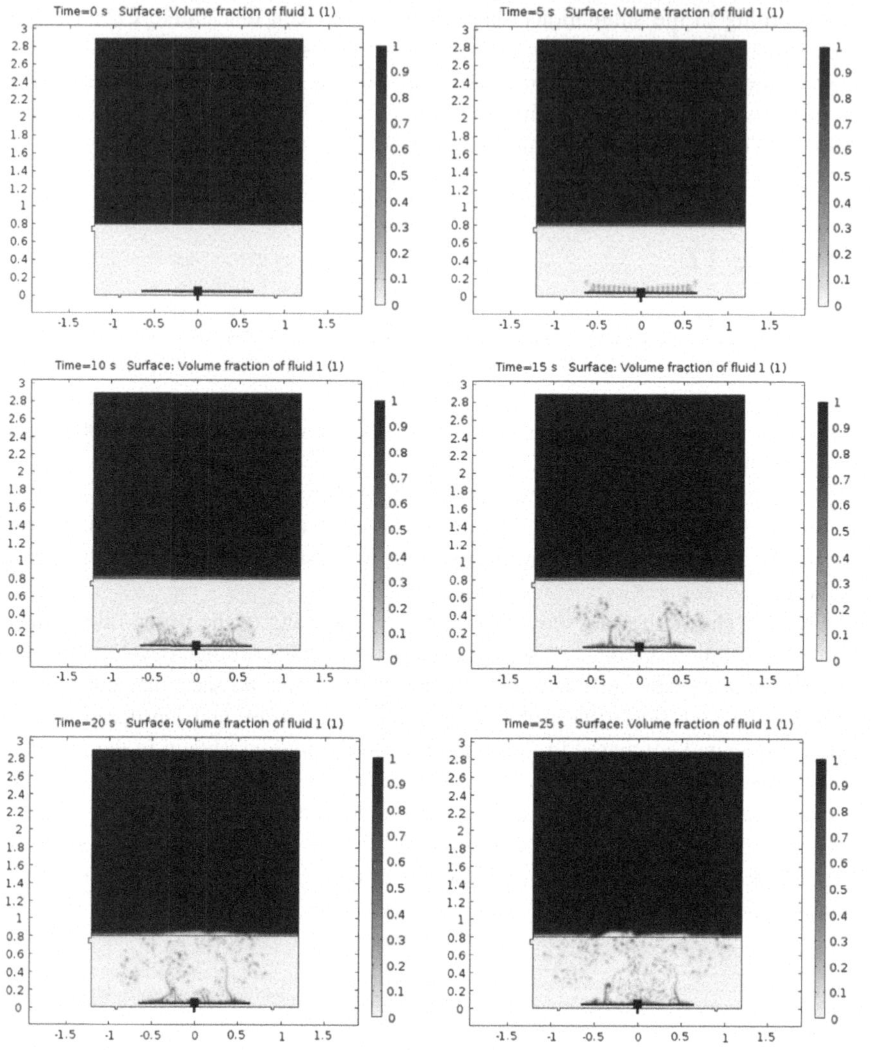

Fig. 4 Snapshots of oil–water interface obtained in Case VI

4 Conclusions

In this work we report the result of improving a hydraulic design of the distributor of a tank separator of a heavy crude called the Gunbarrel tank. Different simulations were carried out to find the right type of meshing, through which there is a resolution of the interfaces using short computation times. The modifications of the number and distribution of holes at the distributor were also carried out in order to improve the efficiency of the washing and dehydration processes.

Six case studies were carried out; the last one indicating an increase of efficiency in the distribution of flows when the configuration of the distributor has 10 holes, and the diameter of the first is 0.0006 m, with increases considering this last value. The mesh used for this simulation was 154,913 triangular elements.

Results allowed us to analyze that the use of computational fluid dynamics is a useful tool to study mainly the optimal behavior of a process, finding the design variables that allow experimentally performing this process in a pilot plant.

References

1. Bassane, J.F.P., Sad, C.M.S., Neto, D.M.C., Santos, F.D., Silva, M., Tozzi, F.C., Filgueiras, P. R., de Castro, E.V.R., Romão, W., Santos, M.F.P., da Silva, J.O.R., Lacerda Jr., V.: Study of the effect of temperature and gas condensate addition on the viscosity of heavy oils. J. Petrol. Sci. Eng. **142**, 163–169 (2016)
2. Liu, Y., Cheng, Q., Zhang, B., Tian, F.: Three-phase hydrocyclone separator-A review. Chem. Eng. Res. Des. **100**, 554–560 (2015)
3. Martínez, P.R., Mosqueira, M.L., Zapata, R.B., Mar, J.E., Bernal, H.C., Clavel, L.J.C., Aburto, J.: Transportation of heavy and extra-heavy crude oil by pipeline: A review. J. Petrol. Sci. Eng. **75**, 274–282 (2011)
4. Payne, J.R., Kirstein, B.E., McNabb, G.D., Lambach Jr., J.L., De Oliveira, C., Jordan, R.E., Hom, W.: Multivariate analysis of petroleum hydrocarbon weathering in the subarctic marine environment. Int. Oil Spill Conf. IOSC **2005**, 4932 (2005)
5. Qu, H., Wang, X., Liu, J., Zhou, X., Zhu, H., Liu, D.: The research on optimization design for export pipeline of watercut crude oil with high viscosity. J. Petrochem. Univ. **28**(4), 81–85 (2015)
6. Yi, P., Long, W., Feng, L., Wang, W., Liu, C.: An experimental and numerical study of the evaporation and pyrolysis characteristics of lubricating oil droplets in the natural gas engine conditions. Int. J. Heat Mass Transf. **103**, 646–660 (2016)

Challenges and Opportunities of the Transition to a New Management Model in a Shipbuilding and Design Company

Natasha Águila Valdés and Yasmin C. Brey Fornaguera

Abstract In the process of integrating management systems in compliance with international regulations in a shipbuilding design and construction company, an important challenge was raised relating to the maintenance and certification of its quality management system: the transition of the system with views to comply with the new requirements of ISO 9001: 2015, without losing the recognition of accredited third parties. The work focuses on the diagnosis carried out, the planning of the transit from one model to another, the adequacy of the processes, the training of the personnel, and the modification and preparation of documentation of the management systems from an integrating vision. The first results of the transition process are set forth, where not only challenges are identified, but also opportunities to improve the management systems and the ship design, construction, and repair processes.

Keywords Quality management · ISO 9001: 2015 transition · Shipbuilding

1 Introduction

1.1 The Quality Management System (QMS) in a Shipbuilding and Design Company

The Empresa de Proyectos Navales, CEPRONA, was founded as a state-run (public) company for the development of ship designs in 1976, subsequently expanding its performance not only to new platform fishing vessels, with the use of several construction materials (wood, steel, ferrocement, and plastic), but also to

N. Águila Valdés (✉) · Y. C. Brey Fornaguera
Empresa de Proyectos, Construcciones y Servicios Navales CEPRONA,
Calle 26 No. 56 e/ 11 y 13, Vedado, Plaza de la Revolución, La Habana, Cuba
e-mail: valdesnaty06@gmail.com

Y. C. Brey Fornaguera
e-mail: yasminbrey64@gmail.com

A. Vega Sáenz et al. (eds.), *Proceedings of the 25th Pan-American Conference of Naval Engineering—COPINAVAL*, https://doi.org/10.1007/978-3-319-89812-4_15

boats for transport, cargo, and recreation. The technical services were diversified, extending from the shipbuilding sector to civil works, mainly hydrotechnical ones, and to the industrial sector.

The implementation of a quality assurance system began in 1998, achieving its certification in 2000 by Lloyd's Register Quality Assurance (LRQA), the Havana Office and the National Standardization Office (ONN), the national authority for these matters in Cuba. The certificate was granted by ISO 9001: 1994, the model applicable to the design of ships and naval technical services; subsequently, along with the diversification of services, this field of certification has been extended to others. The current services cover design, construction, repair and maintenance of vessels, recreational floating and nautical means; design, construction, installation and repair of equipment; designs and services of architecture and engineering of hydrotechnical works and non-tourist diving; ship electronics services; engineering services, consultancy and applied research in ship, civil, and industrial works. Likewise, the system itself evolved from the transition to the new models of the standard approved in 2000 and 2008 and maintained the recognition of third parties.

In 2011, the company had an important change caused by a restructuring in state-run companies: it merged with other companies linked to the shipbuilding industry, within the Ministry of the Food Industry (MINAL), expanding its business portfolio to the construction of vessels and other services such as ship electronics. The merger of companies, as a strategy for growth and optimization of resources, was positive as the value of the resulting entity is higher than that of the companies that joined. They became basic production units, two of them being located in services (NAVALEC and PROYECTOS) and two shipyards (ASTIGAL and CHULLIMA) in the capital of the country and three more shipyards in three provinces of the center and east: ASTISUR, ASTINOR, and ASTIGOLF. The new Empresa de Proyectos, Construcciones y Servicios Navales (CEPRONA) has been a national company ever since.

At the time of the merger, the quality management system (QMS) was in the process of integration to other systems according to international and country-specific requirements expressed in various regulations. In addition to the Occupational Health and Safety System according to ISO 18000 and the Environmental Management System according to ISO 14000, the Internal Control System and the Management and Business Administration System were implemented according to national legislation.

The new production units did not have the aforementioned systems implemented at the same level and only one had its QMS certified. This slowed down the integration process of the systems in the Oficina Central and the UEB PROYECTOS e INGENIERÍA, which were areas included in the scope of certification, according to the new structure; the status of implementation in those units had to be diagnosed, and a timetable for the development and integration of the systems in the new conditions was established. The continuous improvement method was applied, planning the changes, training the personnel involved, including the internal auditors, and carrying out the planned actions in a controlled manner, measuring and improving the results in the reviews led by top management; nevertheless, we continued to ensure compliance with the requirements in those business processes engaged in the certified QMS [1].

As a consequence of the applied continuous improvement approach, the results in sales levels, customer satisfaction, and the development of all processes were not affected by the new conditions and the recognition by third parties was maintained. In April 2015, CEPRONA recertified its system by ISO 9001: 2008, with the scope "Naval architecture projects, including the design of vessels up to 65 m in length, and maritime and industrial technical services. Installation and maintenance services for ship electronic systems, equipment, and computer equipment." The certificate is valid until June 2018.

As of 2012, the nationally established requirements for environmental management have been implemented in all production units and in the Oficina Central, based on the diagnosis of environmental impacts, the granting of environmental licenses, and the inclusion of the plans of the units in national projects for the protection of areas or priority bays. The requirements of the Health and Labor Management System were also implemented, although there was already compliance with the current national legislation on these issues.

It is important to highlight that as part of compliance with national regulations in the company an internal control system was implemented, which was regulated by Resolution No. 60/2011 of the General Comptroller of the Republic of Cuba. This established risk management and prevention for the achievement of objectives and their identification by processes, activities, and operations; the risk prevention plan includes the application of corrective and preventive actions, which makes it compatible with the requirements of ISO standards for business management. This regulation is based on the analysis of potential risks and this has been one of the elements taken into account for the integration of business management. The prevention of risks in the processes avoids breaches of those established by the regulations for the quality of the product, for the environment, safety, and health, and also guarantees the reasonable assurance that the information is reliable, of the efficiency, of the business efficiency, and of control of resources. Risk analysis has been the tool to achieve a matrix of preventive actions and prevent nonconformities from occurring in integrated systems. Efforts and resources are not dispersed or wasteful: priorities are established in the actions according to the severity, occurrence, and ease of detecting the risks.

2 Procedure

2.1 The New Version of the Standard: A Challenge for the QMS of a Shipbuilding and Design Company

The ISO 9001 standard has already had 25 years of history since its first appearance in 1987; to 2015 it has successively gone through five revisions by the International

Organization for Standardization (Committee ISO/TC 176 Management and Quality Assurance, Subcommittee SC 2 Quality Systems). Its purpose has been the establishment of generic requirements, which are applicable to all organizations, for the implementation of a management system regularly to provide products and services that meet the client's requirements and applicable legal and regulatory requirements. The International Accreditation Organization (IAF) establishes a three-year term for the transition to the new model, counted to September 2018 [2]. The organizations that have, like CEPRONA, the system certified by the 2008 model must continue receiving the maintenance audits to their management system according to the foreseen schedule, with the accredited bodies and at the same time they must make the necessary changes and adjustments in the system to transit to the new model [3].

This implies a great challenge when the business does not stop and there are established goals to differentiate the company from the rest in the market, to improve performance, and acquire more sales based on the systematic compliance of the expectations of the customers.

Changes to Take into Account

The ISO 9001: 2015 standard incorporates formal and content changes that decisively affect the strategy to be carried out for the transition process. For the purposes of this paper, the most relevant ones are highlighted, taken into account in the strategy applied in the transition process of CEPRONA's QMS.

First, it is formally based on the high-level structure (HLS) provided by the so-called Annex SL, whose objective is the alignment of all ISO management systems. In this way, the formal coherence of the clauses and subclauses between the standards of the systems is maintained and a common language is applied to all new editions. The regulation has 10 clauses, establishing the requirements to be met from No. 4 and are organized according to the cycle of continuous improvement (Plan-Do-Check-Act), hereinafter the PDCA cycle, in such a way that clauses 4, 5, and 6 correspond to the Planning stage, clauses 7 and 8 to that of Do, 9 to Verify, and 10 to Act.

Second, new concepts are included and others are updated. The standard continues to be based on the principles of quality management, but now eliminates the principle of the system approach to management; some other changes in its formulation are maintained with the other principles: customer focus, leadership, commitment of people, approach to processes, improvement, decision making based on evidence, and relationship management. It specifically emphasizes the approach to processes as a way to understand and control the interrelations and interdependencies between the processes of the management system, and improves overall performance (efficiency and effectiveness of the organization). It is also added that process and system management in general can be achieved using the PHVA cycle with a global approach of thinking at risk.

It is notable that a key component is the risk-based approach; although preventive thinking based on risks was implicit in the previous model, this is now explicitly stated as an important pillar that allows determining those factors that may cause the planned results not to be achieved. Risk management involves identifying, evaluating, and treating them through actions that eliminate, mitigate, or reduce their impact. The purpose of this is to improve confidence and security in satisfying the expectations of customers and the interested parties, as well as to generalize in the organization a proactive prevention culture. It is worth noting that the standard clarifies that organizations must also address opportunities to increase efficiency, achieve better results, and prevent negative effects; it understands the opportunities from two points of view, those generated from favorable events or the associated positive effects of a risk.

New concepts are included: the context of the organization and the internal and external issues, positive or negative, that the organization must identify and follow up due to their relevance and effects on the processes, products, and services and the management system in general; the necessary identification and treatment of the interested parties such as those persons or groups of people, natural or legal, that may affect, be affected or be perceived as such by a decision or activity of the organization; the actual approach of the risk associated with threats and opportunities, where the planning replaces the preventive action, for it no longer appears in the body of the standard accompanying the corrective actions.

Other concepts are maintained but contain in themselves more detailed requirements: leadership, with a more committed top management and focused on harmonizing the management system (processes, objectives, policy) with the strategic direction; communication, with more explicit and detailed requirements with the interested parties; the documented information, which replaces the concepts of documents and records, with greater flexibility and seen as the basis for the materialization and preservation of knowledge and its integration into processes; performance evaluation with emphasis on measurement indicators, to guarantee the validity of the results; a more detailed analysis of nonconformities and corrective actions is required, as well as more detailed requirements in the entries for review by management.

2.2 The Transition of the QMS in CEPRONA: First Results

CEPRONA recertified its system in July 2015, and set out the following objectives (Fig. 1).

(a) Maintain the certification of the QMS and extend the scope of the certification to nontourist diving services and shipbuilding and repair services.
(b) Continue integrating management systems.
(c) Transition to ISO 9001: 2015 and recertify the QMS before September 2018.

Taking into account the three pillars stated in the regulations, the process approach, the PHVA improvement cycle, and risk management, the work was deployed based on the following steps.

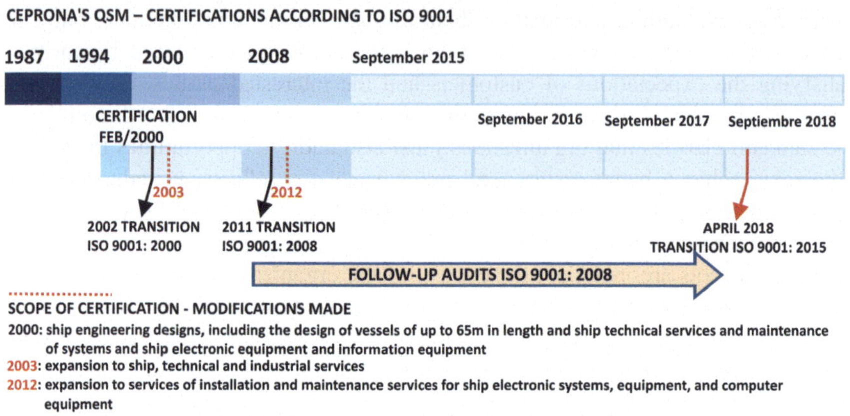

Fig. 1 Evolution of the certification of CEPRONA's QMS with respect to the ISO 9001 standard. *Source* The author

First. Preparation of a diagnosis of the QMS regarding the changes; the DAFO matrix was applied to the challenge of transiting and recertifying the system. Among the identified strengths are:

(a) That the implemented QMS was effective and in a state of progress, with improvement projects that would enable its certification to be extended within three years.
(b) Regarding the progress in the integration with other systems, it went according to plan in all areas.
(c) Work prevention from the risks in the processes in compliance with the aforementioned Resolution No. 60/2011 established a good basis to meet the new requirements; it only remains to harmonize what has been implemented.

Of the weaknesses identified, the following can be related.

- Insufficient knowledge about the new requirements of the standard and its implementation at different levels of the organization.
- Insufficient analysis and monitoring of the context of the organization (internal and external factors) that affect the processes, as well as the needs and expectations of all interested parties.
- Incomplete identification of risks by processes and their management and nonconformities and corrective actions are not always taken into account; opportunities are not always taken advantage of and risk management has been focused more on the negative aspect.
- Documentary system still extensive, susceptible to optimization.

Second. The risks and opportunities in the transition process were identified; although three years for the transition seems a lot, there are multiple tasks to be accomplished with a system underway; improvement plans in process, internal audits, and maintenance to be received periodically. Among the identified risks it was foreseen that the review of the documentary system required a long time, despite the flexibility offered by the new standard, giving greater relevance to the information than to the document; due to the characteristics of the company (cultural, human resources, etc.) and specificities in the provision of services throughout the national territory, optimization would not lead to a notable decrease in procedures and instructions.

The proposed goal of extending the scope of certification could be postponed for specific reasons in infrastructure issues in dry docks and hydrotechnical works after the system transition, due to the need to prioritize the certification thereof.

On the other hand, the transition offers CEPRONA, in particular, the opportunity to complete risk management and make it a fundamental part of the system. With the integration of the management systems and the risk prevention plan associated with the processes, it was taken into account in the audits and reviews by management as part of compliance with national and international regulations. The very high-level structure of ISO 9001: 2015 facilitates the integration process with the management systems regulated by ISO. This, in addition, allows a greater flexibility in the documentary support of the systems; that of CEPRONA can be further simplified by optimizing and improving the quality of the documents that are kept. Although the regulations no longer require a quality manual or mandatory procedures, it was decided to maintain the integrated management manual and part of the procedures in the system for their usefulness regarding the purposes of audits, personnel training, and commercial purposes. Of course, exceptions are those imposed by national regulations.

Continuing in the planning stage of the improvement, the strategic direction was revised and adjusted in terms of mission, vision, and strategy, based on the strategic analysis (DAFO Matrix) after five years of the merger of the companies. This served as the basis for the revision and adjustment of the quality policy and the objectives, and for the analysis of the context of the organization.

During the planning stage, the transition plan was prepared, which was checked in internal audits and maintenance by third parties as part of the proposed changes to the QMS. The improvements to processes and the capacity of the necessary resources were planned, the preparation of the personnel to face the changes, and the adjustments in the documentation, among others.

Third. At the stage of the PHVA cycle, the activities planned and carried out or in the process of execution focused on:

Training of personnel involved in the transition (*top management, internal auditors, and process managers.*) Top management included all the heads and officers at all levels and in all areas. The team of internal auditors was composed of specialists from all units; two of these auditors qualified as leading auditors.

Conducting the analysis and revision of the Organization's Context, with the identification of the issues that affect it internally and externally, at the local, national, or international level. The result was documented in the management manual, including its follow-up and revision in the revisions by the management.

Identification of interested parties, their needs and expectations, at the level of the organization and in each basic unit, recording the result in the management manual and its follow-up and revision, including in the revisions by management.

Revision and adaptation of the process map; the cards of each process were modified and completed according to the updated map. The risks and opportunities to achieve the expected results were addressed.

Revision and adaptation of the documentary system, opting to keep the manual referred to as a means of documenting relevant information. General procedures and instructions were revised and are being revised, as well as those specific to the production units, related to the services they provide.

Planning of internal audits including as criteria the requirements of the new standard. The risk approach in the planning of each audit activity has been a working tool to guarantee its execution.

Fourth. Although the planned activities have not yet been completed, the first measurements have been made, corresponding to the verification phase of the PHVA cycle, with the aim of catalyzing the process through internal audits and conducting revisions by management, with the incorporation in its agenda of the new requirements in the entry and exit information and those other requirements that are established in other clauses (e.g., those contained in clause 4 "Context of the Organization."

First Results of the Transition

At the end of June 2017, a self-evaluation was conducted with a view to the revision by management; from an applied questionnaire Table 1 was obtained.

Taking into account that the planned date to transit in the QMS certification is April 2018, it was considered feasible to execute the total of the activities still in process and complete a cycle of internal audits to the areas involved in the current certification scope and another revision in January 2018. Shortage were identified in the maintenance audit carried out in October 2017 by the certifying bodies.

Parallel to this effort, the QMS is certified by the ISO 9001: 2008 standard, and the planned improvement plans are maintained. There is a team of internal auditors

Table 1 Results of the diagnosis of the transition of CEPRONA's QMS, June 2017

Clause ISO 9001: 2015	% Implementation	Actions
4. Context of the Organization	80	KEEP
5. Leadership	92	KEEP
6. Planning	60	IMPROVE
7. Support	95	KEEP
8. Operation	90	KEEP
9. Evaluation of performance	95	KEEP
10. Improvement	89	KEEP
Total measurement result		86%
Global qualification in quality management		HIGH

Source The author

with updated training in the requirements of ISO 9001: 2015, headed by two leading auditors, and the top management and technical personnel involved have been updated.

Among the results, it is highlighted that the company won a gold medal for Product Quality at the International Fair of Havana in 2015, for the 18 m "ARGUS" tuna fishing boat, greatly influencing the maintenance of a certified QMS since 2000.

The business results have not been affected by changes and adjustments in the system, because sales have been rising steadily since 2011, a time of great transformations for the organization, so that at the end of 2016 there was an increase of 18.9%, even though it has been affected by a depression in the portfolio of shipbuilding contracts. The client portfolio has been renewed year after year, but the organization maintains a high loyalty index. Customer satisfaction has remained above 97% globally and the level of customer complaints has not exceeded three in a year.

The top management has identified joint business opportunities with foreign investors in the provision of ship electronics services, which are included in the scope of certification and with the possibility of exporting them.

The strategic planning of the company has been aligned with the policy and the objectives and identified the external and external factors to the organization that interact, influence, or affect its performance from the legal, technological, market, cultural, economic, and social point of view. The monitoring of these conditions is already subject to analysis in the revisions by the management of the QMS.

3 Conclusions

The three-year term to move to the new model of the ISO 9001: 2015 standard constitutes a challenge for organizations that, like CEPRONA, already have a certified QMS, while the business must continue, improve its operational performance, reduce errors, become a benchmark competitor in the market, and provide constant service to its customers, motivate and engage staff with the outlined strategies, and, in short, expand opportunities for the entity. It could mean, as in the case study, to establish new priorities and postpone goals such as the extension of the scope of certification to other services at a time when the construction of new vessels for platform fishing is depressed.

However, the ISO 9001: 2015 standard itself recognizes three important elements on which its implementation is based: the application of the PHVA cycle to the processes and to the QMS as a whole, the approach to processes, and risk-based thinking [4]. Organizations in the transition period can optimize time and resources if they approach the task from this perspective; CEPRONA has verified the feasibility of doing so, especially when the personnel involved are the same that guarantee and follow up on the management systems implemented. In this way, the challenge of transiting has become the opportunity to fit and optimize the QMS certified from the modifications or new requirements, to integrate the quality management more easily with that of other systems, and, in addition, to improve and deepen the management of the risks in which it had been working from the perspective of preventive actions and partially in compliance with other regulations.

The consideration of the context of the organization and the follow-up of internal and external issues that could affect or favor the opportunities identified for business management gain relevance, and its follow-up enables decisions to be made when faced with market effects, crisis, favorable or not economic conjunctures, and so on.

Finally, the transition is the only option to maintain recognition by third parties of quality management, as a stable framework to achieve projected goals, systematically meet the expectations of customers and interested parties, improve performance, and continue to make a difference in the current market and that of the future.

4 Recommendations

Both for organizations in the sea and maritime sector or other sectors of the economy, whether large or small, with management systems implemented or not, it is considered advisable to:

- Assume the transition as an improvement to management with the PHVA cycle, using all the time, resources, and efforts necessary in the planning stage as a guarantee to achieve the expected results.
- Identify and manage the risks and opportunities in the transition or implementation of ISO 9001: 2015 in the planning stage.
- Even when all the planned activities have not been completed, carry out, when appropriate, measurements through internal audits and revisions by management to verify and act in a timely manner and leave documented information on these activities.

References

1. Valdés, A.. Natasha "Integration of Management Systems: from Quality to Environment in a Shipbuilding and Ship Design Company", CD XXIV COPINAVAL Uruguay 2015, Annex of the Book of Magisterial Lectures and Free Papers, pp. 24–40
2. CUBAINDUSTRIA Homepage. www.nc.cubaindustria.cu/, last accessed 14 July 2017
3. BSI British Standards Institution Homepage. http://www.bsigroup.com/es-ES/, last accessed 14 July 2017
4. International Organization for Standardization, ISO 9001:2015, Quality Management System Requirements ISO/TC 176 SC 2

Industry 4.0. An Opportunity for the Relationship Between University and Shipbuilding in the Future

Salvador Naya

Abstract The Industry 4.0 concept, also called the fourth industrial revolution, represents a new era for the organization of production. One of the objectives of this fourth industrial revolution is the implementation of the so-called "smart factory," capable of greater adaptability to the needs and production processes, as well as a more efficient allocation of resources, opening the way to a new industrial era. In the context of naval engineering it presents a great opportunity to improve both the shipyard and the construction process itself of a ship.

Keywords Industry 4.0 · Shipyard 4.0 · Big data · Quality control Digital twin

1 Industry 4.0

1.1 Introduction

Industry 4.0 represents an industrial revolution that will undoubtedly mark, and is already marking, important social changes in the coming years. Its characteristics are to make an intensive use of process simulation, to use the so-called Internet of Things, big data, simulation, and other cutting-edge technologies, with the primary goal of developing industrial plants with much better interconnected production chains and obviously much more competiveness.

In the case of shipbuilding, Industry 4.0 is already being implemented in different shipyards with the aim of reducing production times and planning construction tasks more efficiently. In this paper some cases of success of application of these new tools are exposed, especially those of the statistical area, such as

S. Naya (✉)
MODES, CITIC and ITMATI Groups, Department of Mathematics,
Escuela Politécnica Superior, Universidade da Coruña, A Coruña, Spain
e-mail: salva@udc.es

A. Vega Sáenz et al. (eds.), *Proceedings of the 25th Pan-American Conference of Naval Engineering—COPINAVAL*, https://doi.org/10.1007/978-3-319-89812-4_16

dimensional quality control, naval reliability, or simulation. Specifically, some of the topics addressed in the university–business transfer project, called "Shipbuilding 4.0" or "Shipyard 4.0," are analyzed, specifically the "Shipyard 4.0. The Shipyard of the Future," which is a Mixed Research Unit between the University of A Coruña and the Navantia Company, and which has been operating successfully since the end of 2015.

The concept of Industry 4.0 is motivated by the last major crisis of the 1980s when Germany, concerned about the lack of competitiveness of its companies, began to look objectively at its most direct competitors, the Asian market; this benchmarking spotlights the need to turn the industry around to make it more competitive.

The other European countries with greater or lesser speed will follow this first approach to the problem, which was raised at the Hannover Summit in 2011. In the case of Spain it has been taken as a matter of state, until the current moment in which it is difficult to fit any industrial research project where the concept of Industry 4.0 does not appear. Some other countries, as is the case of the United States, also have their own initiatives to boost Industry 4.0, such as the "Smart Manufacturing Leadership Coalition" project, which also focuses on the industrial manufacturing modalities of the future and some multinationals including General Electric are working on their own projects such as the industrial Internet [1].

The paper ends with a description of the fundamentals and basic tools of Industry 4.0 in the second section, then a consideration of the opportunity that represents this fourth industrial revolution for the necessary transfer between the university and the company, to present a successful story of this transference, as is the case of the Mixed Research Unit project between the University of A Coruña and the Navantia Shipyard.

1.2 Historical Evolution of Industry 4.0

As can be seen in Fig. 1, the evolution of the industry began with the dawn of the steam engine. What could be called the first industrial revolution (or Industry 1.0), with the incorporation of chain (assembly line) work, which many establish in the first automotive industries, as in the case of Ford, a substantial change would be made that together with the organization of the work was a milestone that would mark what we could call a second industrial revolution or Industry 2.0. Following this thread of the industry, the biggest change in manufacturing has been without a doubt the massive implementation of electronics and computing, which could be defined as Industry 3.0, hence the current situation, which we call Industry 4.0, has been in place for very few years, because these evolutions have constantly taken shorter intervals of time (see Fig. 1).

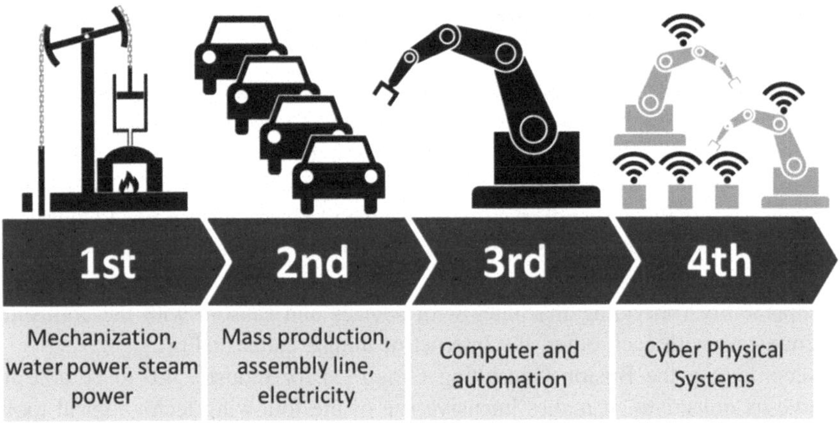

Fig. 1 The complexity of the industry methodological changes as a function of time

1.3 Basic Pillars of Industry 4.0

Classically, Industry 4.0 is based on five basic pillars, which are the following.

1. Digitization: Given that Industry 4.0 implies a high degree of automation and digitalization of factories, it is necessary to use virtual machines (Internet and cyber–physical systems) to control physical objects, which will facilitate instant intercommunication between different workstations.
2. Flexibility and customization of production: Communication enters the different stages of production, together with the monitoring of the stock of raw materials and semi-finished products, allows great flexibility in the production process, and great adaptability to fortuitous situations, all of which can contribute to the increase and improvement of production.
3. Logistics: The Industry 4.0 system is capable of generating a regular flow of information, which can be exchanged quickly, both internally and externally. This would allow adaptations to changing situations, both internally and at a general level, which requires a great logistical development that will reduce costs.
4. Simulation: The use of simulation allows reproducing a virtual replica of all or part of the production chain, which also makes it possible to generate simulations of procedures or products and optimize the processes without expense.
5. Efficiency in energy and raw materials: Industry 4.0 also seeks energy savings as in the management of natural and human resources. With a well-organized organized system, it will be possible to address issues of energy efficiency that is also one of the pillars of this philosophy.

1.4 Industry 4.0 Technology Tools

The five pillars described in the previous section and characterize this new industry, are developed through 10 tools, which are commonly associated with this fourth industrial revolution. The objective of this new industrial philosophy is the optimization of production processes for which new concepts based on 10 key issues are used. Among these new elements that make up Industry 4.0 we would highlight the use of massive data thanks to the digitization and the use of tools for big data and the concept of culture marked DIY (do it yourself) and the possibility of interoperability (providing machines with devices and sensors with the ability to communicate with each other, the Internet of things, called IoT) [2].

According to the Boston Consulting Group [3] for Industry 4.0 to be able to achieve its objectives, it makes intensive use of the following technological tools [4]. In addition to these new tools, I would add one more that integrates part of the previous tools, the concept of digital twin [5].

The following 10 tools are:

1. Big data: The ability to collect, store, and analyze large amounts of data through sensors and actuators distributed by the company. This will allow adjusting and optimizing production, identifying in particular its inefficiencies and bottlenecks. The sensorization of the processes generates a massive volume of data that will need to be analyzed in detail for future exploitation.
2. Robotics: Autonomous technology allows machinery and robots to think, act, and react autonomously. Under this paradigm the control systems will act at a distance to outline these behaviors. This can help contribute to competitiveness, productivity, and, of course, the profitability of the company.
3. Universal integration of the system: The integration of horizontal and vertical information technology systems. This is done both within the company (vertical) and beyond its borders through integration (horizontal) with suppliers of raw materials and suppliers of equipment and services.
4. Internet of things (IoT): Thanks to the use of servers and their connection to objects it is possible that unfinished devices and products will be embedded with intelligent sensors connected to the company's data system. This will allow autonomous robotic processes to interlace and adjust production allowing real-time tracking.
5. Cybersecurity: The increase in computer attacks makes it necessary to implement cybersecurity tools to protect industrial systems and network manufacturing lines.
6. The Cloud: The use of spaces in the Cloud allows supporting a greater interchange of data between different locations of the company and beyond its limits.
7. Processes of additive manufacturing: The new concept of additive manufacturing driven by the possibility of using 3D printing to produce customized products or rapid design changes allows reductions in transport costs through onsite manufacturing and having parts in minimum time.

8. Augmented reality (AR): The use of tools that combine information with the user's environment or live video in real-time through technological interfaces such as mobile devices and smart glasses, will provide workers with real-time instructions to improve decision making.
9. Simulation: It will allow operators and system designers to model and optimize production lines, supply chains, material, or people flows and, in general, all the processes involved in the process production, reducing the actual configuration time to a fraction of what was previously thought possible.
10. Digital twin: Digital twins integrate artificial intelligence, machine learning and software analytics with data to create living digital simulation models that update and change as their physical counterparts change. A digital twin continuously learns and updates itself from multiple sources to represent its near real-time status, working condition, or position [5].

2 Opportunity for University Transfer

The universities of today have to develop the transfer of research results to the society that sustains them; in this transfer policy one of the missions of the university is centered on teaching and research, which constitute the objectives of higher education. Within this interest to intensify the collaboration between the company and the university, Industry 4.0 represents a great opportunity that will depend on our institutions to know how to take advantage of it for the benefit of both, in a culture of collaboration that should promote innovation and technological development for a modern and advanced society, based on a "win to win" philosophy.

For this, it is necessary to create models that allow the university to adopt the necessary collaboration with the company. Universities must pay attention to the technological needs of companies by contracting R&D to favor agreements and transfer contracts.

This third mission of the universities should promote the necessary boost of business "culture" in universities, which allows access to knowledge, technology, and innovation of universities by companies. The generation of collaboration of companies in the university environment is a reality that this fourth industrial revolution is facilitating and we should take advantage of, sooner rather than later.

3 Shipyard 4.0. The Shipyard of the Future

The project of the Joint Research Unit 4.0 (UMI) started in 2016. At the moment it has 15 lines of action, which are part of 12 research groups of the University of A Coruña (UDC) as well as several teams of Navantia. It is undoubtedly a good example of a transfer between the university and the company, which has a total budget of 3.5 million euros [6].

The UMI project is the result of previous benchmarking work carried out by the Navantia Shipyard and the good relationship, which already exists with the UDC, where several research groups have been collaborating with units of this company. From this study, which is developed worldwide, the need to undertake a clear line of optimization of the processes necessary for the construction of ships is detected. The path to follow is set in this UMI project, in which different research groups of the University of A Coruña collaborate under the name of Shipbuilding 4.0.

The fundamental objective of the UDC-NAVANTIA Mixed Unit (UMI) is the development of new techniques and technologies that increase the competitiveness of the shipyard, in order to face the industrial technical challenge of the F110 frigate program, through the improvement of existing production processes and the development of new ones, in such a way that productivity increases and time and costs are reduced.

Initially, the lines of research of this UMI focused on the optimization of processes and the development of new tools that would bring digitalization, robotization, and new technologies to the manufacture of ships. In order to adapt these lines of action to the new context of Industry 4.0, a series of actions have been proposed, such as the following.

- Transformation of production methods to improve shipbuilding efficiency
- Development of new logistic schemes for shipbuilding
- Automation of production processes
- Development of specific ICT tools and devices for engineering, production, and knowledge management
- Transfer of new developments to the supply chain.

3.1 Statistical Quality Control (Example of a UMI Line)

Among the different actions, the optimization of processes has been considered as one of the most essential, including different actions such as modeling and simulation of shipbuilding processes within the context of statistical quality control (SQC), using robotization techniques, and dimensional control by images. To do this, powerful statistical programs will be applied, such as the free software libraries of Software R [7], and using massive data manipulation techniques, such as the case of functional data analysis [8, 9] or the treatment of images [10].

To carry out this initiative, techniques linked to the Industry 4.0 concept, such as those described in the previous section, will be applied to the statistical control of processes in steel fabrication and pipelines in the shipyard. For this, the following phases have been developed (see https://umi.udc.es).

- Determining quality indicators. Concerning dimensional control and alignment, establishing indicators (based on measurable variables on the components or blocks) is important to measure manufacturing quality. The production process stage from steel plate reception to block completion consists of a large number of individual processes: cutting, bending, and panel manufacturing, subblock and block assembly, pipeline manufacturing, and the like. This first objective will allow the construction of indicators (at the level of the specific production processes) of the dimensional quality of the process and will identify the most critical processes.

- Process capability analysis. An important objective is to determine the capability that the different processes have to comply with the established specifications. To this aim, an important task is to analyze the probability distribution of the variables to be measured. For this capability analysis, as well as for the application of other statistical methods, it will be necessary to verify some fundamental assumptions, among others, the assumptions that the process is stable and that the characteristics studied are normally distributed. Therefore, analyzing the distribution of these measures will allow the indices to be adapted to analyze their capability. Apart from serving to quantify each process, using adequate capability indices will allow assessing whether Six Sigma criteria are being met [11].
- Process control. Once the quality indicators have been identified, the next objective is to know the internal manufacturing standards. In other words, the quality control thresholds for each process are found. These are the normal manufacturing parameters. It is often the case that these thresholds do not coincide with the standard applicable to the project or with the needs of an internal customer. This implies reprocessing in order to achieve the desired control. Analyzing the collected data, the manufacturing tolerances for each component will be established [12].
- Process improvement. Once the deviation parameters are known and analyzed, the data will identify improvement needs. This is important both to reach the minimum standard required and to improve productivity by avoiding reprocessing. The more specific these objectives are and the lower the number of variables initially handled, the easier the problems will be identified and solved. For example, if the initial objective is to build without oversizing panels, a lot of overlapping processes will come into play. So it will be difficult to draw conclusions and to propose improvement actions. If the aim focuses on the improvement of a specific process (e.g., cutting, bending, panel making, etc.) a smaller number of parameters will be analyzed and it will be easier to identify the sources of variation. Adding all these improvements in specific processes will result in complete process improvement.

Many of these actions will also seek the further study of the reliability of these processes based on RAM techniques (reliability, availability, maintainability) [12, 13].

4 Conclusions

In this paper, a reflection is made on the need for adaptation of the industry, and especially of the naval industry, to the technological changes of the fourth industrial revolution. The opportunity offered by the transfer of the productive fabric with the research centers, and especially with the universities, emphasizes that this collaboration is something that facilitates companies that can compete more easily in this globalized society.

Digitalization and personalization, which promote the new concept of Industry 4.0, allow the majority of this new methodology to be implemented in the shipbuilding processes. The proposed model of a mixed research unit, under the name of "Shipyard 4.0," can be an example to be followed by other universities and shipyards to carry out the necessary updating of the environment of this new industrial era.

Acknowledgements The author is grateful for the funding received from MINECO grants MTM2014-52876-R and MTM2017-82724-R, and by the Xunta de Galicia (Grupos de Referencia Competitiva ED431C-2016-015 and Centro Singular de Investigación de Galicia ED431G/01), all of them through the ERDF.

References

1. GE homepage. https://www.ge.com/digital/industrial-internet. Last accessed 21 Dec 2017
2. SMLC homepage. https://smartmanufacturingcoalition.org/. Last accessed 21 Dec 2017
3. Industry 4.0 homepage. https://en.wikipedia.org/wiki/Industry_4.0. Last accessed 21 Dec 2017
4. Industry 4.0: the future of productivity and growth in manufacturing industries, Boston Consulting Group's BCG perspectives. Available at: https://www.bcgperspectives.com/content/articles/engineered_products_project_business_industry_40_future_productivity_growth_manufacturing_industries/. Last accessed 21 Dec 2017
5. Digital Twin: Introduction to digital twin: simple, but detailed. https://www.youtube.com/watch?v=RaOejcczPas. YouTube. IBM Watson Internet of Things. Last accessed 21 Dec 2017
6. UMI Shipbuilding 4.0. homepage. https://umi.udc.es. Last accessed 21 Dec 2017
7. Flores, M., Naya, S., Fernández-Casal, R., Tarrío-Saavedra, J.: In qcr: quality control review, R package version 1.0. https://CRAN.R-project.org/package=qcr (2016)
8. Naya, S., Francisco-Fernández, M., Tarrío-Saavedra, J., López-Beceiro, J., Artiaga, R.: Application of Functional Data Analysis to Material Science. Recent Advances in Functional Data Analysis And Related Topics. Springer, Berlin (2011)
9. Naya, S., Tarrío-Saavedra, J., López-Beceiro, J., Francisco-Fernández, M., Flores, M., Artiaga, R.: Statistical functional approach for interlaboratory studies with thermal data. J. Therm. Anal. Calorim. **118**(2), 1229–1243 (2014)
10. Mallik, A., Tarrío-Saavedra, J., Francisco-Fernández, M., Naya, S.: Classification of wood micrographs by image segmentation. Chemometr. Intell. Lab. Syst. **107**, 351–362 (2011)
11. Naya, S.: Fiabilidad en la era del Big Data. Oportunidades y retos para el Astillero 4.0. Exponav2015. La construcción naval. Estado actual y evolución futura. En http://www.exponav.org (2015). Last accessed 21 Dec 2017

12. Naya, S., Cao, R., Jácome, A., López de Ullibarri, I., Oliveira, M., Muñoz. S.: Estimation of the availability, Reliability and Maintainability (RAM) of a submarine using the Monte Carlo simulation. In: Developments in Maritime Transportation and Exploitation of Sea Resources. Taylor and Francis Group, London (2014)
13. Grieves, M.: Virtually Perfect: Driving Innovative and Lean Products through Product Lifecycle Management. McGraw-Hill, New York (2011)

Identification of the Operating Conditions of the Main Engines of a Fast Support Vessel (FSV)

Ricardo H. R. Gutiérrez, Ulisses A. Monteiro, Luiz Vaz Pinto and Severino Fonseca Da Silva Neto

Abstract To avoid operational failures in the four main propulsion engines of a fast support vessel, which could place at risk its structure, and consequently also the crew on board, measurements were taken during the vessel's sea trial aiming at identifying excessive vibration and resonance in the engines. Data were acquired in two stages: free navigation, with 80% MPE load and during the MPE start-up period, considering a 600–1800 rpm variation in rotational speed. The measuring equipment and the points to be monitored were chosen following international standards. The data obtained were analyzed using specialized signal-processing techniques and the results were evaluated as follows: in the free navigation condition, the overall values of the vibration spectra were compared to the limits established by classification societies; for the MPE start-up period, waterfall plots were analyzed to identify the rotation harmonics that excite the engines' natural frequencies. The final results showed that two MPEs had excessive vibration, and that the main orders of excitation were 1× rpm, 1.5× rpm, 2× rpm, 4× rpm, and 4.5× rpm and the transient analysis identified that the 2× rpm, 4× rpm, and 4.5× rpm orders of excitation caused resonance in the heads and bases of the MPE cylinders.

Keywords Vibration analysis · Fast fourier transform · Waterfall plot
Main propulsion engines

1 Introduction

Offshore production facilities are located increasingly farther from the coast (as a result of deepwater exploration), complicating the logistics involved in transporting the supplies needed for the operation of each unit and for subsistence of the staff on

R. H. R. Gutiérrez (✉) · U. A. Monteiro · L. V. Pinto · S. F. Da Silva Neto
Programa de Engenharia Oceânica, Centro de Tecnologia, Universidade Federal do Rio de Janeiro, Bloco I-108, Cidade Universitária, Ilha do Fundão, Rio de Janeiro 21945-970, Brazil
e-mail: RHRAMIREZ@oceanica.ufrj.br

A. Vega Sáenz et al. (eds.), *Proceedings of the 25th Pan-American Conference of Naval Engineering—COPINAVAL*, https://doi.org/10.1007/978-3-319-89812-4_17

board. Support vessels are therefore necessary to supply offshore production facilities as efficiently as possible, reducing operational costs and ensuring safe continuous production.

Fast supply vessels (FSV) stand out among other support vessels (e.g., PSV, AHTS, etc.) for their high performance, small size, lightness, and quickness, and are therefore used to carry very urgent loads.

To ensure safe navigation, the design of FSVs must anticipate possible causes that might place the structure or the crew on board at risk. In this context, good structural integration between the hull and the main propulsion engines (MPEs) is essential to avoid excessive vibration levels and to ensure that the system does not operate in resonance [1].

MPEs burn fuel in cylinders to generate the power required by the ship. This causes vibration in the engine blocks, and this vibration is transmitted to the ship through the main engine foundation. Classification societies define maximum limits for this vibration to ensure crew comfort, structural safety, and proper function of the MPEs [2, 3].

In this framework, the objective of the present study is to assess the operational conditions of a fast supply vessel's MPEs, through an analysis of the vibration signatures collected during the vessel's sea trial using advanced signal-processing techniques.

2 Experimental Procedure

The vibration of the MPEs on board a FSV vessel was measured. The measurements were carried out at the top and base of the noncoupled sides of the MPEs that presented excessive vibration, following recommendations of international standards [4]. Table 1 shows the vessel's main characteristics.

Figure 1 shows the arrangement of the four MPEs in the vessel's engine room, and Fig. 2 illustrates the measuring points and the reference axes, which apply to all four MPEs.

Table 1 Main characteristics of the FSV

Type of hull	Fast supply vessel
Extreme length (m)	48.00
Length between perpendiculars (m)	46.10
Molded breadth (m)	9.50
Molded depth (m)	4.25
Service speed (knots)	21.0
Number of main propulsion engines	4
Gearbox reduction ratio	2.93:1
Number of generators	2
Number of bow thrusters	2

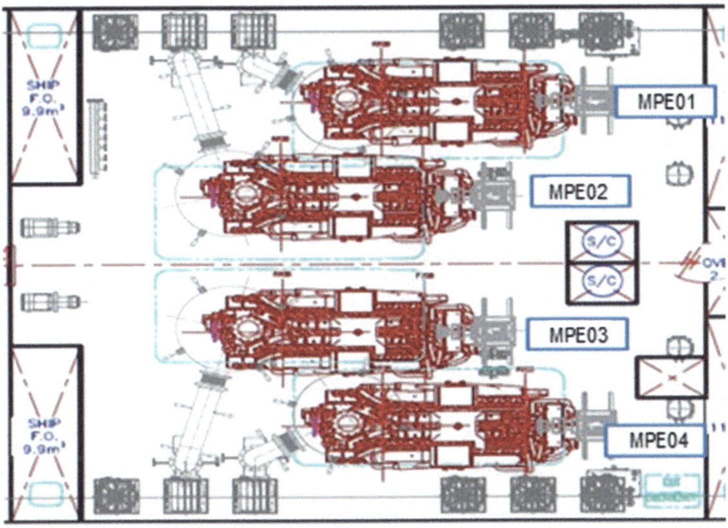

Fig. 1 Arrangement of the four MPEs in the engine room

Fig. 2 Measuring points and reference axes shown in one of the MPEs

2.1 Measuring Conditions

The measurements were carried out under two conditions.

Steady-State Regime Condition

Vibration measurements in the MPEs followed the recommendations in [4], which describes the ideal conditions for taking measurements in the steady-state regime, all of which were met for this study:

1. Ship at constant speed and with rudder angle below 2° on each side
2. Constant propulsion power
3. Sea state not more than 3
4. Propellers completely under water
5. Depth of water at least five times the draft of the ship.

The vibration levels of MPEs in the steady-state regime were measured in transit, with the engines operating at 80% power. Environmental conditions were good and did not influence the results.

Transient Regime Condition

The vibration responses of MPEs 01, 03, and 04 were measured with the engines decoupled from the propelling shafts as engine rotation increased from 600 to 1800 RPM.

3 Background

Several signal-processing methods are described in the literature (in the time, frequency, time-frequency domain, etc.). This study used a method known as FFT (fast Fourier transform) for signals obtained in steady state, and STFT (short-time Fourier transform) for the analysis of signals in transient regime.

3.1 Fast Fourier Transform (FFT)

The FFT is a tool used to obtain the distribution of vibration energy in terms of the frequency components present in the signal. It is one of the most commonly used techniques for the analysis of stationary signals and can be expressed as follows.

$$X(\omega) = \sum_{n=-\infty}^{+\infty} x[n]w[n]\mathrm{e}^{-i\omega n} \tag{1}$$

where $x[n]$ is the steady-state vibration, $w[n]$ is the window function (e.g., Hann) used in the analysis, and $X(\omega)$ is the discrete Fourier transform.

3.2 Short-Time Fourier Transform (STFT)

The STFT is a variation of the FFT for analyzing nonstationary signals, that is, when the statistical properties of the signal change over time, as is the case for the vibration signals acquired for the MPEs under a transient regime. Use of the FFT assumes that excitation frequencies do not vary during the time interval considered. However, when the rotation of an MPE varies, both its fundamental frequency and its excitation harmonics vary over time, making an adaptation to the FFT method necessary.

This adaptation consists in considering a time interval such that, in that period, the excitation frequencies remain constant. This leads to the representation of the vibration signal in the time-frequency or rotation-frequency domain, whichever is more convenient for the analysis being carried out.

Mathematically, adapting the FFT to analyze transient signals changes Eq. (1) into:

$$X(m,\omega) = \sum_{n=-\infty}^{+\infty} x[n]w[n-m]\mathrm{e}^{-i\omega n} \tag{2}$$

where m is the size of the window used.

It is important to highlight that the squared magnitude of the STFT ($|X(\tau,\omega)|^2$) is known as the spectrogram of the analyzed function.

4 Results

Vibration results were acquired and processed using the Portable BR-ADX software, developed by LEDAV/COPPE/UFRJ in partnership with Petrobras. This software is used to monitor the conditions of equipment and structures. Figure 3 shows the main screen of the software.

4.1 MPE01

Figure 4a shows that, in the steady-state regime, MPE01 is operating at about 1695 rpm (28.26 Hz), and also that, within the frequency range presented, the largest amplitude occurs near 113 Hz, which is equivalent to the engine's

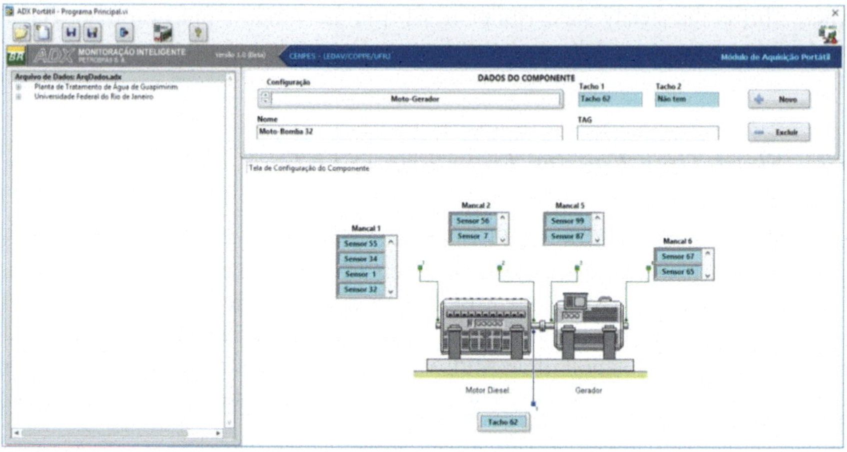

Fig. 3 Main screen of the Portable BR ADX software

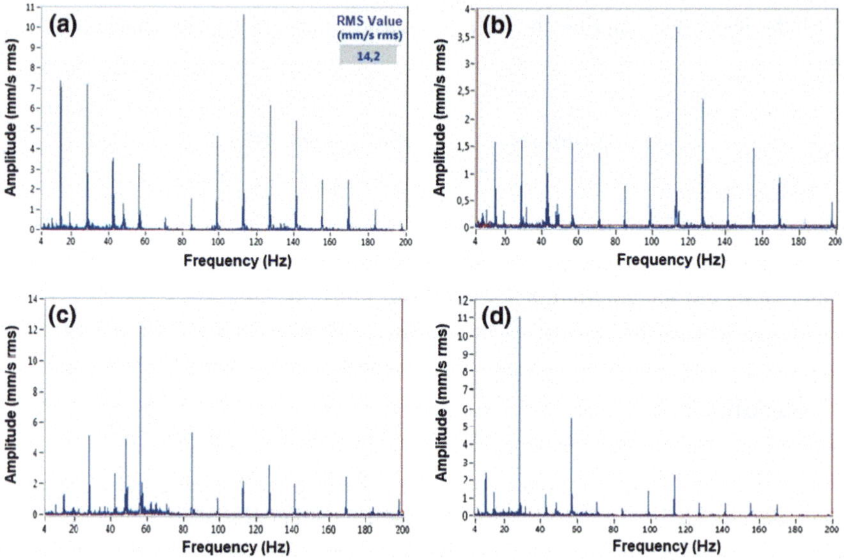

Fig. 4 Vibration spectra at all MPE01 measuring points

fourth-order harmonic (4× rpm). It is also possible to observe that there are vibration amplitudes excited by the seventh harmonic of the engine's rotation (7× rpm), around 198 Hz.

Figure 4b shows the vibration spectrum at the base of MPE01 in the transverse direction. The largest amplitudes occur at 42 Hz, equivalent to 1.5× rpm, and 113 Hz which is equivalent to the fourth-order harmonic (4× rpm). The influence of the rotation harmonics in the vibration responses at the top of MPE01 for the other directions (vertical and axial) differ. For the axial direction, the largest amplitude occurs close to 56.5 Hz, which is equivalent to the engine's second harmonic (2× rpm), as shown in Fig. 4c. Finally, for the vertical direction at the top of the engine, the largest amplitude occurs at 28.26 Hz, equivalent to the first harmonic (1× rpm), as seen in the vibration spectrum in Fig. 4d.

In the transient vibration analysis of MPE01, using STFT, the resonance regions and the rotation harmonics responsible for the excitation of the set MPE01 + FOUNDATION were identified.

Figure 5a shows the variation of vibration amplitudes in the transverse direction at the top of MPE01 as rotation increases from 600 to 1800 rpm, with MPE01 running unloaded. The waterfall plot shows several resonance regions, excited by the harmonics 2× rpm, 3.5× rpm, 4× rpm, and 4.5× rpm.

Figure 5b shows the transient vibration response in the transverse direction at the base of MPE01 in the same conditions described above. Once again, there are several resonance regions excited by the harmonics 2× rpm, 3.5× rpm, 4× rpm, and 4.5× rpm.

The waterfall plot of MPE01base vibrations in the axial direction, Fig. 5c, shows that, despite an increase in vibration levels at 2× rpm, there is no indication of resonance due to this harmonic. Finally, the waterfall plot for the measurements at

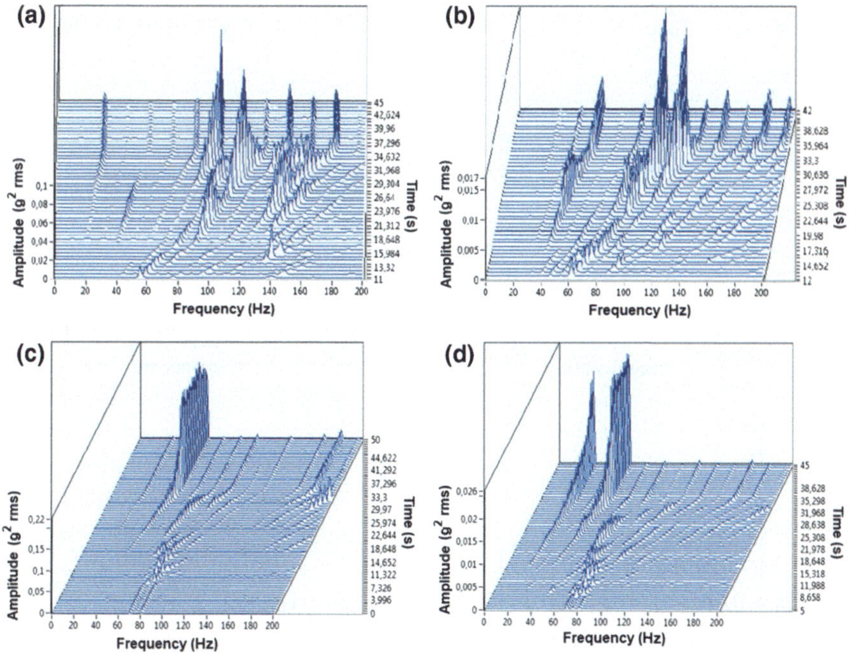

Fig. 5 Waterfall plots for the points measured in MPE01

the base of MPE01 in the vertical direction (Fig. 5d) indicates an increase in the vibration levels at 1× rpm and 2× rpm, but no resonance due to those harmonics.

4.2 MPE02

Vibration measurements in MPE02 were taken only in the steady-state regime during navigation. Under this condition, the highest vibration levels were measured at the top of the engine in the axial direction (Fig. 6a), and the vibration spectrum shows that, when MPE02 is running in steady state, vibration amplitudes up to the 6.5× rpm harmonic (around 186 Hz) can be identified.

The vibration spectra also show that the harmonics that lead to the highest vibration levels at the top of MPE02 depend on the measuring direction (Fig. 6b, c, transverse and vertical direction, respectively). On the other hand, the measurements at the base of MPE02 showed the two main harmonics responsible for excitation: 1.5× rpm and 4× rpm (Fig. 6b).

4.3 MPE03

In this case, the highest vibration levels were measured at the top of the engine, in the transverse direction. Figure 7a shows the vibration spectrum for MPE03

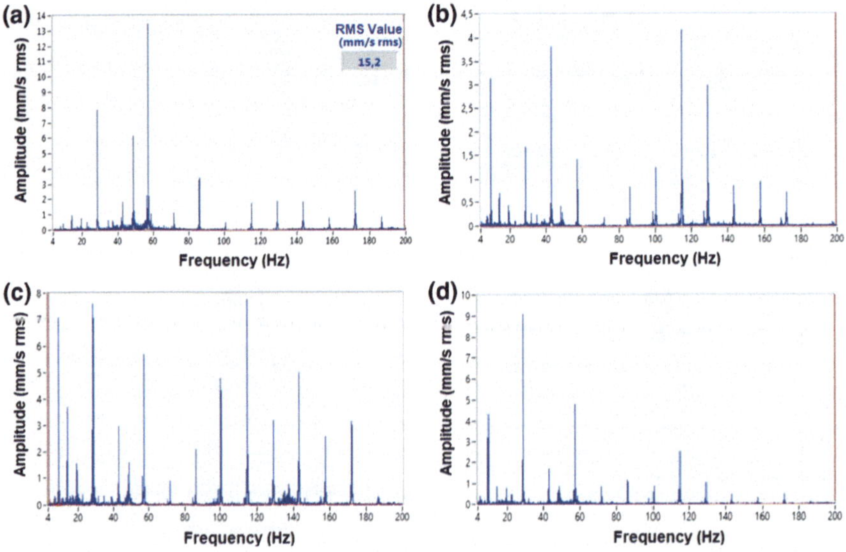

Fig. 6 Vibration spectra at the top (**a**) and base (**b**) of MPE02

operating in the steady-state regime. Figure 7b–d show the vibration spectra measured at the base in the transverse direction and at the top in the axial and vertical directions, respectively. It can be observed that, although the spectra in Fig. 7c, d present higher amplitudes than the spectrum in Fig. 7a, the energy contained in the spectrum shown in Fig. 7a is greater for the frequency range considered.

During the transient analysis of MPE03, resonance conditions were identified at the top of the engine in the transverse, axial, and vertical directions (Fig. 8a, c, d, respectively), excited by the second-order harmonic (2× rpm), whereas the first-order rotation, 1× rpm, excites the top of the engine in the vertical direction (Fig. 8d). Furthermore, 4× rpm and 4.5× rpm excite the top of the engine in the transverse direction (Fig. 8a). Finally, the base of MPE03 is excited in the transverse direction (Fig. 8b) by the frequencies 2× rpm, 3.5× rpm, 4× rpm, and 4.5× rpm.

4.4 *MPE04*

In the steady-state regime, it is possible to identify that the base of MPE04 is excited in the transverse direction (Fig. 9b) by 1× rpm, 4× rpm, and 4.5× rpm frequencies. Additionally, the top of the engine is excited in the transverse, axial, and vertical directions by 4× rpm, 2× rpm, and 1× rpm frequencies, respectively (Fig. 9a, c, d).

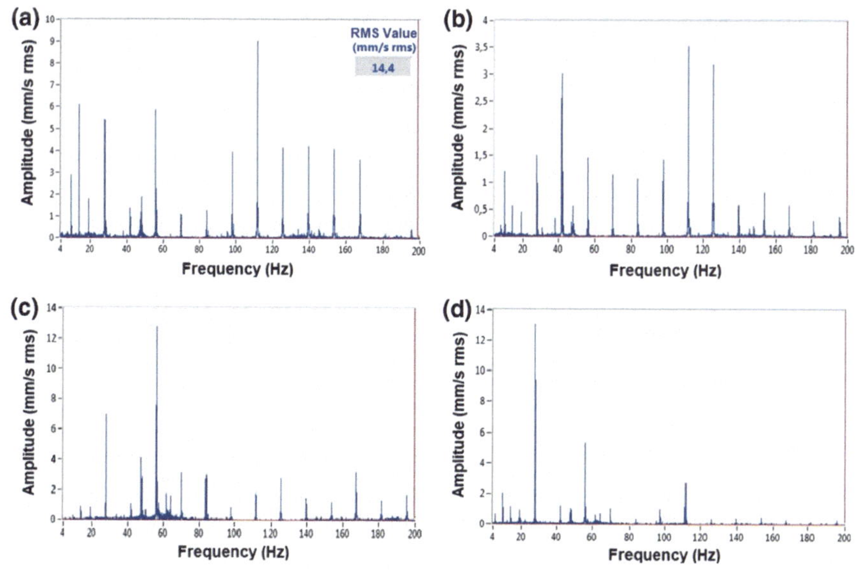

Fig. 7 Vibration spectra at all MPE03 measuring points

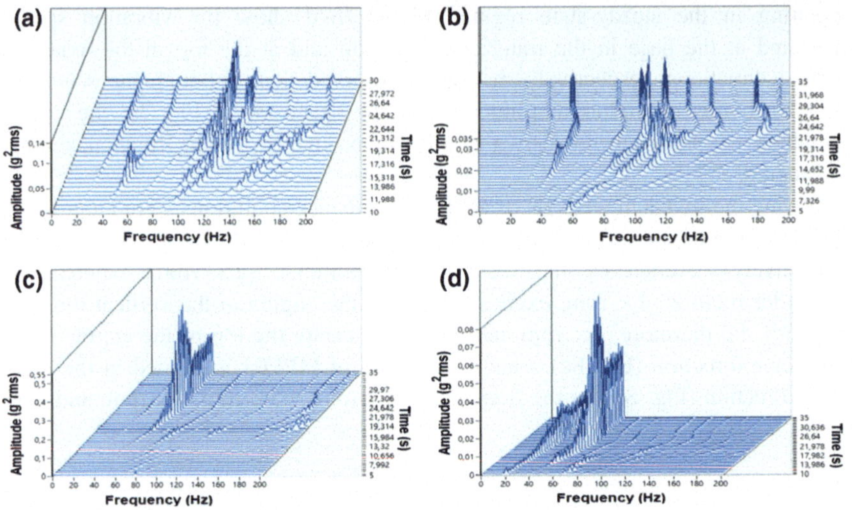

Fig. 8 Waterfall plots for the points measured in MPE03

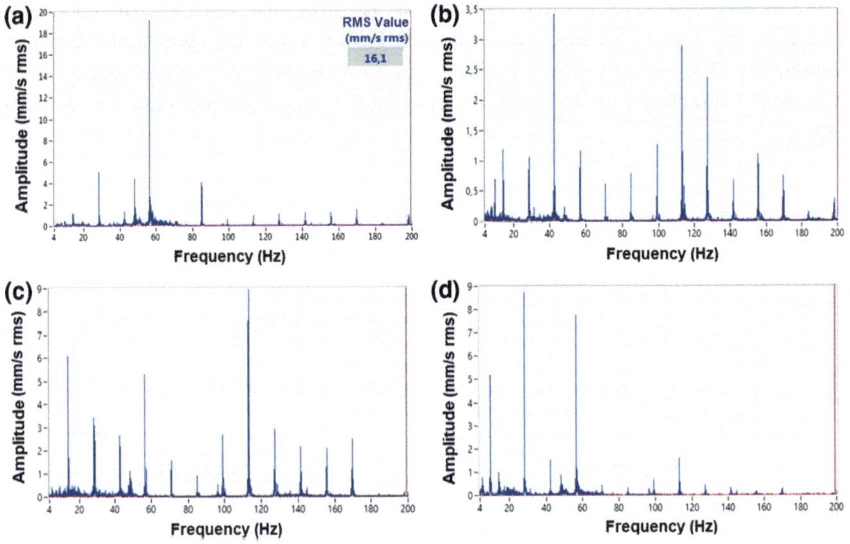

Fig. 9 Vibration spectra at all MPE03 measuring points

In the analysis of MPE04 unloaded in the transient regime, resonance regions can be identified at the base of the engine in the vertical direction, excited by 2× rpm, 3.5× rpm, 4× rpm, and 4.5× rpm (Fig. 10b), as well as at the top of the engine in the transverse direction (Fig. 10a), excited by 2× rpm, 4× rpm, and 4.5× rpm.

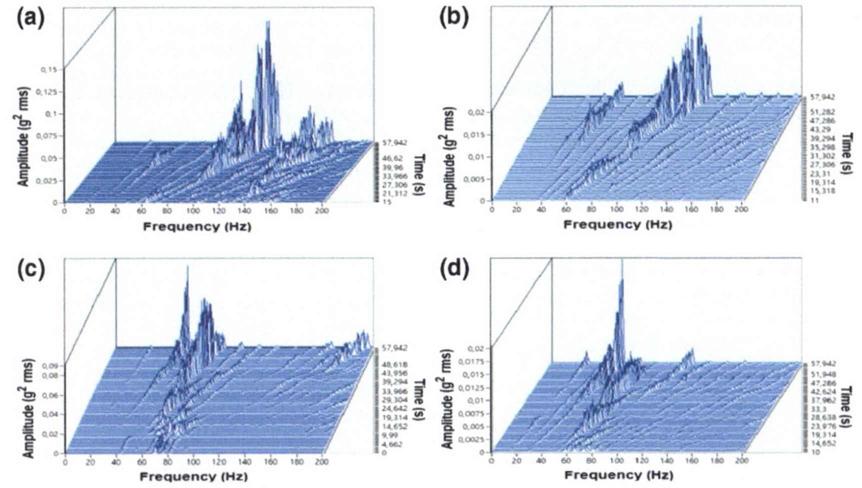

Fig. 10 Waterfall plots for the points measured in MPE03

Finally, resonance regions in the transient regime for MPE04 are also observed at the top of the engine in the axial direction, excited by 1× rpm and 3.5× rpm frequencies (Fig. 10d).

5 Conclusions

The greatest vibration amplitudes in the steady-state regime at the top of all four MPEs are excited by the rotation harmonics 1× rpm, 2× rpm, and 4× rpm, and the transient analysis has identified that the harmonics 2× rpm, 4× rpm, and 4.5× rpm cause resonance at the top and base of all four MPEs.

The results have shown that the engines were operating with excessive vibration levels, which could lead to a reduction of the service life of components of the propelling system.

Two measures are suggested to solve this problem: reinforcing the base of the main engine foundation and replacing the rigid wedges used in the MPEs by flexible wedges.

References

1. Monteiro, U.A., Vaz, L.A., Gutiérrez, R.H.R., Mattos, J.H.V.: 25° Congresso Nacional Transporte Aquaviário, Construção Naval E Offshore, p. 9 (2014)
2. DNV-GL: Rules for Classification of Ships—Confort Class0 (2012)
3. DNV-GL: Rules for Classification of Ships—Newbuildings—Vibration Class (2010)
4. ISO 20283-4: Mechanical Vibration—Measurement of Vibration on Ships—Part 4: Measurement and Evaluation of Vibration of the Ship Propulsion Machinery (2011)

Part II
Maritime Transportation and Logistic

Alternative Route with Integrated Road–Water Transportation System Between Ananindeua/Marituba to the Federal University of Para

Rodolpho Soares, Adib Maués, Rita Moraes, Hito Moraes and Alan Borges

Abstract This paper presents a proposal of an alternative route between Ananindeua and Marituba to the Federal University of Para through an integrated system, road–fluvial, where the students can have a cheaper, shorter, and more confortable alternative. To develop the new system, an analysis of the current public transport model was made and, from that, the integrated system was elaborated and analyzed regarding its technical and economic viability, showing it to be a possible solution for public transport, reducing the traveling time to Ananindeua and Marituba 34 and 20%, respectively. In the economic aspect, the proposed model has shown to be unviable regarding the price, with a ticket price almost doubled when compared with the current ticket value, however, some measures can be made to diminish this and make it possible to operate more inexpensively. As to the technical feasibility, the waterway has shown to be navigable, having a minimum depth twice the required value and a minimum width 4 m longer than necessary to operate the proposed system.

Keywords Water transport · Urban mobility · Public transport
Intermodal transport

1 Introduction

The transport matrix in Brazil is extremely unbalanced, focusing on the road system as its main option for moving people and loads. Brazil is a continental-sized country, with extensive and navigable rivers, where millions of people, products, and wealth travel through, making it an important asset for national integration, especially in the Amazon region [1].

R. Soares (✉) · A. Maués · R. Moraes · H. Moraes · A. Borges
Federal University of Pará, Belém, Brazil
e-mail: rodolphosoares_@hotmail.com

A. Vega Sáenz et al. (eds.), *Proceedings of the 25th Pan-American Conference of Naval Engineering—COPINAVAL*, https://doi.org/10.1007/978-3-319-89812-4_18

Furthermore, the water transportation system is crucial in the Amazon due to places of difficult access that can only be reached by water, giving the local population no other choice [2].

Belém, located on the western side of the Guajará Bay and to the south of the Guamá River, has water transport as the main system to connect the 39 islands and the municipalities on land. Throughout the 28 km of waterfront, there are small docks, mostly private, that are used for small boats, operating in an unsafe and poor manner [1].

With the weak infrastructure and huge potential, Belém does not differ from other cities in Brazil, where water transportation is sidelined and the population endures an outdated, saturated, and not functional public road transport system.

The area of study of this paper is the metropolitan region of Belém (MRB), that is made up of seven municipalities, including Ananindeua and Marituba, which, together with Belém, are the three biggest cities of the metropolitan region. The road route will be done in two ways: internal and express. The first is an internal route within the neighborhoods of the municipalities to collect the passengers and bring them to the express terminal, and the other will be in the municipality of Ananindeua, through Osvaldo Cruz Street, going through the old garbage depot of Belém, and arriving at the Aurá Stream, where the water route of the integrated system will operate.

Aiming to reduce the saturation of the public road transportation system, this paper presents an integrated road–water alternative to connect the municipalities of Ananindeua and Marituba with the Federal University of Pará (UFPA), having the water system as its focus, assigning to it the proper parameters to make the project economically and technically viable.

2 Methodology

This research was done with a quantitative approach, where all the data were collected in a neutral and standardized way, and assuming that they represent in a realistic manner the problematic situation.

As to the nature of the research, it can be classified as applied, once it is in practical application, evolving local interest.

Regarding its procedures, both bibliographic and field studies were used. Several articles, dissertations, theses, and papers were analyzed to support the methods and to guide the parameters for the new system, such as the water fare, the choice of boats, and the location of terminals.

To determine the locations of the terminals, the literature has shown that it is crucial that they are in a place with easy access to the population, and in an area where they can be influenced and can change the dynamics of the region, bringing development to it.

The type of boat was determined following the terms revealed in the bibliographical studies, where the most important feature is its compatibility with the waterway dimensions. With that, the boat used in the federal government's program "Caminho da Escola" was chosen, since the model is already in use in the project area, having only some features of the boat altered to improve its performance and bring it into conformance with the needs of the integrated project.

As for the fare, the method used is the one developed by the Instituto de Pesquisas Tecnológicas de São Paulo (IPT), where it considered for the calculation of the fare the following annual costs: investment amortization and recovery, repair and maintenance of the fleet, insurance of the fleet, salaries and social taxes, fuel and lubricants, and administration. Once these costs were properly identified, a spreadsheet was created with the assist of Microsoft Excel software to add these costs and generate an annual cost for the system. To achieve the final value of the fare, the time for round trip (TRT), the number of travels (Nt), and the cost per travel (Ct) and the fleet were also calculated.

The field research was conducted in two separate stages, with the first a boat trip to the Aurá Stream to verify if it could technically support the project. The technical viability of the stream was also verified with a depth study by the Companhia de Recursos Minerais (CPRM), where field research was done to verify the critical depth points in the stream.

The second stage was a study conducted to diagnose the current public road transportation system, identifying and analyzing its characteristics.

The first of these was demand: it was necessary to make sure that the proposed system was a viable investment, necessitating a study to verify that this route has a reasonable demand of people that would justify its implementation. Research was done together with the Centro de Tecnologia da Informação e Comunicação (CTIC) and the operating companies on the existing road route to identify this demand.

Other characteristics were: route, average distance, average speed, average traveling time, cost, fleets, and terminal, which also helped to guide the parameters of the new integrated system.

3 Results

The location of a passenger terminal must be defined considering studies of the movement of demand (origin and destination) [1]. The results found for demand showed that this route has a big flux of people and would benefit several users. Matching the data in Fig. 1, where the amount of students who live in the two municipalities and use the current system to go to the university is shown, and the data collected with the companies, it was estimated that 1190 students could benefit from the implementation of this new system.

From the field research it was possible to establish the parameters of the current public transportation system. From those results, the integrated road–water model was developed in order for it to be economically and technically better than the current model.

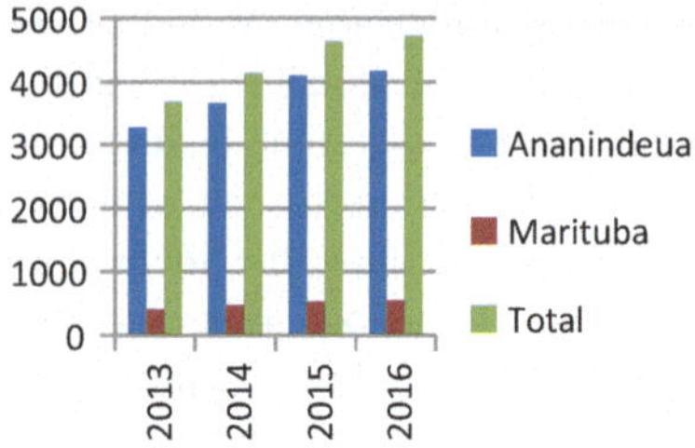

Fig. 1 Students who live in Ananindeua and Marituba and study at UFPA

3.1 Current Transportation System

The route for the current transportation system consists of leaving the UFPA bus terminal, going through Bernardo Sayão Avenue to Almirante Barroso Avenue until it reaches Marituba center, and afterwards goes back to the university through a similar route. This route takes an average distance of 24.5 km each way, totalling 49 km.

The average traveling time varies according to the time of the day. From 5 AM through 5:30 PM, the average traveling time is 1:10 h for each way of the route. From 6 PM to 8 PM, the traveling time can be 2–2:30 h each way, depending on the traffic.

In the MRB, the cost of a bus ticket is R$2.70, and in spite of being one of the lowest prices for public transportation in a Brazilian capital, does not have any integration with any other bus line or mode of transportation.

The average speed of the buses on this route varies from 30 to 40 km/h. During the most congested hours of traffic, this average speed can drop to 15–20 km/h.

Regarding the fleet, the buses that operate on this route are owned by the city office and, by public bidding, leased to the operating companys. In this case, there are two companies acting on this route, and each of them has eight vehicles; due to maintenance, repairs, and other operational problems, not all vehicles are available for use. The buses can run around 8 km per liter of diesel and they have an average capacity of 40 seated passengers. The vehicles have a medium to good general condition according to the users. Inside, the buses are often dirty, but the seats are in good condition and the external part of the vehicles is well repaired and maintained. In addition, by municipal law, all buses must have accessibility elevators for disabled passengers.

As to the terminal, the main and most important one is the UFPA Bus Terminal, located next to the campus of the university, with over 4.000 m^2 and cost of over 22 million reais. The terminal is rated by the users as degraded and in poor condition despite the still good infrastructure. The terminal has no accessibility for disabled people whatsoever.

3.2 Integrated Road–Water System

The integrated system has three different routes: the internal neighborhood road route, made by internal buses, the express road route, and the water route. The internal road route will be executed inside the neighborhoods of the municipalities, in a 3 km range. The second one will start from the Br-316, going through Osvaldo Cruz Street, until the old garbage depot and reaching, through an extension of the road, the Aura Stream. The water route begins on the stream and stays there until it reaches the federal university. Figure 2 shows the express route and the water route; the neighborhood routes need to be elaborated with a more detailed study on the demand inside the cities.

The medium distance for this course is 3 km for the neighborhood route, 6.8 km for the express route, and 8.27 km for the water route.

The average speed chosen for the calculation of the average traveling time was 40 km/h for the buses on either road route and 25 knots (46.30 km/h).

Therefore, the average traveling time is different for each neighborhood, as shown in Table 1, and they depend on the distance to the express bus terminal. The loading and unloading time considered was 2 min for each mode of transport. In the municipality of Marituba, due to the small sizes of the neighborhoods, they were divided into two groups, the first comprised the neighborhoods of Pedreirinha, Dom Aristides, Nova União, São Francisco, and Bairro Novo e Bela Vista (Group 1) and

Fig. 2 Express route and water route

Table 1 Average traveling time

Municipality	Neighborhood	Internal buses (min)	Express bus (min)	Water transport (min)	One way (min)	Total
Ananindeua	Cidade Nova	18	7	18	43	1 h 26 min
	Maguari	18	7	18	43	1 h 26 min
	Distrito Ind.	24	7	18	49	1 h 38 min
Marituba	Group 1	27	7	18	52	1 h 44 min
	Group 2	27	7	18	52	1 h 44 min

the other is the neighborhoods of Decouville, Parque das Palmeiras, and Centro e Boa Vista (Group 2).

The cost for the integrated system was calculated considering the possible costs that would be faced on the implementation of this route, and it was found the value of R$3.92 per passenger, using an occupational rate of 75%.

As to the fleet, the road part of the system will use the same type of bus as the ones that are in use currently, both for the internal neighborhood route as for the express one. As for the water fleet, it will use the same boat as the federal government's program "Caminho da Escola," which has a 22 seated passengers capacity, although the original project has an engine of 80 hp, which will be replaced for a 250 hp engine in order to achieve the necessary speed of 25 knots.

The integration terminal is an element of support to the system, which represents an intermediate point of transfer between models of transport on a trip [1]. The proposed system will have three different terminals. One, for the neighborhood routes, will follow the currently used models of the city's public transportation system; a terminal for the express bus, which will have space for the internal buses and, in that way, integrate the ticket; and the last a road–water integrated terminal, where the passenger can enter the water system to go to the university.

3.3 Technical and Economic Feasibility

Before the implementation of the water transport, it is necessary to know the characteristics of the waterway such as depth and width, among other factors that can compromise the operational performance of the boat [2]. Therefore, three analyses were done to guarantee the technical feasibility of the project: bathymetry, minimum depth, and minimum width. The first was done with the help of CPRM, on a field research to the stream to elaborate a spreadsheet with the depths of the stream. Figure 3 and Table 2 show the lowest points of the stream, highlighting that the stream is viable for water transportation.

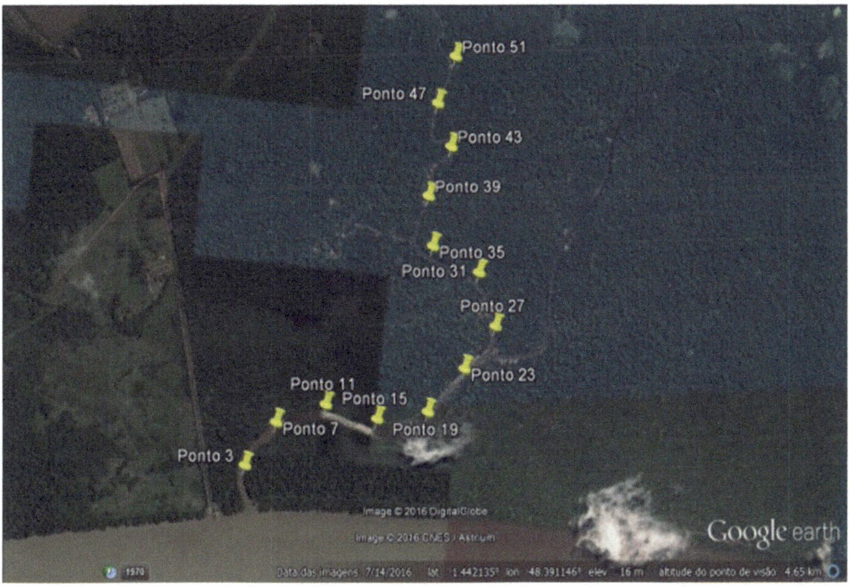

Fig. 3 Depth points of the Aura Stream

Table 2 Depth of the points

Point	Depth
3	4.77
7	8.65
11	6.96
15	9.01
19	8.76
23	8.73
27	7.17
31	7.53
35	7.33
39	7.21
43	4.80
47	4.42
51	3.12

According to the Agência Nacional de Transporte Aquaviários (ANTAQ), the minimum depth is determined by:

$$H\text{min} = 1.5 \times D$$

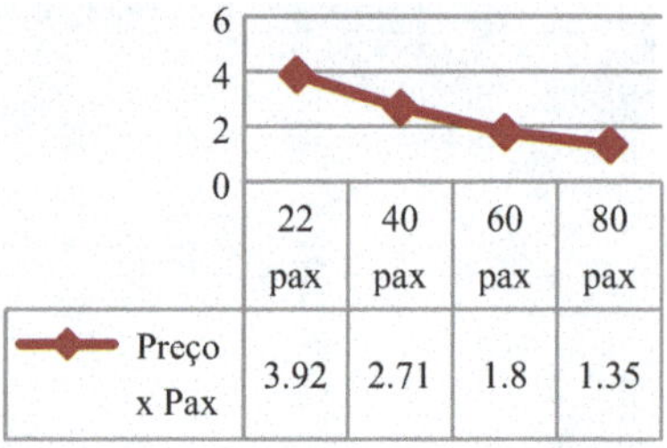

Fig. 4 Graphic price times number of passengers

where D is the vessel draught of the ship. The boat that will be used on the water route is a vessel of 0.67 m, making the stream a possible waterway for the proposed model.

Also according to ANTAQ, the minimum width for a waterway is:

$$W\min = 4. \times B$$

where B is the breadth of the ship. The chosen boat has a breadth of 2.53 m, making a width of 11.53 necessary. The Aurá Stream has a width of 15 m at its narrowest point, which makes it viable in this aspect to navigate.

As to the economic aspect, the project has shown to be infeasible due to the high price of the ticket. However, there are some measures that can be made to enable the proposed system to be executed. The main one is regarding the number of passengers. The original boat has the capacity of carrying 22 passengers with a ticket price of R$3.92, but if the number of seats is increased, the price of the ticket decreases, making the system viable. Figure 4 shows the possible values for an increased number of passengers.

4 Conclusion

In spite of the necessity of further technical, environmental, and economic analyses, the project has shown to be viable and advantageous.

Even though the project has different impacts in the neighborhoods of each municipality (Fig. 5), the integrated system has proven to be faster when compared to the current transportation system, decreasing the traveling time 34% for the neighborhoods in Ananindeua and 20% in Marituba as shown in Table 3.

The proposed system can also have a cheaper ticket than the current one, once the passenger capacity is raised or a subsidy is implemented by the three municipalities. In addition, the environmental benefits will be enormous, with the decrease of pollutant gases with fewer vehicles in the urban centers of the cities, improving traffic and quality of life for the citizens.

The integrated route also has a social impact due to the traditional community that lives on the Aurá Stream. With the implementation of the new route, they will have access to better services and integration with the urban centers.

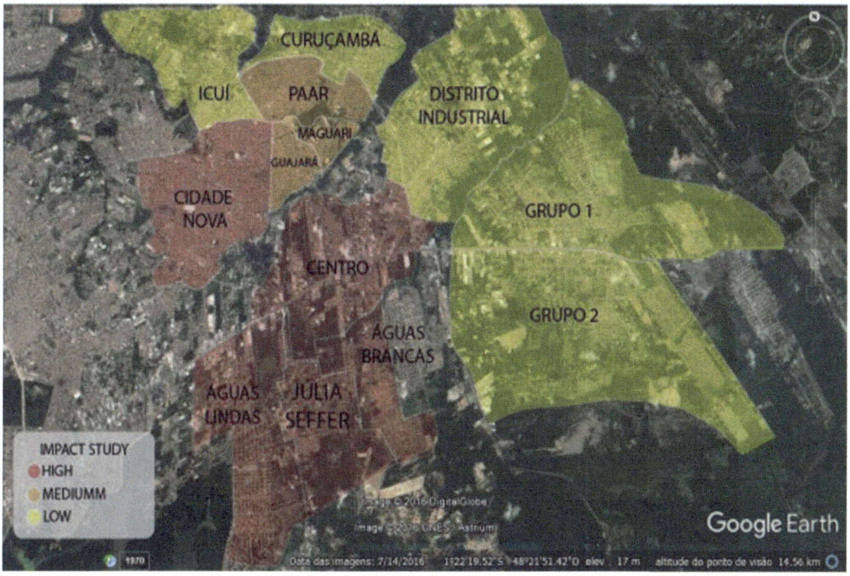

Fig. 5 Map of impact

Table 3 Average travelling time

System	Municipality	Average traveling time
Road	Ananindeua	2 h 10 min
	Marituba	2 h 10 min
Road–water	Ananindeua	1 h 26 min
	Marituba	1 h 44 min

Finally, the road–water integrated transportation system from Ananindeua and Marituba to the UFPA has proven to be a viable and beneficial solution for urban mobility in Belém, benefiting the citizens with shorter, faster, more comfortable, and affordable travels to their destinations.

References

1. SOUZA, M.H.: Contribuição metodológica para localizar terminal de integração de passageiros do transporte hidro-rodoviário urbano. 157f. Tese de Doutorado. Universidade Federal do Rio de Janeiro, Rio de Janeiro, 2009.
2. Moraes, R.D.C.M.D.: Ferramenta de avaliação de transporte hidroviário de passageiros da região Amazônica. Belém: CDD-23. ed. 629.048 (2013)
3. Márquez, L., Cantilo, V., Arellana, J.: How are comfort and safety perceived by inland waterway transport passengers? Transp. Policy, **36**, 46–52 (2014) (Elsevier, Amsterdam)

Study of Sustainable Maritime Transport on the Brazilian Coast—Green Corridor

Delmo Alves de Moura and Rui Carlos Botter

Abstract Brazil has the potential to develop Green corridor systems using sustainable transport and innovative technology in its freight operations. Due to the territorial extension of the country, water transportation would be the most appropriate. This work analyzes the characteristics of the systems of Green corridors of Europe and the perspectives for implantation of these technologies in Brazil. Increasing the efficiency of transport modes through the creation of corridors and consequent improvement of resources would reduce the need for investments to increase the capacity of transport networks in a scenario of budgetary caution and opposition to large projects of infrastructure in regions close to urban centers. The use of short sea shipping, which is maritime transport on the coast of a country, is a means that can increase Brazil's logistics competitiveness, and allow the reduction of operational costs and the number of public ports along the Brazilian coast.

Keywords Short sea shipping · Green corridors · Intermodality

1 Introduction

1.1 Europe as a Reference

The SuperGreen project is a model that guides sustainable logistical decision making for the region. The objective of the project is to provide support and recommendations for the implementation of a Green freight corridor in the European Union. The project involves several important players to the success of new logistics concepts. It uses intermodality in an optimized and sustainable way

D. A. de Moura (✉)
Federal University of ABC, Santo André, São Paulo, Brazil
e-mail: delmo.moura@ufabc.edu.br

R. C. Botter
University of São Paulo, São Paulo, Brazil
e-mail: rcbotter@usp.br

A. Vega Sáenz et al. (eds.), *Proceedings of the 25th Pan-American Conference of Naval Engineering—COPINAVAL*, https://doi.org/10.1007/978-3-319-89812-4_19

and targets innovations in operations through the well-structured use of information technology. The program involves a network of players from the public and private sectors (stakeholders).

The project involves 22 partners from 13 European countries. It covers transport, logistics operators, infrastructure, shipping companies, environmental organizations and authorities responsible for urban social and territorial planning, consultants, universities, research, and technological development centers.

The project encompasses transport modes such as road, rail, coastal shipping, long-haul maritime transport (between two or more countries), and inland waterway transport.

For more than 10 years, the European Union has been running a transport program among its member countries to eliminate or reduce environmental damage such as the emission of gaseous pollutants, atmospheric particulate matter, noise pollution, contaminated solid and liquid wastes, and congestion.

In general, the project guides the use of sustainable fuels, innovation in transportation operations, the use of information technology in the management of processes, and permeates information in all logistics systems to minimize actions that do not add value to the whole transport chain [1].

It is a complex and audacious project that has several actors committed to the well-being of the European logistics system, with private companies, federal governments, local governments (state or municipal), universities, research centers, and laboratories.

2 Bibliographic Review

There is a growing need for the use of renewable energies in logistic operations to obtain a transport system that does not emit pollutants and other particles in the environment [1–5].

A Green corridor is characterized by transporting a considerable volume of goods onto its tracks. In its course, there is the transshipment of products, which also deserves full attention to the use of renewable and sustainable energy, considered clean. The use of technology and innovation is essential in this process to obtain a Green transport [2].

The European Union has defined specific rules for the amount of sulfur emission in the fuel used in the maritime sector. That rate is 0.1% [2, 5].

There are international rules aimed at limiting the emission of nitrogen oxide (NO_x) from marine diesel engines and other pollutant gases found in road vehicle fuels. Also, when it concerns transport modes, speed reduction is an important factor for the reduction of gaseous pollutants [2, 5–11].

The European Union has defined specific rules regarding the destination of waste from transport systems, regardless of the modalities used for operations. These same rules apply to the ballast water of ships and all the garbage they produce [2].

Noise from transport operations is another factor that deserves particular attention. It has to be controlled and monitored to avoid injury to the local and regional population [2].

The European Union has established the European Rail Traffic Monitoring System (ERTMS) with a focus on reducing or eliminating traffic congestion. It means that all new projects must contemplate this new system to avoid congestion [2].

Transportation must be efficient, using information technology, sustainable technologies, and innovation in operations and equipment. It should always aim at renewable energies that do not cause environmental damage and problems for the health of the population. A sustainable and efficient logistics system is one that contributes to the progress of the region, without causing problems such as noise, environmental pollution, congestion, and damage to fauna and flora. It should harmonize with its surroundings and with the planet [2, 4, 5, 7].

Green corridors are implemented to improve the transportation system through innovation in operations and the use of sustainable renewable energy. Another purpose is to optimize road, rail, air, sea, and waterway transportation systems so that there is substantial integration among them, as well as efficiency in transshipment operations with the use of clean and sustainable technology [2, 11].

3 Methodology

From nature, this work can be considered basic research. The approach is qualitative. From its objectives and procedures, it is considered exploratory research, involving a bibliographical survey of the subject [12].

The research instruments of this work were to make a survey of the scientific articles about the theme and to evaluate its essential points that contributed to this study. Researchers from Sweden who are committed to the implementation of a system of European Green corridors were interviewed; technical visits to European ports that also implemented the Green corridor system were carried out. In addition, interviews were conducted with specialists who implemented sustainable port management in Europe.

The advantage of this methodology is to be able to understand a system theoretically through the opportunity to verify its characteristics in loco, the details of all aspects of the project, and to know closely the innovative technologies used in the logistics operations and, consequently, to understand the role of each actor in this process.

4 Characteristics of Transportation in Brazil

The Brazilian navigable coast is 7491 km long. The states with the highest contribution to the national GDP are located on its maritime coast. Most of the Brazilian population is concentrated in the main hinterlands covered by the more significant domestic ports. These statements are enough for the planning of a transport matrix focused on the use of waterway transportation. Intermodality with other means of transport would contribute to an optimized and sustainable system. An innovation study on transport, handling, and storage logistics operations should address the use of clean technologies, the reduction of pollutants in the atmosphere, and noise pollution at minimum levels. Also, there should be a focus on reducing traffic congestion in the major cities of the country [13].

Along the Brazilian coast, there are 37 public ports. The Port of Santos, in São Paulo, is the largest port in Latin America.

The state of Rio Grande do Sul (RS) located in the extreme south of the country is at a distance of approximately 5400 km from the state of Roraima (RR), which is in the north. That reveals the continental extension of Brazil.

Table 1 shows the composition of Brazil's transport matrix, with the percentages of each mode of transportation.

Table 2 presents the average speed of Brazilian railroads, compared to US railroads, which interferes with a country's logistics costs and competitiveness.

Table 3 shows a summary of the characteristics of the waterways in Brazil.

Table 1 Matrix of transport in Brazil

Matrix of cargo transportation		
Modal	Million (TKU)	Participation (%)
Road	485.625	61.1
Rail	164.809	20.7
Water	108.000	13.6
Pipeline	33.300	4.2
Air	3.169	0.4
Total	794.903	100.0

Source Ref. [14]

Table 2 Average speed on railways

Operational average speed—railways	
Brazil	25 km/h
United States	80 km/h

Source Ref. [14]

Table 3 Waterway transport characteristics

Infrastructure—waterways	
Private terminals	156
Freight transshipment station	25
Organized ports	37
Waterways—length (in km)	
Navigable ways	41.635
Economically navigable ways	22.037

Source Ref. [14]

5 Short Sea Shipping on Brazilian Coast

Despite having 37 ports, 156 terminals, and a majority of its population living about 200 km from the coast, short sea shipping is still little explored. It is via this means of transport that Brazil has the potential to reduce the current road transport index (61.1%) and contribute to a more sustainable logistics system.

Integration in freight transport between the north and south of the country should increase through short sea shipping. The road modal should be used for distances up to 400 km between origin and destination.

Currently, only three shipping companies operate in the short sea shipping segment in Brazil. If there were reliable and sustainable planning with public and private policies to encourage the use of that modal, more companies would participate in this transportation segment. Due to a likely increase in the supply of services, investments in port infrastructure and neighborhoods might be needed. There would be a need for legislation to eliminate beadledom and, above all, to optimize operations to integrate all types of transportation (waterway, road, rail, and pipeline).

The "motorways of the sea," a concept that aims to introduce new intermodal maritime-based logistics chains in Europe, is a sustainable way of contributing to the environment, emphasizing water transport, especially maritime traffic, and the others as part of an integrated transport chain.

Concerning freight transport in Brazil, it is still very early to think and act sustainably. The Brazilian Plan of Logistics and Transport aims to increase the use of water transport in the transport matrix. However, the actions and policies implemented by the federal government and public–private partnerships (3P) are still practically insignificant for the segment.

Brazil has enormous potential to exploit the use of short sea shipping interconnecting the northern and southern regions of the country. Several products could stop crossing the highways of the country if the companies used the maritime and waterway modal. Intermodality would improve the quality of service offered to the market and reduce the costs of transport operations.

The transportation of soybeans, whose production is concentrated mainly in the center-west region of Brazil, could obtain logistical gains with the appropriate use of the railroads and, then, the waterway through barges.

Given that Brazil is the second largest soybean exporter in the world, behind only the United States, and that this product contributes considerably to the domestic balance of trade, encouraging Green corridors would help reduce logistics costs and help the growth of sustainable transport.

The automobile industry could take advantage of the potential of short sea shipping to transport its production to car dealerships through the use of Ro-Ro (roll-on/roll-off) ships.

These vehicles could be transported by rail or road from the ports of the country to their final destination, contributing to the intelligent and sustainable use of transport in general.

Figure 1 shows the division of road cargo by the group in Brazil in 2015. The highest percentage is related to general cargo, which could be transported between the main ports of Brazil through the use of short sea shipping.

General cargo consists mainly of processed food and beverages, cellulose and paper, farming products, rubber, plastic and nonmetallic commodities, forest products, and manufactured goods.

Figure 2 shows the cargo distribution through short sea shipping and by groups of goods in Brazil in 2015. The most transported type of product is liquid bulk inasmuch as the largest energy company in Brazil is Petrobras, and its subsidiary that manages the transport activities is Transpetro, which uses short sea shipping to transport its products.

There is, therefore, a need to increase the use of short sea shipping for general cargo in Brazil. It is time to define sustainable Green corridors, as has the European Union applying these concepts and technologies in handling, storage, and transshipment. Innovative forms of operations contribute to the environment and the planet.

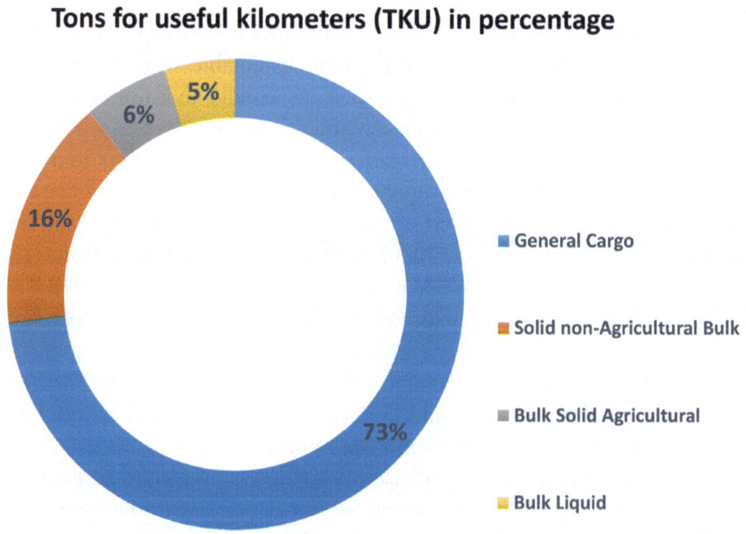

Fig. 1 Division of road cargo by commodity group, 2015 *Source* Ref. [15]

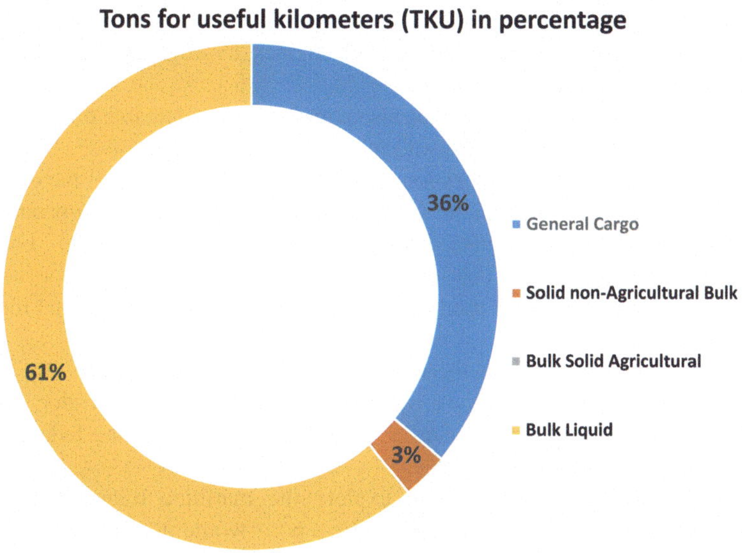

Fig. 2 Division of cargo by short sea shipping and by cargo groups, 2015 *Source* Ref. [15]

6 Conclusion

Brazil has unrealized potential to implement Green corridor systems along its coast, aiming at the optimization of waterway transportation, focusing on short sea shipping (maritime transport on the Brazilian coast). However, it is necessary to implement public policies to increase the matrix of waterway transport and encourage companies to use this means of transportation.

A Green and sustainable logistics system demands a call for business, universities, research centers, and federal, state, and municipal governments to discuss the benefits of implementing a robust and long-term system.

The outflow of soybean production in Brazil would already be a corridor that deserves particular attention from the actors involved in this industrial segment, aiming at reducing costs of operation and improving services and management of intermodal transport operations.

The adoption of transparent medium- and long-term public policies would favor the allocation of public and private resources for the development of state-of-the-art technologies that would contribute to an integrated, optimized, and above all environmentally sustainable transport system.

Conventional transportation must be rethought and innovated. New alternatives should be put to use, including hybrid systems, use of renewable energy, liquefied petroleum gas, biomethane, compressed natural gas, and so on. Therefore, for urban areas, electric motors or compressed natural gas could be used in transport vehicles. On the highways, compressed natural gas could also be used, as long as there were several points of supply for the trucks along their routes. An electric network for fluvial and maritime transport could be used in port operations, for example, the use of wind-powered energy in ports; for vehicle engines, hydrogen gas; and for ships' engines, LNG gas [16].

The ballast water treatment of ships also corroborates sustainable transport. Care must be taken to avoid the transfer of pathogenic aquatic organisms.

Avoiding or minimizing congestion in transport is another critical factor for the proper development of sustainable Green corridors. Monitoring all operations to avoid congestion will contribute to the sustainable logistics system. For this, information technology and communications systems are essential for success. Limiting the speed of vehicles used in transport modes is also an important measure for the entire logistics system.

Improving the logistical flow between the south and north of Brazil, by intense use of short sea shipping would be a way to start a Green logistics corridor. There is potential for its development. However, government support at the federal, state, and municipal levels is indispensable.

Brazil has enormous potential to explore the maritime transport of goods interconnecting the main Brazilian ports. These ports would use intermodality with the rail, road, and waterway systems to supply the entire national market, because a large part of the Brazilian population, about 80%, lives, on average, 200 km off the country's coast.

Acknowledgements Project sponsored by the following Brazilian agencies: FINEP and CNPQ.

References

1. Presbitero, A., Roxas, B., Chadee, D.: Sustaining innovation of information technology service providers—Focus on the role of organisational collectivism. Int. J. Phys. Distrib. Logistics Manage. **47**(2/3), 56–174 (2017)
2. Psaraftis, H.N. (ed.): Green transportation logistics. The quest for win-win solutions. In: International Series in Operations Research and Management Science. Springer Publisher, International Publishing Switzerland (2016)
3. Kontovas, C.A.: The green ship routing and scheduling problem (GSRSP): a conceptual approach. Transp. Res. Part D **31**, 61–69 (2014)
4. Guerrero, S.E., Madanat, S.M., Leachman, R.C.: The trucking sector optimization model: a tool for predicting carrier and shipper responses to policies aiming to reduce GHG emissions. Transp. Res. Part E **59**, 85–107 (2013)
5. Zis, T., Psaraftis, H.N.: The implications of the new sulphur limits on the European Ro-Ro sector. Transp. Res. Part D **52**, 185–201 (2017)
6. Bel, G., Rosell, J.: Effects of the 80 km/h and variable speed limits on air pollution in the metropolitan area of Barcelona. Transp. Res. Part D **23**, 90–97 (2013)
7. Hoen, K.M.R., Tan, T., Fransoo, J.C., van Houtum, G.-J.: Switching transport modes to meet voluntary carbon emission targets. Transp. Sci. **48**(4), 592–608 (2014)
8. Mohajeri, N., Gudmundsson, A., French, J.R.: CO2 emissions in relation to street-network configuration and city size. Transp. Res. Part D **35**, 116–129 (2015)
9. Duran, V., Uriondo, Z., Gutiérrez, J.M.: The impact of marine engine operation and maintenance on emissions. Transp. Res. Part D **17**, 54–60 (2012)

10. Lee, T.-C., Lam, J.S.L., Lee, P.T.-W.: Asian economic integration and maritime CO_2 emissions. Transp. Res. Part D **43**, 226–237 (2016)
11. Swiftly Green. https://www.swiftlygreen.eu/en (2015). Last accessed 20 Apr 2017
12. Vergara, S.C.: Projetos e Relatórios de Pesquisa em Administração, 12th edn. Atlas, São Paulo (2010)
13. Confederação Nacional de Transporte. Pesquisa CNT de rodovias 2016. http://pesquisarodovias.cnt.org.br/. Last accessed 19 Apr 2017
14. Confederação Nacional de Transporte. Boletim Estatístico—CNT—fevereiro de 2017. http://www.cnt.org.br/Boletim/boletim-estatistico-cnt. Last accessed 03 May 2017
15. Empresa de Planejamento e Logística. Transporte Inter-regional de carga no Brasil—Panorama 2015. Carregamento com a matriz origem-destino (2015). http://www.epl.gov.br/matrizes-do-transporte-inter-regional-de-carga-no-brasil. Last accessed 06 June 2017
16. Chan, S., Moreno, L.F.M., Alam, A., Hatzopoulou, M.: Assessing the impact of bus technology on greenhouse gas emissions along a major corridor: a lifecycle analysis. Transp. Res. Part D **20**, 7–11 (2013)

Analysis of Sustainable Logistics Operations: An Example to Follow for Port of Santos in Brazil

Delmo Alves de Moura, Davi Goulart de Andrade and Rui Carlos Botter

Abstract This paper analyzes new technologies and innovations implemented in major European ports to improve their logistics operations and their relevance in the study of sustainability in the operational processes of the Port of Santos, Brazil. The study focuses on the reduction of emissions of pollutant gases such as sulfur oxides (SO_x), nitrogen oxides (NO_x), atmospheric particulate matter (PM), and carbon dioxide (CO_2). This research studies the port operations of world reference and analyzes the current situation of port logistics operations in Brazil's largest port, the Port of Santos, São Paulo, and evaluates the processes and technologies currently implemented that contribute to sustainable port logistics operations. The paper deals with possible innovations in the logistics operations of transport, storage, and handling of goods. It analyzes the resources used in the port management of the Port of Santos. It also examines the use of renewable technologies that reduce the emission of polluting gases into the atmosphere and improve the efficiency of operations, such as:

- Transport technologies: Focused on projects related to transport technologies that involve different types of equipment
- Transport and logistics solutions: Solutions that integrate different parts to promote efficiency and, at the same time, reduce environmental impacts.

Keywords Green port · Sustainable port operations · Clean technologies

---

D. A. de Moura (✉) · D. G. de Andrade
Federal University of ABC, Santo André, São Paulo, Brazil
e-mail: delmo.moura@ufabc.edu.br

D. G. de Andrade
e-mail: davigoulartandrade@gmail.com

R. C. Botter
University of São Paulo, São Paulo, Brazil
e-mail: rcbotter@usp.br

A. Vega Sáenz et al. (eds.), *Proceedings of the 25th Pan-American Conference of Naval Engineering—COPINAVAL*, https://doi.org/10.1007/978-3-319-89812-4_20

1 Introduction

1.1 Focus of Work

Some of the world's leading ports are operating with a focus on activities using clean and renewable energy. Increasingly, there are incentives for the use of technologies and innovations that contribute to eliminating or reducing the emission levels of polluting gases in the environment, the reduction of noise pollution, and the emission of chemicals caused by the port operations to their surroundings.

Promoting an environmental management system for each port is a necessary condition for the implementation of effective measures that contribute to Green ports. One differential that can positively contribute to the sustainability and use of renewable energies in port operations is to have public policies focused on that area.

One of the challenges of humankind is to have ports integrated with their surroundings, which can interact harmoniously with the city around them, generating wealth, and not denigrating the environment.

Some of the major world ports in Europe and the United States can serve as models for other ports in the world. Those ports use renewable energy in their transport, warehousing, and handling. Their differentiated operations include the use of renewable energies in ship loading and unloading movements, the reduction or elimination of the use of fossil fuels, and the management of every chain that involves the access of vessels to its terminals. These ports manage the other model integrated into the ports, regarding the use of clean energies, treatment of residues, elimination of noise pollution, and so on.

This paper analyzes the legal aspects, physical characteristics, and current public policies that rule port activities in Brazil and the ongoing logistics operations of the Port of Santos, which is the largest port in Latin America. It addresses what possible processes, technologies, and innovations could be implemented in the medium- and long-term to make it a capable and integrated port with the other international ports regarding sustainable logistic operations.

2 Bibliographic Review

The use of liquefied natural gas (LNG) in port operations can contribute to gas reduction when compared to diesel-powered ships' engines. LNG is the cleanest fossil fuel and has the potential to reduce the emission of the following gases: NO_x, 92%; CO_2, 23%; SO_2, 100% e atmospheric particulate matter, 100% [1].

The Association of the Baltic Ports has developed an infrastructure for LNG to be the fuel for ships in some ports in the Baltic Sea region, with the support of leading industry organizations including the ESPO (European Sea Ports Organization) [1].

The ports that have participated in the studies are Helsingborg, located in the region of Skåne in southern Sweden and specialized in the transport of containers; Trelleborg, also located in the south of Sweden and specialized in RoRo transportation; Sundsvall, in northern Sweden, concentrated in the transport of paper, cellulose, and sawn wood; and Rostock, in Germany, which is a port with combined load terminals for forest products and RoRo cargo [1].

Easy access from both landside and seafront, as well as major investments in infrastructure, contributed to the port of Rostock on the Warnow River, becoming the number two port among the German Baltic Sea ports [1].

Long-term exposure to atmospheric particulate matter (PM) and nitrogen oxides (NO_x) has a devastating effect on the health of human workers who work directly in port operations, as well as in the surrounding population.

Since 2010, the European Union has determined that the maximum sulfur limit on marine fuel be 0.1% (against 1% previously) for docked ships. Vessels may choose to use alternative fuels while they are docked or connect at specific points to obtain electricity to power their operations [1].

The use of electric power from the port, called onshore power supply (OPS), reduces the harmful effects of noise and air pollution in the environment as they allow the auxiliary engines of the ships to be switched off. The implementation of the OPS is an opportunity to reduce the emission of carbon dioxide (CO_2), one of the contributors to the greenhouse effect [1].

OPS can be profitable for regular waterway transport lines when the vessel always docks in the same port and terminals. For ships carrying containers, which dock at different piers, such ports must have several terminals with adequate infrastructure to connect the OPS system to ships.

Some ports in the Baltic Sea are already prepared to offer the OPS system, mainly for water transport ships, auxiliary fleets, tugs, boats, barges, and icebreakers. They are usually low-voltage electricity points, with substations generating from 20–100 kV, with connection points of 6–20 kV and vessels with a transformer of 6–20 kV, of 400 V (the variation is from 230 to 400 V with a frequency of 50 Hz) [1].

The traditional equipment used in the port operations, such as RTG (rubber tire gantry), RMG (rail mounted gantry), ALV (automated lifting vehicle), ASC (automated stacking crane), automated transfer crane, forklifts, and other equipment used in ports, are usually large consumers of energy. When it comes to the topic of energy efficiency in port operations, it is not easy to separate this work philosophy from the investments that will be needed in port equipment. Therefore, it is necessary to generate renewable and clean energy to use in this port equipment, with a focus on reducing costs as well, as innovation and technology imply financial investments [2, 3]. The implementation and development of the use of renewable and sustainable energy are not simple to accomplish in ports that use, for example, fossil fuel. An essential condition is to have public policies and to involve all actors in port activities. It is necessary to understand the current reality, to evaluate the possible alternatives of the market for the use of renewable energies, and to be willing to face the technological challenges at a level that requires much work, planning, and the commitment of several areas of port management.

Having all the information of the current processes and practices of a port is an essential condition for the beginning of the implantation of a Green port. It is fundamental to know in detail the use of energy from the port, to have key performance indicators of current energy consumption in operations, and to have environmental management indicators. Having information, for example, of the energy consumption of each operation, will allow the team to gather real conditions to analyze possible Green logistics solutions for a port [2–5].

Therefore, the port administrators must be committed to this system of sustainable logistic operations to implement a Green port. Otherwise, it will practically go to failure [2–4, 6–13].

There are good examples of ports with the implementation of sustainable operations systems. Among them is the port of Koper, Slovenia, which has created a system to replace fossil fuel (diesel) water heating by biomass and solar energy for the next few years [2]. The port of Koper is also committed to encouraging and increasingly supporting initiatives in the introduction of more modern technologies focused exclusively on clean sustainable systems with a focus on environmental protection [2–8, 10–13].

This type of sustainable initiative in the port of Koper has opened doors for a pilot project to develop sustainable and clean projects with a strong appeal for the reduction or elimination of emission gases that harm the environment. The initial project had the endorsement and commitment of the board of directors. This attitude, therefore, served as an example for all personnel [2, 3, 6, 8, 11].

The RTG electrification at the port of Koper was a process that helped immensely reduce the use of diesel and thereby eliminate the emission of pollutants into the atmosphere. It has also reduced the emission of noise pollution in port operations [2, 3, 7, 10, 13].

The port authority has a key role to play in implementing a sustainable operations system in a port. The commitment of the port authority to the use of clean technologies in the operations of handling, storage, loading, and unloading is essential for the success of a Green port [3, 9, 12, 13].

3 Methodology

The methodology employed in this work was the bibliographic review of the researched topic. The bibliographic review is a stage that provides support for the research, based on articles and scientific material on the theme Green Port, bringing knowledge subsidies. The theoretical framework was based on international journals and documents. This study also used fieldwork through which data were collected from people who contributed to the understanding of the application of sustainable operations in the Port of Santos, Brazil. Therefore, an exploratory research was used in this study, involving bibliographical review, interviews with people with practical experiences of the subject, and analysis of examples that contribute to the understanding of the problem.

4 Green Port Operations

The Maasvlakte II terminal in Rotterdam is considered the major civil engineering project in the Netherlands. It includes the construction of a new port and backup infrastructure on a brownfield site next to the Maasvlakte terminal. The size of the site is approximately 167 ha, with 2800 m of the wharf and an additional 500 m for the operation of barges, with a capacity of 2.7 million TEUs. The Maasvlakte II terminal has the first Cargotec remote control operated porteiners, as well as cranes for the terminal of barges, Kuenz RMG cranes for the rail tracks on the quay line, Gottwald automated guided vehicles (AGVs) with Gottwald lifting battery, Gottwald robotic battery changing stations and management of terminal operating systems (TOS), and equipment control systems (ECS) by Navis (Cargotec) and TBA (Gottwald) simultaneously.

The terminal construction is being carried out according to the environmental performance certification rules of the Building Research Establishment Environmental Assessment Method (BREEAM) adopted by the Dutch Green Building Council. The use of electrification in all equipment allows the terminal to be free of emissions of CO_2, NO_x, or particulates. In 2009, the Maasvlakte I terminal became the world's first container terminal powered by wind power.

A new generation of battery-powered AGVs emerged in 2011. It was adopted in terminals such as the CTA in Germany and LBCT in the United States. The objective was reducing energy and maintenance costs, zero emission of pollutants, noise reduction, and use of renewable energy.

There is equipment used in automated systems called OHBC (overhead bridge crane). It can be found at PSA's Pasir Payang Terminal (Singapore) and Antwerp's Churchilldock (a 200 m long pilot project). Such equipment requires high initial investment due to the structure of the floor and supporting pillars of the crane, but has the potential for automation, is faster than other equipment of its category, and has precise positioning, in addition to being electrically powered equipment.

The electrification of equipment is an important item for a Green port. Several terminals have electrified their RTGs, which have been renamed the E-RTG, such as Modern Terminal Limited (Hong Kong) and Lazaro Cardenas Terminal (Mexico). These terminals collaborate with the environment in the reduction of the Greenhouse effect.

5 The Port of Santos—Brazil

The Port of Santos handled 2,358,220 container units out of a total of 5,652,587 in Brazilian ports in 2016, which corresponds to 41.72% of the total.

The Port of Santos is of enormous importance for Brazilian imports and exports. All types of cargo—solid, bulk, and liquid—are transported through it. According to the World Shipping Council, the Port of Santos appears in 39th position in the 2015 Top Fifty World Container Ports.

Most of the exports are carried out through the Port of Santos and the main products exported are orange juice, red meat, coffee beans, ethyl alcohol, carded cotton, cane sugar, and corn.

The Port of Santos has a vast area of influence. The primary hinterland that it serves involves five Brazilian states: Mato Grosso (MT), Mato Grosso do Sul (MS), Goiás (GO), Minas Gerais (MG), and São Paulo (SP). Together, these cover approximately 75 million people and 67% of Brazilian GDP. The Port of Santos handled a total of US$116.3 billion (FOB–US$) of Brazilian international trade in 2014.

The Port of Santos is strategically located in São Paulo and has intermodality with the railway, road, and pipeline modes. Some products are transported from the central-western region through a waterway and then transshipped to the road mode to supply the Port of Santos, as in the case of products such as soy and sugar.

When compared to international ports that have implemented or are implementing sustainable operations systems, the Port of Santos is still a long way from being among these privileged ports. Its logistics operations are not yet considered Green. Isolated movements are working to fit this system, although very incipient.

The automation systems of transport operations, handling, and storage using renewable energies are limited in the Port of Santos. Its equipment in the terminals is far behind those used in the Green ports.

The vehicles used in port operations are diesel-powered, which increases the emission of polluting gases in the atmosphere. There is still no electrical system to supply equipment and ships, such as the onshore power supply scheme.

Many aspects contribute to the port failing its operations considered sustainable, and all of them should receive particular attention because they degrade the environment. They include ship-generated garbage, handling, and storage of chemicals, leakage of fuel from ships, inappropriate dredging operations, and the like. The Port of Santos controls most activities directly related to environmental management. However, national legislation is still not appropriate to ensure sustainable logistics operations.

The Port of Santos is studying the feasibility of having a bulk terminal for solids operating sustainably, reducing truck traffic, and using the railroad. The goal is to reduce truck movements through the region by more than 600 a day and thereby reduce the emission of polluting gases and particulate matter in the atmosphere.

Other improvements include the use of solar panels for energy generation and enclosed systems for the loading and unloading of ships. The goal is to reduce the consumption of electrical energy demand through LED illumination. The collection and use of rainwater is also being considered.

Another current project provides for the use of electric vehicles combined with renewable energy for handling, transportation, and storage operations at the terminal. Once more, the goal is to obtain international certification as a Green port.

6 Environmental Management of the Port of Santos

The Docks Company of the State of São Paulo (CODESP) has been seeking, as a strategic decision, to implement an environmental management system in its administrative units. This system will deal with the Environmental Audit in Organized Ports, as well as with obtaining certification to the international standard ISO14001. The implementation of this system is already in progress. One of the most critical steps in this process is the establishment of CODESP's Environmental Policy. This policy is CODESP's reference for the sustainable development of its activities, and due to its importance, it should be widely disseminated to the staff and outsourced personnel of the company.

CODESP, the Administrative Authority of the Organized Port of Santos, in its quest for excellence in environmental performance, has the following objectives.

- Fully comply with current environmental legislation and the standards defined by CODESP.
- Promote efficiency in the provision of port services, having as controlling principles conservation and environmental control.
- Prevent pollution in all its forms.
- Preserve the historical and cultural heritage of the region.
- Stimulate technological innovation in the search for opportunities for business growth and continuous improvement of the environmental management system.
- Promote the constant training and qualification of its collaborators.

The supply of energy for CODESP's self-consumption and to the many tenants served by the system, comes primarily from the Itatinga hydroelectric plant. It has a capacity of 15 MW. An exception is the Alemoa Liquid Bulk Terminal, which is powered by CPFL (Companhia Paulista de Força e Luz), an energy supply company in the state of São Paulo.

The Port of Santos's solid waste management plan occurs through companies outsourced by CODESP (in the public port areas) and by the tenants (in the areas under lease), with specific rules in the contracts for each type of service and waste.

The solid waste from all buildings located on the right and left bank of the Port of Santos estuary is deposited in defined places following current environmental standards.

In the Right Margin, the waste is collected and transported to the transshipment area, approved by CETESB (Environmental Company of the State of São Paulo). At the Left Bank, the collection is not performed daily due to low demand.

The Solid Waste Management Plan of the Port of Santos is the instrument of compliance and a guide for all actors within the port, regarding the adequate management of solid waste generated in all activities carried out in the Organized Port of Santos. The Environmental Ports Management (GPM), together with the Occupational Health and Safety Management, are the management and oversight units of this plan within the Organized Port of Santos.

One of the achievements of CODESP that brought several benefits to the region of the Port of Santos was the implementation of its potable water supply and sanitary sewerage systems, which have been in operation since 2007.

The Environmental Policy of the Port of Santos is the protocol of intentions to be observed for the creation of the Environmental Management Systems of the Organized Port of Santos. In this sense, the environmental policies of each tenant must be consistent with this policy, of the Port of Santos, as well as the Environmental Management Systems of each tenant should acknowledge the synergy between them and the Environmental Management System of the Port Administration.

7 Conclusion

The Port of Santos needs a master plan aimed at implementing public policies with sustainable objectives, such as a focus on innovation and the use of clean technologies to carry out port operations at its public and private terminals.

The role of the federal, state, and municipal public authorities is essential to define sustainable policies for the national port authorities and especially the Port of Santos. Involving public and private partnerships in these decisions is critical to securing resources for the implementation of a long-term project focused on sustainable logistics indicators for the port and the application of financial resources in projects with technologies and innovation in clean and sustainable energies.

The international ports that have implemented sustainable projects will serve as benchmarks for the Port of Santos to focus on their needs.

Several operations carried out in the port activities of Santos that use fossil fuel must be eliminated or reduced to the maximum. That would help the port to lead a process of clean and sustainable operations. At the same time, other clean and renewable energy alternatives should be implemented. Other possible improvements include the approach to national research centers, universities, and other innovation centers so that new technologies are discovered and implemented in the current operations of the Port of Santos.

Acknowledgements Project 2015/00277-8, sponsored by FAPESP, a Foundation for Research Support of the State of São Paulo.

References

1. Clean Baltic Sea Shipping: Cleanship report. http://www.clean-baltic-sea-shipping.com/uploads/files/CLEANSHIP_final_report_for_download.pdf (2013). Last accessed 14 June 2017
2. Pavlic, B., Cepak, F., Sucis, B., Peckaj, M., Kandus, B.: Sustainable port infrastructure, practical Implementation of the green port concept. Therm. Sci. **18**(3), 935–948 (2014)
3. Peris-Mora, E., Orejas, J.M.D., Subirats, A., Ibánez, S., Alvares, P.: Development of a system of indicators for sustainable port management. Mar. Pollut. Bull. **50**, 1649–1660 (2005)
4. Puig, M., Wooldridge, C., Casal, J., Darbra, R.M.: Tool for the identification and assessment of environmental aspects in ports (TEAP). Ocean Coast. Manage. **113**, 8–17 (2015)
5. Chiu, R.-H., Lin, L.-H., Ting, S.-C.: Evaluation of Green Port Factors and Performance: A Fuzzy AHP Analysis. Mathematical Problems in Engineering, vol. 2014, article ID 802976, 12 pages. http://dx.doi.org/10.1155/2014/802976 (2014)
6. Davarzani, H., Fahimnia, B., Bell, M., Sarkis, J.: Greening ports and maritime logistics: a review. Transp. Res. Part D **48**, 473–487 (2016)
7. Lirn, T.-C., Wu, Y.-C., Chen, Y.J.: Green performance criteria for sustainable ports in Asia. Int. J. Phys. Distrib. Logistics Manage. **43**(5/6), 427–451 (2013)
8. Puig, M., Wooldridge, C., Darbra, R.M.: Identification and selection of environmental performance indicators for sustainable port development. Mar. Pollut. Bull. **81**, 124–130 (2014)
9. Acciaro, M., Ghiara, H., Cusano, M.I.: Energy management in seaports: a new role for port authorities. Energy Policy **71**, 4–12 (2014)
10. Roh, S., Thai, V.V., Wong, Y.D.: Towards sustainable Asean port development: challenges and opportunities for Vietnamese ports. Asian J. Shipping Logistics **32**(2), 107–118 (2016)
11. Sciberras, E.A., Zahawi, B., Atkinson, D.J.: Reducing shipboard emissions—assessment of the role of electrical technologies. Transp. Res. Part D **51**, 227–239 (2017)
12. Tang, J., McNabola, A., Misstear, B., Caulfield, B.: An evaluation of the impact of the Dublin port tunnel and HGV management strategy on air pollution emissions. Transp. Res. Part D **52**, 1–14 (2017)
13. Walker, T.R.: Green Marine: An environmental program to establish sustainability in marine transportation. Mar. Pollut. Bull. **105**, 199–207 (2016)

Competitiveness Criteria in Ports

Cassia Lima Machado, Ilton Curty Leal Junior, Vanessa de Almeida Guimarães and Newton Narciso Pereira

Abstract An organization gains competitive advantage when it develops or performs actions that allow it to surpass its competitors. Within the competitive market environment, ports, as important links in the international supply chain, should focus on their performance. With the increasing importance of seaports in the economy, port competitiveness has become a necessary topic of study. However, many are the criteria used to analyze competitiveness in organizations. In view of this, what are the criteria cited in the literature applicable to the analysis of the competitiveness of seaports in the world? To answer this question, this study identifies the main criteria used to analyze the competitiveness of international seaports reported in the literature. The importance of this study is given by the participation of port activity in the global economy. About 90% of the volume of trade around the world is traded through the ports. As ports are also organizations that need to be competitive, studies on this topic can be of help to professionals and academics in the area. Based on a bibliographical review, it was concluded that among the criteria found, the most important, according to the authors, in hierarchical order were: costs, efficiency, handling capacity, infrastructure, and port connectivity. During the research, no studies were found that address criteria related to port competitiveness specifically for the Brazilian scenario, and this is a suitable topic for future research.

Keywords Competitiveness · Ports · Port competitive criteria

C. L. Machado (✉) · I. C. Leal Junior
Federal Fluminense University - Administration Postgraduate Program - PPGA, Volta Redonda, Brazil
e-mail: cassia.lima09@hotmail.com

V. de Almeida Guimarães
Federal Center for Technological Education Celso Suckow da Fonseca, Rio de Janeiro, Brazil

N. N. Pereira
School of Industrial Engineering Metallurgical at Volta Redonda, Center for Sustainable Systems Studies—CESS, Rio de Janeiro, Brazil

A. Vega Sáenz et al. (eds.), *Proceedings of the 25th Pan-American Conference of Naval Engineering—COPINAVAL*, https://doi.org/10.1007/978-3-319-89812-4_21

1 Introduction

In the global competitive environment, an organization can gain competitive advantage by developing or executing actions that enable it to outperform its competitors. This search for competitive advantage is a central theme in the field of strategic management [1].

Regardless of the view that is used as a basis for analyzing the competitive environment, resources, or market, the fact is that, in some way, the search for competitiveness is present in most organizations. In this way, companies make use not only of their differentials vis-à-vis the external market, but also of their internal resources, considered by some authors as the main source of competitive advantage of organizations [3, 29].

With organizational strategies aimed at competitiveness, the main objective of companies has become the creation of competitive advantages within the competitive market environment, and seaports, as important links in the international supply chain, are part of this environment.

With the increasing importance of seaports in the economy due to their participation in the international market, port competitiveness has become a topic studied by several authors around the world [15, 23, 26, 45].

Low costs, quality of service, and productivity are criteria cited in the literature to analyze competitiveness in organizations in general, but do these criteria also apply to ports? What are the criteria cited in the literature that allow us to analyze the competitiveness of seaports in the world?

To answer these questions, this study identifies the main criteria used to analyze the competitiveness of international seaports reported in the literature. The specific objectives include: understanding the competitiveness theme and its applicability to ports and verifying the importance of these criteria according to the literature.

This work is relevant given the importance of port performance in the global economy. Seaports are part of the global supply chain and their activities are essential for the economic development of countries, as they account for around 90% of the volume of trade around the world. As ports are also organizations that need to be competitive, studies of their competitiveness as well as the classification of their criteria can contribute to the literature of the theme.

2 Port Competitiveness

The market-driven competitive positioning through the generic strategies of leadership by cost, focus, and differentiation as well as the competitive strategy of the "five forces" model were ideas presented by Porter [29, 30, 31]. Initially based on the SCP structure (structure–conduct–performance), these ideas introduced the concept of competitive advantage within the academic field of strategic management.

According to Baumann [4], in a globalized system the various organizational units become components of the same integrated value generation structure, whose overall strategies lead to the search for cost reduction and the specialization of production lines, among other attributes that aim at increasing efficiency and a greater degree of competitiveness. The organizational strategies turn to these objectives, seeking to combine and apply the existing attributes directed to the achievement of the goal, that is, to create competitive advantages within this competitive environment.

With the evolution of competitiveness studies, its basic concepts related to the market [29, 30, 32] also evolved and became linked with other factors such as innovation, sustainability, business strategy, and others [1, 8, 19, 28].

Every decade, a specific feature within organizations has been most desired by companies in search of competitiveness. In 1970 it was costs, in 1980 the quality of products, in 1990 it was the speed, and in 2000 flexibility, until reaching the years 2010–2015 where innovation is considered the "engine of development" and has weight in the survival of enterprises in a competitive environment [1, 13].

Seaports are also organizations that compete with each other. Throughout the twentieth and twenty-first centuries, they have evolved, increasing their importance in the economy and becoming key elements in the management of the global supply chain in conjunction with the development of international trade [23].

Competitiveness in seaports is important for the performance of organizations operating in the international market, not only because of their effects on the economy but also because of the large volume of goods involved in the process [26].

Due to their economic function, which is essentially to benefit those whose trade passes through them, seaports basically require three elements: maritime access with sufficient ballast, capacity to handle goods, and distributive capacity, including proper connections with the interior [23].

With increasing world trade cargo volumes, gaining competitive advantage requires an effective strategy in managing the global supply chain. To be competitive, a port needs to be part of an efficient supply chain. This thinking in terms of supply chains presupposes the existence of different types of competition, where each member of the chain has specific objectives. Efficient interaction between members of the supply chain makes every chain successful. Hence comes the importance of choosing a competitive port to be part of the chain [23].

Due to the fierce international competition between ports, it is essential to identify the determinants of port competitiveness, because they influence the decision to choose the port.

3 Methodological Procedures

As to its nature, this research is classified as basic, inasmuch as it aims to build knowledge without an immediate application using a qualitative approach.

Regarding the objectives, it is classified as descriptive, due to the concern with analysis, registration, classification, and interpretation of the data, without interference of the researcher. The technical procedure used was bibliographic research related to ports, competitiveness, port competitiveness, and competitiveness criteria, based on books and scientific articles taken mainly from the bases Webofscience, Periódicos Capes, and Scielo, as well as material taken from the Internet [33, 44].

The bibliographic research on port competitiveness was based on 35 articles found in the databases consulted. Each criterion used to evaluate port competitiveness was analyzed in the articles of origin and a term was found that, through its broad meaning, summarized its meaning. Criteria that had similar meanings were grouped and for each group of criteria a new name was assigned in order to facilitate their understanding.

The importance of the competitiveness criteria applied to ports is given in this study according to the frequency with which each criterion was cited by the authors of the articles surveyed.

4 Results

The result of consolidating the determinant criteria for port competitiveness was 17 criteria, and this consolidation was important due to the fact of assigning unique adapted nomenclature and conceptualization to criteria that had the same meaning, as shown in Table 1.

After analyzing the literature review, as shown in Table 2, five more cited criteria were found, which follow, in order of importance: costs (cited in 22 articles), efficiency (quoted in 14 articles), capacity (quoted in 13 articles), infrastructure (cited in 13 articles), and port connectivity (cited in 11 articles). The least-cited criteria were politics and economy, flexibility, availability of resources, and productivity.

Still based on the analysis of the bibliographic survey, it is possible to verify that there is predominance of some criteria over the years.

From 1985 to 1999, the most frequently cited criteria in the literature for determining international port competitiveness were costs, port management, capacity, and efficiency. From 2000 to 2010, the cost criterion remained predominant, but followed by infrastructure, frequency, port location, and loading and unloading time.

In the most recent literature, from 2011 to 2015, cost no longer remains the most cited criterion, giving rise to capacity, a criterion that is related to the volume of cargo handling and the possibility of the port loading and unloading larger volumes of goods, and connectivity of the port, a criterion related to the proximity of the ports of origin with the ports of destination, as well as their links with the interior for loading and unloading of goods. This shows that cost has lost importance in the face of other criteria that when not fully met, can cost more for organizations that depend on port services. Obviously, it means that with the complexity of the

Table 1 Consolidation and conceptualization of the criteria on port competitiveness

Criteria	Conceptualization	Examples found in literature
Capacity	It is related to the volume of cargo handling and the possibility of a port loading or unloading a larger volume of goods	Capacity, cargo handling capacity, volume of cargo handled, container handling volume, TEUS loaded by a crane and total TEUS moved in the port, static storage capacity, gates capacity, and patio equipment
Port connectivity	It is related to the proximity of the origin/destination ports, as well as their links of connection with the interior of the country for loading and unloading	Connectivity of the port, proximity to the ports of origin and destination, connections to the country interior, alternative routes for transportation, connections with the interior, quality of connections with the interior, port connection capacity, and intermodal connection links
Cost	Disbursements are generally made to allow the use of a particular port	Costs, travel cost, generalized costs, transportation costs, sea costs, port costs, port expenses, port taxes, amounts paid, price charged, and cost of chargers per port call
Information availability	It is related to how easy it is to get information about loading and unloading of goods	Availability of information, information technology, port information services, availability of port communication systems and cargo information
Resource availability	Equipment that is available for use inside the port at the time of loading and unloading the ships	Availability of equipment machines and services
Efficiency	Ports performance, related to who carries out activities, moving high quantities of goods with minimum failure rate	Efficiency and performance of port operations (involving cargo handling, berth occupancy rate, average number of vessels in queue, vessel queue size, congestion, and pier productivity) performance of losses and damages (including frequency), port efficiency
Flexibility	Possibility to deal with different types of cargo, including special cargoes (that are not accepted in all ports)	Flexibility, special cargo handling ability, flexibility in meeting large freight, flexibility in service, and suitability for special loads and special projects
Frequency	It is the number of times the ship visits the port; how many times the ship can load and unload at that port	Frequency of vessel visits, frequency of call of ships, frequency of trips, frequency of loading and unloading, number of routes available, and number of routes offered

(continued)

Table 1 (continued)

Criteria	Conceptualization	Examples found in literature
Port management	It is related to port management, from labor problems to reputation for cargo malfunctions	Capacity for claims management, port reputation, port management, labor issues, good reputation for damages and delays and port reputation for cargo breakdowns
Infrastructure	It is the physical structure and facilities that the port makes available to operationalize its activities	Adequate infrastructure, multiple transport infrastructure, superstructure, port facility, port conditions for cargo reception, available facilities, and physical and technical port infrastructure
Port location	Place where the port is geographically installed, close to freight roads or to other markets	Port location, geographical location, distance from the port to competitors, proximity to other markets, and main routes served
Internal logistic	It is related to the organization and management of activities within the port	Container handling logistics, empty container distribution, in-port freight consolidation and consolidation, cargo allocation, and cargo storage availability
Economy and policy	It is related to the state of politics and economy of the country where the port is located	Economic stability, political stability, political evolution, economic liberalization, and social policy
Customer concern	It is related to how the managers deal with customer needs, as well as how quickly they address the identified issues	Customer focus, ability to adapt to customer requirements, customer convenience, and rapid response to the needs of port users
Quality	Related to the quality of the services offered by the port	Quality of customs services, service level, reliability, quality of services, quality of terminal operators, quality of equipment, quality of the whole port, quality of work, quality of port services, and quality of services rendered
Time for loading and unloading	Related to the time spent in loading and unloading activities. It ranges from the moment the ship docks in the port until it leaves	Time of loading and unloading, transit time of the cargo in the port, ship turnaround time, loading time, and time performance
Others	These are several criteria found in the literature that do not fit into any criteria already mentioned	Free enterprise spirit, maritime company revenue, faster global route, and port security

Source Adapted by the authors (2016)

Table 2 Port competitiveness criteria found in the literature in chronological order

	Capacity	Port connectivity	Cost	Information availability	Resource availability	Efficiency	Flexibility	Frequency	Port management
Slack [35]	X		X			X		X	
Bird and Bland [5]			X					X	X
Frankel [9]	X		X						
Murphy et al. [26]	X			X	X	X	X		X
Murphy and Daley [25]	X		X	X	X	X	X		X
Tongzon [39]			X			X		X	X
Malchow and Kanafani [20, 21]	X							X	
Mangan et al. [22]		X	X	X	X	X	X	X	
Ha [12]			X	X					X
Tiwari et al. [38]	X								
Lirn et al. [18]			X						
Song and Yeo [36]	X		X						
Teng et al. [37]			X						
Cullinane et al. [6]			X						
Tongzon and Heng [42]						X			
Guy and Urli [11]			X						X
Ugboma et al. [43]						X		X	
Acosta et al. [1]				X		X			X

(continued)

Table 2 (continued)

	Capacity	Port connectivity	Cost	Information availability	Resource availability	Efficiency	Flexibility	Frequency	Port management
Shintani et al. [34]			X						
Tongzon and Sawant [40]			X			X		X	X
De Martino and Morvillo [7]									
Grosso and Monteiro [10]		X	X	X					
Leachman [16]		X	X						
Meersman et al. [24]		X							
Wiegmans et al. [45]		X		X		X			X
Tongzon [41]			X			X		X	X
Aronietis et al. [2]	X	X	X				X	X	
Yeo et al. [46]		X	X		X				
Oportus and Echeverría [33]		X							
Jafari and Noshadi [15]	X		X			X			
Hyuksoo and Sangkyun [14]	X					X			
Lee and Lam [17]	X	X		X					
Nazemzadeh and Vanelslander [27]	X	X	X						
Van Dyck and Ismael [44]	X	X	X			X			
Frequencies	13	11	22	8	4	14	4	9	10

(continued)

Table 2 (continued)

	Infrastructure	Port localization	Internal logistic	Economy and policy	Customer concern	Productivity	Quality	Time for charge and discharge	Others
Slack [35]	Infrastructure	Port localization	Internal logistic	Economy and policy	Customer concern	Productivity	Quality	Time for charge and discharge	Others
Bird and Bland [5]			X					X	X
Frankel [9]			X						X
Murphy et al. [26]	X		X						
Murphy and Daley [25]			X						
Tongzon [39]	X				X				
Malchow and Kanafani [20, 21]		X						X	
Mangan et al. [22]								X	X
Ha [12]	X	X			X			X	
Tiwari et al. [38]								X	
Lirn et al. [18]	X	X						X	
Song and Yeo [36]	X	X					X		
Teng et al. [37]				X		X	X	X	
Cullinane et al. [6]				X			X		
Tongzon and Heng [42]					X				
Guy and Urli [11]	X	X							
Ugboma et al. [43]	X								
Acosta et al. [1]	X								
Shintani et al. [34]			X						

(continued)

Table 2 (continued)

	Infrastructure	Port localization	Internal logistic	Economy and policy	Customer concern	Productivity	Quality	Time for charge and discharge	Others
Tongzon and Sawant [40]	X				X				
De Martino and Morvillo [7]					X		X		
Grosso and Monteiro [10]			X						
Leachman [16]								X	
Meersman et al. [24]									
Wiegmans et al. [45]	X	X	X				X		X
Tongzon [41]	X				X				
Aronietis et al. [2]		X					X		X
Yeo et al. [46]							X		
Oportus and Echeverría [33]									
Jafari and Noshadi [15]	X	X					X	X	
Hyuksoo and Sangkyun [14]							X		
Lee and Lam [17]							X		X
Nazemzadeh and Vanelslander [27]		X				X			
Van Dyck and Ismael [44]	X	X		X					
Frequencies	13	10	7	3	6	2	10	9	6

Source Adapted by the authors (2016)

operations associated with the ports, the criteria of operational efficiency have become decisive for customers to select the ports that will operate. This leads to a comprehensive analysis of the system, which involves complex decision factors that will ultimately determine the best port to operate. However, it should be noted that in the face of competition, better services and efficiency in the activity influence the decision, but the cost in Brazil still has a strong influence on the final decision of the owner of the cargo.

5 Final Considerations

The term competitiveness has a broad sense, both in market positioning and resource-based view [3, 30]. When applying the concept of competitiveness to ports, the criteria that assess competitiveness are similar, even as a port is also a complex and heterogeneous organization. After a bibliographic analysis where the criteria of competitiveness applied to ports were drawn, it was concluded that among the criteria found, the most important according to the authors in order of importance were the costs, efficiency, capacity, infrastructure, and connectivity of the port.

This paper presented the criteria used to evaluate the competitiveness of the port in the world, reaching its proposed objective. In Tables 1 and 2 the criteria used by the authors were drawn up, exemplifying each criterion.

No articles were found in the literature that discuss the criteria used to evaluate port competitiveness in the Brazilian scenario, therefore, this theme can be approached in future research. In this way, the present paper is a contribution to the state of the art of Brazilian port literature.

References

1. Acosta, M., Coronado, D., Mar Cerban, M.: Port competitiveness in container traffic from an internal point of view: the experience of the Port of Algeciras Bay. Marit. Policy Manage. **34** (5), 501–520 (2007)
2. Aronietis, R., et al.: Some effects of hinterland infrastructure pricing on port competitiveness: case of Antwerp. In: 12th World Conference on Transport Research, Lisbon (2010)
3. Barney, J.: Firm resources and sustained competitive advantage. J. Manag. **17**(1), 99–120 (1991)
4. Baumann, R.: O Brasil e a economia global. Editora Campus/SOBEET, Rio de Janeiro (1996)
5. Bird, J., Bland, G.: Freight forwarders speak: the perception of route competition via seaports in the European Communities Research Project. Part 1. Marit. Policy Manage **15**(1), 35–55 (1988)
6. Cullinane, K., Teng, Y., Wang, T.: Port competition between Shanghai and Ningbo. Marit. Policy Manage. **32**(4), 331–346 (2005)
7. De Martino, M., Morvillo, A.: Activities, resources and inter-organizational relationships: key factors in port competitiveness. Marit. Policy Manage. **35**(6), 571–589 (2008)

8. Ferraz, J.C., Kupfer, D., Haguenauer, L.: Made in Brazil: desafios competitivos para a indústria. Campus, Rio deJaneiro (1997)
9. Frankel, E.G.: Hierarchical logic in shipping policy and decision-making. Marit. Policy Manage. **19**(3), 211–221 (1992)
10. Grosso, M., Monteiro, F.: Relevant strategic criteria when choosing a container port-the case of the port of Genoa. Res. Transp. Logistics, 299 (2009)
11. Guy, E., Urli, B.: Port selection and multicriteria analysis: An application to the Montreal-New York alternative. Maritime Economics & Logistics **8**(2), 169–186 (2006)
12. HA, M.-.S A comparison of service quality at major container ports: implications for Korean ports. **Journal of transport geography**, v. 11, n. 2, p. 131-137, 2003.
13. HASHIMOTO, M. **Espírito empreendedor nas organizações:** aumentando a competitividade através do intra-empreendedorismo. Saraiva, 2006.
14. Hyuksoo, C.H.O., Sangkyun, K.I.M.: Examining Container Port Resources and Environments to Enhance Competitiveness: A Cross-Country Study from Resource-Based and Institutional Perspectives. The Asian Journal of Shipping and Logistics **31**(3), 341–362 (2015)
15. Jafari, H., Noshadi, E.: Ranking Ports Based on Competitive Indicators by Using ORESTE Method. International Research Journal of Applied and Basic Sciences **4**(6), 1492–1498 (2013)
16. Leachman, R.C.: Port and modal allocation of waterborne containerized imports from Asia to the United States. Transp. Res. Part E Logistics Transp. Rev. **44**(2), 313–331 (2008)
17. Lee, P.T.-W., Lam, J.S.L.: Container port competition and competitiveness analysis: Asian major ports. In: Handbook of Ocean Container Transport Logistics, pp. 97–136. Springer International Publishing, New York (2015)
18. Lirn, T.C., et al.: An application of AHP on transhipment port selection: a global perspective. Marit. Econ. Logistics **6**(1), 70–91 (2004)
19. Machado-Da-silva, C.L., DA Fonseca, V.S.: Competitividade organizacional: uma tentativa de reconstrução analítica. Organizações Soc. **4**(7), 97–114 (1996)
20. Malchow, M., Kanafani, A.: A disaggregate analysis of factors influencing port selection. Marit. Policy Manage. **28**(3), 265–277 (2001)
21. Malchow, M.B., Kanafani, A.: A disaggregate analysis of port selection. Transp. Res. Part E Logistics Transp. Rev. **40**(4), 317–337 (2004)
22. Mangan, J., Lalwani, C., Gardner, B.: Modelling port/ferry choice in Ro-Ro freight transportation. Int. J. Transp. Manage. **1**(1), 15–28 (2002)
23. Meersman, H., et al.: Port competition revisited. J. Pediatr. Matern. Fam. Health Chiropractic **55**(2), 210 (2010)
24. Meersman, H., et al.: The relation between port competition and hinterland connections: the case of the Iron Rhine and the Betuweroute. In: International Forum on Shipping, Ports and Airports (IFSPA 2008)-Trade-Based Global Supply Chain and Transport Logistics Hubs: Trends and Future Development (2008)
25. Murphy, P.R., Daley, J.M., Dalenberg, D.R.: Port selection criteria: an application of a transportation. Logistics Transp. Rev. **28**(3), 237 (1992)
26. Murphy, P.R., Daley, J.M., Dalenberg, D.R.: Port selection criteria: an application of a transportation. Logistics and Transportation Review **28**(3), 237 (1992)
27. Nazemzadeh, M., Vanelslander, T.: The container transport system: selection criteria and business attractiveness for North-European ports. Marit. Econ. Logistics **17**(2), 221–245 (2015)
28. Nir, A.S., Lin, K., Liang, G.S.: Port choice behaviour–from the perspective of the shipper. Marit. Policy Manage. **30**(2), 165–173 (2003)
29. Pizzoloto, N.D., Scarvarda, L.F., Paiva, R.: Zonas de influência portuárias-hinterlands: conceituação e metodologias para sua delimitação. Gestão Produção **17**(3), 553–566 (2010)
30. Porter, M.: Competitive strategy: techniques for analyzing industries and competitors. The Free Press, New York (1980)
31. Porter, M.: Vantagem competitiva. Editora Campus, São Paulo (1986)
32. Porter, M.E.: A vantagem competitiva das nações (1993)

33. SALGADO OPORTUS, O.; CEA ECHEVERRÍA, P. Análisis de la conectividad externa de los puertos de Chile como un factor de competitividad. **Ingeniare. Revista chilena de ingeniería,** v. 20, n. 1, p. 25-39, 2012.
34. Shintani, K., et al.: The container shipping network design problem with empty container repositioning. Transportation Research Part E: Logistics and Transportation Review **43**(1), 39–59 (2007)
35. Slack, B.: Containerization, inter-port competition, and port selection. Maritime policy and management **12**(4), 293–303 (1985)
36. Song, D.-W., Yeo, K.: A competitive analysis of Chinese container ports using the analytic hierarchy process. Maritime Economics & Logistics **6**(1), 34–52 (2004)
37. Teng, J., Huang, W., Huang, M.: Multicriteria evaluation for port competitiveness of eight East Asian container ports. Journal of Marine Science and Technology **12**(4), 256–264 (2004)
38. Tiwari, P., et al.: Shippers' port and carrier selection behaviour in China: a discrete choice analysis. Maritime Economics & Logistics **5**(1), 23–39 (2003)
39. Tongzon, J.L.: Determinants of port performance and efficiency. Transportation Research Part A: Policy and Practice **29**(3), 245–252 (1995)
40. Tongzon, J.L.: SAWANT, Lavina. Port choice in a competitive environment: from the shipping lines' perspective. Appl. Econ. **39**(4), 477–492 (2007)
41. Tongzon, J.L.: Port choice and freight forwarders. Transp. Res. Part E Logistics Transp. Rev. **45**(1), 186–195 (2009)
42. Tongzon, J., Heng, W.: Port privatization, efficiency and competitiveness: Some empirical evidence from container ports (terminals). Transp. Res. Part A Policy Pract. **39**(5), 405–424 (2005)
43. Ugboma, C., Ugboma, O., Ogwude, I.C.: An analytic hierarchy process (AHP) approach to port selection decisions–empirical evidence from Nigerian ports. Marit. Econ. Logistics **8**(3), 251–266 (2006)
44. van Dyck, G.K., Ismael, H.M.: Multi-criteria evaluation of port competitiveness in West Africa using analytic hierarchy process (AHP). Am. J. Ind. Bus. Manage. **5**(06), 432 (2015)
45. Wiegmans, B.W., Hoest, A., Notteboom, T.E.: Port and terminal selection by deep-sea container operators. Maritime Policy & Management **35**(6), 517–534 (2008)
46. Yeo, G., Roe, M., Dinwoodie, J.: Measuring the competitiveness of container ports: logisticians' perspectives. European Journal of Marketing **45**(3), 455–470 (2011)

Maturity Analysis of Processes for the Implementation of the SCOR Model in Companies in the Northern Colombian Zone

Fredy Armando Cuervo Lara

Abstract This research aims to determine the level of maturation of processes for the implementation of the SCOR model in the processes associated with the supply chain. The sample set included 35 companies from the industrial, commercial, and service sectors, which are located in four departments of the northern Colombian zone. The selected departments were Magdalena, Atlántico, Bolívar, and Cesar and Guajira. The methodology applied included the analysis of the strategic planning model, based on an analysis of mission, vision, strategic objectives, projects, and foresight of the organization; in addition, the balance of the business strategy was analyzed based on the breakdown of the strategic map. Afterwards, a business process maturity analysis was carried out applying the Gartner and CMMI models to identify the correlation with the processes associated with the SCOR methodology, among them supply chain planning, procurement, manufacturing, distribution, returns from distribution, and returns from procurement. The information was compiled in an exhaustive document review of each company and surveys were applied to members of the organization to identify elements that contribute to the implementation of the SCOR methodology, whose structure is attached in an annex. The analysis, processing, and contextualization of the information of the strategic planning model resulted in 55% of the companies presenting a moderate alignment

F. A. Cuervo Lara (✉)
Universidad Cooperativa de Colombia, Santa Marta, Colombia
e-mail: fredycuer@gmail.com

F. A. Cuervo Lara
Universidad Internacional Iberoamericana de Puerto Rico, Puerto Rico, USA

F. A. Cuervo Lara
Universidad Sergio Arboleda, Santa Marta, Colombia

F. A. Cuervo Lara
Universidad del Magdalena, Santa Marta, Colombia

F. A. Cuervo Lara
Universidad Pedagógica y Tecnológica de Colombia, Tunja, Colombia

F. A. Cuervo Lara
Universidad Jorge Tadeo Lozano, Bogotá, Colombia

A. Vega Sáenz et al. (eds.), *Proceedings of the 25th Pan-American Conference of Naval Engineering—COPINAVAL*, https://doi.org/10.1007/978-3-319-89812-4_22

in terms of their mission objective, their visionary perspective, and their strategic objectives formulated.

Keywords Processes · Strategic planning · Supply chain · Procurement Manufacturing · Distribution · Return

1 Introduction

This study focuses on the variable business processes, specifically in the processes related to the supply chain in the SCOR model (supply chain operation reference), analyzing the level of maturity of its processes in the light of a model of maturity that has taken elements of different models that fit the supply chain analysis in companies of different sectors and different sizes in the Colombian North Zone. The figures of Colombian logistic context show indicators of considerable importance when evaluating the real competitiveness of the business logistics, among them the results of the national logistics survey made by the national planning department, the guilds, business, and the academy in 2015 are observed. The logistic indicators showed the following results [1].

A small company uses 1.4 times more time to load or unload than a large company; a carrier before loading or unloading must wait an average of 3–5 h with the vehicle still. The three most important items in the cost of logistics present the following distribution: 37% transportation, 20% storage, and 17% purchases and management of suppliers. In terms of logistics efficiency, it is observed that 83.8% of deliveries are made on time, 2 out of 10 orders arrive late, 92% of orders are delivered full, and 77% of deliveries are made on time and complete. The national logistics survey shows that the main barriers that impact the logistics of users of logistics services are high transportation costs with 32%, inadequate infrastructure 21%, lack of logistics information systems 19%, and complexity in customs procedures with 12%.

2 Context of Maturity Models

The maturity reached in the processes determines the business results. An organization is a scenario of transformation of inputs of various kinds, whose exit from the business environment to a competitive business environment is reflected in the reception of the goods or services produced by the organization [2].

Measurement of the maturity of processes arises as an evolutionary process that seeks to find determining elements in the improvement of the process and whose need is manifested in the measure that the organization pursues better results; a maturity model allows describing the characteristics of the process in light of elements defined in the model and compared with standards that show the performance of the process. The model must present the following components: number of levels, descriptive name of each level, generic description of each level, number of dimensions or areas of processes (PA), number of elements or activities of each PA, description for each activity, and the detail of how it can be carried out [3].

2.1 Review of Applied Maturity Models

The first ones were formulated for organizational behavior; however, from the 1970s its use was extended to the understanding of business phenomena. In 1969 Peinado, one of the first authors, mentioned that the company was developed in stages organized with uniform patterns, and since then the organizations have been typified in different numbers of stages of development.

CMM Capability Maturity Model

The technology and its incorporation in business processes, seeking to improve results in processes, generated the need to monitor the reliability and efficiency of the processes. From there comes the CMM-Capability Maturity Model, developed in 1986 at Carnegie Mellon University, by request of the Department of Defense of the United States; this model was applied to software development organizations [4].

The model presents five levels of maturity in its project management processes [5] (Fig. 1).

Subsequent to this model, other models that include common elements were integrated into CMM and the Capability Maturity Model Integration CMMI originated to develop a framework to improve processes [7].

Maturity Level	Feature
Level 5 Optimized	Practices based on continuous improvement
Level 4 Predictable	Measurable Practices
Level 3 Defined	Practices based on competencies
Level 2 Managed	Repeated Practices
Level 1	Initial Chaotic practices

Fig. 1 Capability Maturity Model levels [6]

Gartner Maturity Model

According to Gartner, a greater maturity of the infrastructure and operations (I&O) processes is essential for business processes to function more efficiently and to be oriented towards customer satisfaction [8]. The model was initially oriented to IT and operations, but it has been demonstrated that the impact on other business processes leads to an improvement and evolution in business management and results [9]. The maturity of processes formulated by Gartner initiates technological changes and involves people, processes, and organizational culture.

The levels of maturity of Gartner are shown in Table 1.

SCOR Process Model

The SCOR model is a reference framework developed for supply chain management in 1996 by the supply chain council (SCC); this model has a deployment of first-level processes in five processes: planning, source, manufacturing, logistics, and return, which cover the entire supply chain from source to final customer [11].

The levels of the SCOR model are: higher level: types of processes, configuration level: categories of processes, and level of elements of processes, decomposition of each process, to arrive at metrics based on key performance indicators. These indicators are divided into categories related to reliability in compliance (reliability), speed of attention, cost, assets, and flexibility [12].

Table 1 Maturity levels of Gartner models [10]

Maturity level of the process	Characteristics
0 Survivor	Little or no attention to the IT infrastructure and operations
1 Consciousness	Understanding that infrastructure and operations are critical for the business, beginning to take action (in people, organization, processes, and technologies) to obtain operational control and visibility
2 Commitment	Moving to a managed environment, for example, in day-to-day IT support processes or in improving project management success, to guide customers and increase their satisfaction
3 Proactive	Gaining efficiency and quality of service through standardization, policy development, governance structures, and the implementation of proactive interdepartmental processes such as change management and implementation of versions
4 Aligned to IT	Management service as a business; focused on the client; process quality, competitive and reliable IT service provider
5 Partner	Partner of confidence for the company in order to increase the value and competitiveness of business processes, as well as the business as a whole

2.2 *Business Context of the Region*

The Colombian Caribbean region is located to the north, its main structuring axis is the Caribbean Sea, its extension is 132,270 km^2, equivalent to 11% of the national territory, and it has an estimated population of close to 11 million inhabitants, of whom 74% are located in municipal capitals and the rest in rural areas. It is integrated by seven departments that contribute 107 billion pesos to the national GDP, equivalent to 15% of the national GDP [13].

The sample set of the study included 35 companies from the industrial, commercial, and service sectors, which are located in five departments of the northern Colombian zone; the selected departments were: Magdalena, Atlántico, Bolívar, and Cesar and Guajira (Tables 2 and 3).

Understanding that according to Colombian commercial legislation a company is considered according to the number of employees and the level of assets expressed in minimum legal monthly salaries in force, smmlv.

3 Results Obtained

Regarding the alignment of business strategy, it was inquired about the existence of a strategic platform, about the alignment of mission–vision–strategy, the implementation of integral dashboards, existence of strategic maps and processes, and the result of alignment of business strategy, is based on nine variables [15].

Table 2 Sample composition of the study

Type of companies	Sample	Magdalena	Atlántico	Bolívar	Guajira	Cesar
Micro enterprises	9	4	2	1	1	1
Small	5	2	1	1	1	
Medium	8	3	2	2		1
Large	13	3	3	3	2	2
Total	35	12	8	7	4	4

Table 3 Bancoldex classification of companies in Colombia 2017 [14]

Type of company	Characteristics
Microenterprise	Less than 10 workers
	Total assets less than 500 smmlv
Small	Plant between 11 and 50 workers
	Total assets of 501 to 5,000 smmlv
Medium	Of 51 to 200 workers
	Total assets of 5001–30,000 smmlv
Large	More than 200 workers
	Total assets above 30,000 smmlv

smmlv—legal monthly minimum wages

The analysis, processing, and contextualization of the information of the strategic planning model resulted in 55% of the companies presenting a moderate alignment in terms of their mission object, their vision perspective, and their formulated strategic objectives [16]. An imbalance of the strategy in the perspectives of the Balance Score Card was presented by 45% of the companies, aimed at strengthening and prioritizing financial objectives by 65%, favoring process objectives by 25%, and training and client objectives by 10%.

3.1 Level 1 SCOR Process Characterization

A characterization analysis of processes associated with the SCOR model was carried out, considering the breakdown of activities, inputs, customers, resources, suppliers, leader, objectives, scope, documentation, control parameters, and requirements [17]. The graphic in Fig. 2 summarizes the result of the characterization assessment of SCOR processes in the sample of companies.

3.2 Development Level Analysis of SCOR Indicators

The process development analysis of the first SCOR first-level process indicators was performed, according to the categories established in the model: compliance, flexibility, costs, asset rotation, and speed of responses. The figure shows that in the indicators of compliance in the supply chain, the indicator that shows the greatest development in the sample of companies was the percentage of compliance with perfect orders and the one that showed the least development was delivery performance as well as the service level.

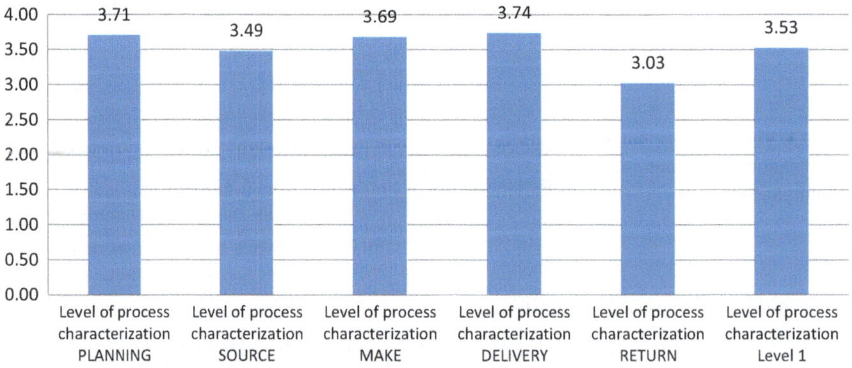

Fig. 2 Level of process characterization level 1 of the supply chain

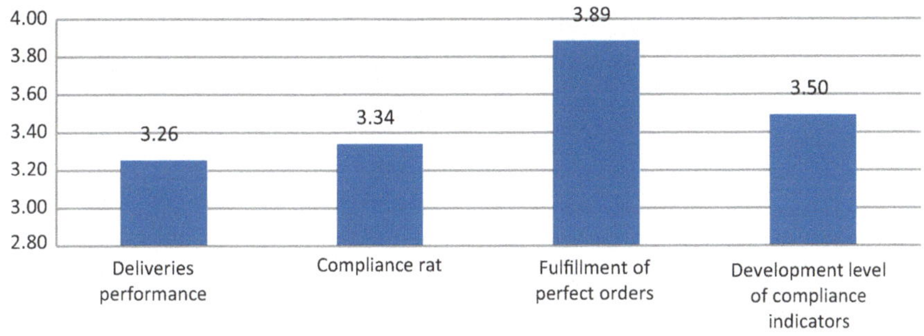

Fig. 3 Level of development of SCOR compliance indicators

Figure 3 shows that in the indicators of flexibility in the supply chain, the indicator that shows the greatest development in the sample of companies was the percentage of flexibility in production and the one that showed the least development was the response time of the supply chain.

3.3 *Gartner Maturity Analysis by Sector in the Sample of Companies*

When evaluating the Gartner level of maturity according to the criteria established by the model, an analysis was performed by the sector to which each company belongs; the study contemplated 15 sectors of economic activity, represented by 35 selected companies. In the table it is shown that the distribution of companies by sector according to the level of maturity that Gartner observed.

It is observed how the sector of large companies presents the highest level of maturity in SCOR processes: aligned level, the other sectors present high percentages of companies at levels of awareness, commitment, and proactive; no company presented the surviving level.

Figure 4 shows that the sectors with the highest level of maturity were automotive, distribution, rail, large surfaces, and maritime; one of the reasons exposed by each sector was the fact of being links of global supply chains with defined standards for SCOR processes.

The distribution of Gartner maturity levels by sector of economic activity in the business sample studied, specifically in companies in the logistics sector, identified that 50% of companies presented commitment level, 25% awareness level, and 25% proactive level.

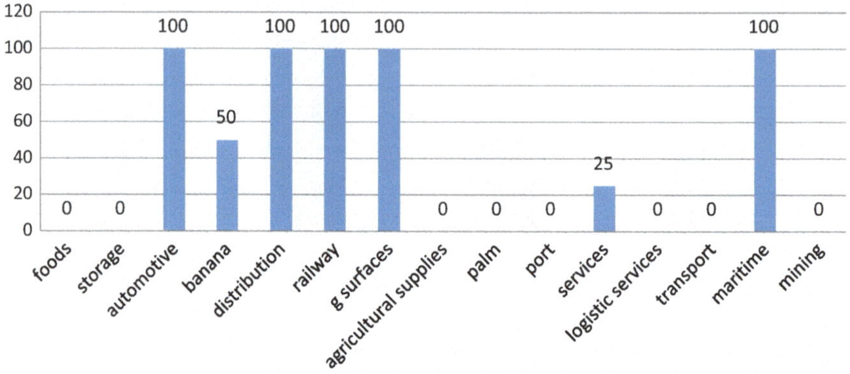

Fig. 4 Sectors with percentage of companies with processes in level of Gartner maturity: aligned

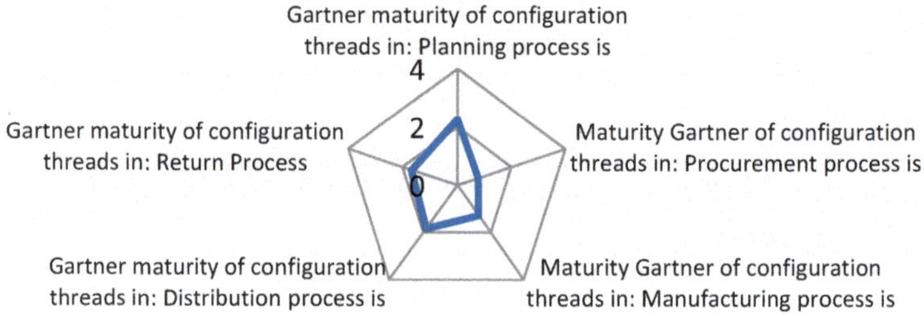

Fig. 5 Gartner maturity of SCOR processes in small companies

Gartner Maturity in SCOR Processes in Small Companies

It can be seen how the planning process in small companies (Fig. 5) was the one that presented a level of maturity 2: commitment, which means that there is a need to move to a state with IT support for project management; the most incipient process in maturation was supplying level 0: survivor, the other processes are in maturity close to level 2; commitment, this situation shows an improvement in maturity in this type of company with a structure in state awareness and commitment status

3.4 *Maturity Analysis CMMI: Capability Maturity Model Integration*

As a complementary contribution, a maturity analysis was applied according to the CMMI. This model is the most widely disseminated worldwide to improve processes in organizations. It is applied through the Software Engineering Institute of Carnegie Mellon University (Pittsburgh, USA). The CMMI can be used in three contexts: development of products and services, establishment of services, management and delivery (CMMI model for services), and acquisition of products and services (CMMI model for acquisition).

An analysis of the maturity of processes was carried out by type of company according to its size: micro, small, medium, and large (Table 4), identifying the

Table 4 CMMI maturity level of SCOR processes by type of company

Process	Average micro enterprises		Average small businesses		Medium companies		Average large companies	
CMM maturity of configuration in: planning process is	1.94	Initial	2.60	Managed	3.22	Defined	4.23	Quantitatively managed
CMM maturity of configuration threads in supply process is	1.89	Initial	2.00	Managed	2.88	Defined	4.46	Optimized
CMM maturity of configuration threads in manufacturing process is	2.00	Managed	2.27	Managed	2.88	Defined	3.77	Quantitatively managed
CMM maturity of configuration in distribution process is	2.00	Initial	2.27	Managed	2.88	Defined	3.77	Quantitatively managed
Maturity CMM of configuration threads in return process	1.96	Initial	2.18	Managed	2.88	Defined	4.00	Quantitatively managed
General maturity CMM SCOR processes	1.96	Initial	2.26	Managed	2.94	Defined	4.05	Quantitatively managed

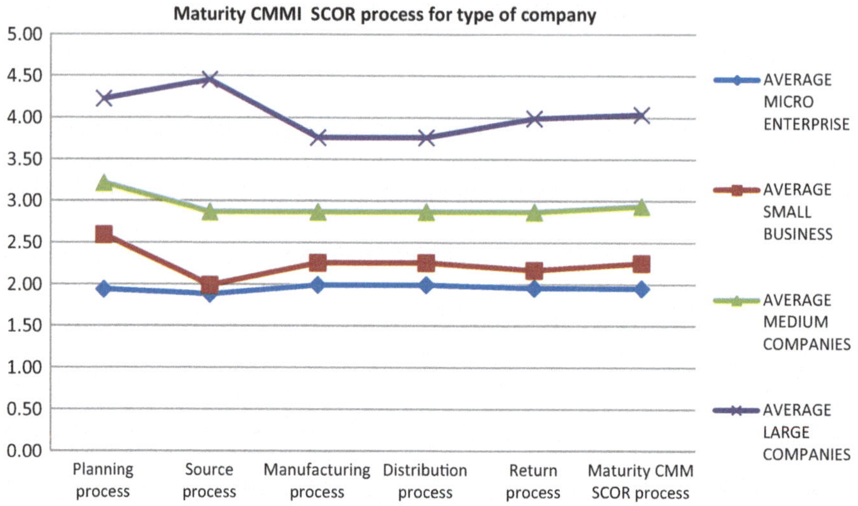

Fig. 6 Maturity process in CMMI model for SCOR process

maturity according to CMMI for each group of processes and their respective subprocesses. The levels of maturity obtained are summarized below.

The CMMI maturity analysis (Fig. 6) was carried out by type of company for each SCOR process, obtaining the initial level of maturity in SCOR processes for micro and small companies, a level defined for medium-sized companies, and quantitatively managed for large companies. Once the maturity analyses were carried out by the Gartner model and the CMMI model, the models were compared by type of company.

4 Conclusions and Recommendations

The study showed, with reference to the application of the Gartner model, that of the 15 sectors sampled, only 5 presented 100% of companies with maturity level aligned in SCOR processes. This shows that there is still a lot of work in inbound logistics or internal process to ensure optimal performance in the supply chain and logistics outbound or external to companies.

The companies of the transport sector presented the lowest levels of maturity of SCOR processes; this represents a high risk for the resilience of the supply chain, because transport activity represents one of the highest costs in the logistics of the region.

The sector of distribution companies presented a level of maturity awareness, which represents a high risk for the logistics chain, given that distribution is one of the most complex processes in the region due to lack of infrastructure and coordination between the links of the logistics chain.

With reference to the application of the CMMI model it is observed how the SCOR processes in micro-enterprises present a maturity in initial level 1, which means that the process is unpredictable, without control and reactive solutions; however, they tend towards level 2 managed. In small companies the SCOR processes presented level 2 maturity managed, which means that the process is oriented to projects and is often controlled, but with reactive solutions. In the medium-sized companies, the SCOR processes presented a defined level of maturity 3, which means that the processes are characterized by organization and are proactive solutions.

In large companies it is observed as the level of maturity 4 managed quantitatively, which means that the processes are measured and controlled with discipline, however, they do not present any level 5 optimized process, which shows an area of opportunity for large companies to guide the processes to continuous improvement.

The large gap in maturity is generated in micro and medium enterprises, in which the problems of planning and supply generate a start of the supply chain with very low performance results. This business line still presents very basic results of vertical integration backwards and forwards in its supply chain, which generates an effect of business isolation that is reflected in the maturity of its processes [18].

It is recommended that the business strengthening plans specifically involving training in the supply chain with training for managers, supervisors, and logistics operators be implemented; given the evident low level of maturity of the supply chain, presented by medium-sized companies, small businesses, and micro companies, it is necessary to act on how the processes of the SCOR model are planned and executed.

It is necessary to perform a maturity analysis of SCOR processes, detailed by sector and in all regions of Colombia; this study opens the door to a national study that allows us to know the national process maturity situation to implement the SCOR model at the national level.

References

1. National Planning Department: National Logistics Survey 2015. Government of Colombia, Bogotá (2016)
2. Tutorialspoint Homepage. https://www.tutorialpoint.com/en/cmmi/cmmi-maturity-level.htm. Last accessed 28 Jan 2017
3. Benavides, R.: Design of a Model of Evaluation and Measurement of the Degree of Organizational Maturity in Project Management of Vehicle Assemblers. Thesis of degree, Repository of the Colombian School of Engineering—Julio Garavito, Bogotá (2015)
4. Riuz, M., Ramos, I.: Evaluation and process improvement models: comparative analysis. In: Ceur-ws.org, vol. 4. Paper4 (2014)
5. Kerzner, H.: Strategic Planning for Project Management Using a Project Maturity Model, 1st edn. Wiley, New York (2001)
6. Quality and software homepage. http://www.calidadysoftware.com/otros/introductioncmmi.php. Last accessed 12 Mar 2017

7. All soft software engineering homepage. http://www.allsoft/recursos/themodelCMMI.pdf. Last accessed 15 Apr 2017
8. Figuerola, N.: Gartner Maturity model of infrastructure and operations. Best practice in goberance. Feb (2014)
9. Perez-Melgarejo, E., Perez-Vergara, E., Rodriguez Ruiz, Y.: Maturity models and their suitability to apply in small companies. Ind. Eng. **35**(2), 146–158 (2016)
10. Diaz-Jaimes, M., Ortiz, N.: Review of models: performance evaluation strategy for manufacturing companies. Ind. Univ. Santander. **11**(1), 55–72 (2012)
11. APICS Homepage. http://www.apics.com/apicsforbusiness/platform/organization. Last accessed 24 June 2017
12. APICS Homepage. http://www.apics.com/productsandservice/apics-scc-frameworks/scor. Last accessed 21 May 2017
13. Observatory of the Colombian Caribbean. The Caribbean Region. Generalities of the Caribbean Region. www.ocaribe.org/caribbeanregion. Last accessed 04 Mar 2017
14. Bancoldex Homepage. http://www.bancoldex.com/Sobre-microempresas/classification-of-companies-inColombia315.aspx
15. Alvarez, M.: Manual of Strategic Planning, 1st edn. Panorama, Mexico (2006)
16. Hijar, G.: Strategic Planning: The Prospective Vision, 1st edn. Editorial Limusa, Bogotá (2011)
17. Chiavenato, I.: Strategic Planning: Fundamentals and Applications, 2nd edn. Ed Mc Graw Hill, Mexico (2000)
18. Sallenave, J.P.: Management and Strategic Planning, 1st edn. Ed Norma, Bogotá (2002)

Evolution of Liquified Gas Transit Through the Panama Canal—An Opportunity for Exploiting the Expanded Canal

Luis Manuel Carral Couce, Laura Castro-Santos, Javier Tarrío Saavedra, Adán Vega Sáenz, Johnny Bogle and Rodolfo Sabonge

Abstract The transit of vessels of liquified natural gas by the Panama Canal is a recent fact derived from the increase in the production and exchanges of natural gas via the sea. The application of new extraction technologies in the United States has as a consequence a surplus in production and an increase in gas exports through the progressive adaptation and entry into service of maritime terminals. At the same time, the consumption in Pacific Asia is rising as a result of the development of emerging economies and Japan's policy of energy diversification after the Fukushima event. This imbalance between production and consumption between the Atlantic and Pacific maritime areas, and the increase in its transport demand, coincides in time with the recent expansion of the Panama Canal. The consequences for shipping seem clear because the Canal opens the possibility of LNG shipments from the Gulf of Mexico in the United States and Trinidad and Tobago to Asia. The LNG transport flows directed to Asian consumers are greatly influenced by the Canal rates. Estimations of minimum and maximum LNG loads corresponding to the years 2020 and 2035 have been used to achieve the objective of this paper: to study the effect of these flows on the sea route that transits Panama and its influence on the exploitation of its Canal.

Keywords Maritime transport · LNG · Panama canal · Canal exploitation

L. M. Carral Couce (✉) · L. Castro-Santos · J. T. Saavedra
Universidade da Coruña, A Coruña, Spain
e-mail: lcarral@udc.es

A. Vega Sáenz · J. Bogle · R. Sabonge
Universidad Marítima Internacional de Panamá, Panama City, Panama

A. Vega Sáenz et al. (eds.), *Proceedings of the 25th Pan-American Conference of Naval Engineering—COPINAVAL*, https://doi.org/10.1007/978-3-319-89812-4_23

1 Introduction

Huebner [1] referred to the two main functions of the Panama Canal (PC) a century ago. The Canal was born because of its strategic value, which allowed the United States to transfer warships between the Atlantic and the Pacific oceans. It offered a shorter route for maritime transport between the two oceans, with the consequent saving of fuel and time. Therefore, it also had an economic value and thus it has an enormous geostrategic transcendence in the military field [2] and in the field of maritime transport.

The PC covers 65 km of distance from the Isthmus of Panama under the principle of a single channel and a double set of three locks (Mirafolores, Gamboa, and Gatún). They are used to save the difference in levels between the Atlantic and Pacific Oceans and Gatún Lake, an artificial navigable waterway located in the center of the country, which is created by the water contributions of the Chagres river basin [3].

Ships have increased in size seeking improvements in the competitiveness of maritime transport under the principle of economies of scale. In this context, the PC has tried to maintain its capacity to comply with the demands of maritime transport [4]. Therefore, the decision to expand the Canal was made in 2006.

A multitude of studies have been published about the changes that in maritime reality will produce the expansion of the Canal during the execution time of the works. Regarding the traffic [2], it indicates that the changes will affect large geographic areas: Asia Pacific, North America, Latin America, and the European Union, and especially the Asia–United States container route. Lindstad et al. [5] note that vessels will avoid long distances with their use (Suez and Cape Horn), with the consequent reduction in miles traveled and emissions of CO_2 and other pollutants. Ungo and Sabonge [6] analyze the competitiveness of the route through the PC for the transport of containers, in comparison with other alternative routes, under different cost scenarios. Pagano and Guevara [4, 7] point out that the expanded Canal will define a new type of vessel, the New-Panamax. Moryadee et al. [8], in analyzing the traffic of liquified natural gas (LNG), pointed out the role the Canal can play as a communicator between two maritime areas, the Atlantic and the Pacific, with very different market realities.

Carral et al. [9–11] have studied the effect of the commissioning of the extension on the maneuverability of the ships after its extension, a statistical proposal for the estimation of the transit time of the New-Panamax vessels, and finally its effects on ships and traffic.

The supply chain of natural gas (NG) through LNG consists of four processes: gas production, processing, liquifaction (conversion to LNG), and transport (Fig. 1). Thus, the development of this technology has been an alternative to transportation through gas pipelines [12] in the case of long distances and the need to cross seas or oceans.

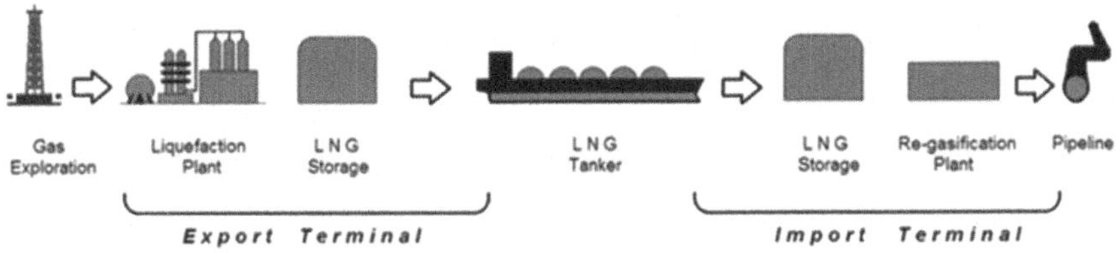

Fig. 1 NG supply chain considering its transformation in LNG. *Source* Own

NG represents approximately a quarter of the global energy demand, representing LNG [13]. It allows consuming countries to diversify their energy supply [14].

The global NG market has changed as a result of the increase in production in the United States and Canada and the application of unconventional extraction procedures [15]. Leather et al. [16] studied the effect on the market of the appearance of new exporters in the case of the United States and the increase in production in Australia. At the same time, LNG consumption has increased, especially in Asian markets [17, 18].

Gas transit between the Atlantic and Pacific marine areas has been scarce due to the fact that before the expansion of the PC only 6% of the LNG transport fleet (LNGC; with a capacity per unit of 32,000 m^3) could use it [19]. Once the expanded PC is in operation, 84% of the LNGC with a load capacity of up to 175,000 m^3 can transit [10].

Therefore, it is possible to incorporate a new type of traffic in the PC. The main objective of this document is the study of this possibility and its influence on the exploitation of the PC.

2 The Global LNG Market and Maritime Routes for LNG Traffic

2.1 LNG as a Global Industry in a Globalized World

The development of LNG allowed access to this key source of energy for the development of countries that, like Japan, Korea, or Spain, do not have their own deposits or connection to the gas pipeline network. On the other side, LNG has become the key piece in the economies of the producers, such as Algeria, Trinidad and Tobago, or Qatar [20].

2.2 Supply and Future Demand

The group of LNG-producing countries has increased with several new exporters, including Australia and the United States, where the latter acted as an importer until now (Fig. 2). The irruption of unconventional gas, not only increases very significantly the amount of recoverable resources, but it also changes the geographic distribution of potential sources of production, because it is more distributed than the conventional one (Fig. 3).

This tendency to diversify LNG production will be maintained [20]. It is important to notice that between the two countries with the largest gas reserves, Russia is the first with marginal production of LNG [21], whereas the second, Iran, is not a producer at present [22].

The future evolution of demand will come from the large emerging countries, China and India, but also from the other emerging countries that need gas for their development [23]. The very gas production itself in Europe will decrease during the century, with gas production in the Netherlands, the United Kingdom, and North America in decline, and without prospects that the barriers to unconventional gas development will disappear, whose resources in Europe are limited [24]. Moreover, it does not seem that the great geopolitical difficulties to bring gas from the Caspian Sea and the republics of Central Asia via gas pipeline will disappear. For all these reasons, if Europe does not want to increase its dependence on Russian gas, it will have to replace the loss of its own production importing LNG. The growth in domestic consumption and the decline in the production of countries that are traditional producers of gas, including LNG producers in the past, can determine these countries to join the group of LNG-importing countries [20].

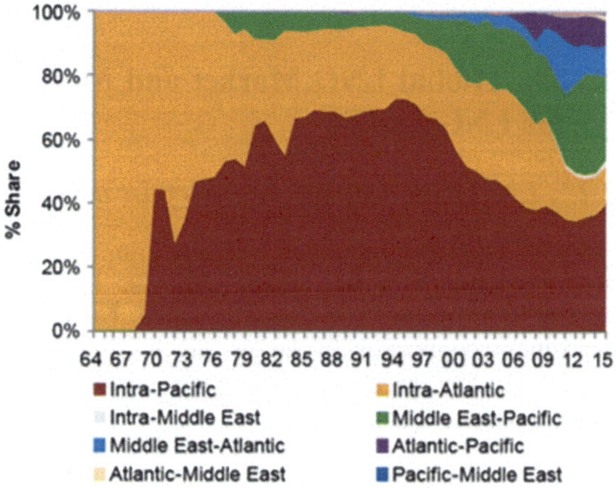

Fig. 2 Transit of NG between regions in the period 1964–2015. The violet color indicates the strong consumption in the Atlantic and Pacific and the reduced trade between them [20]

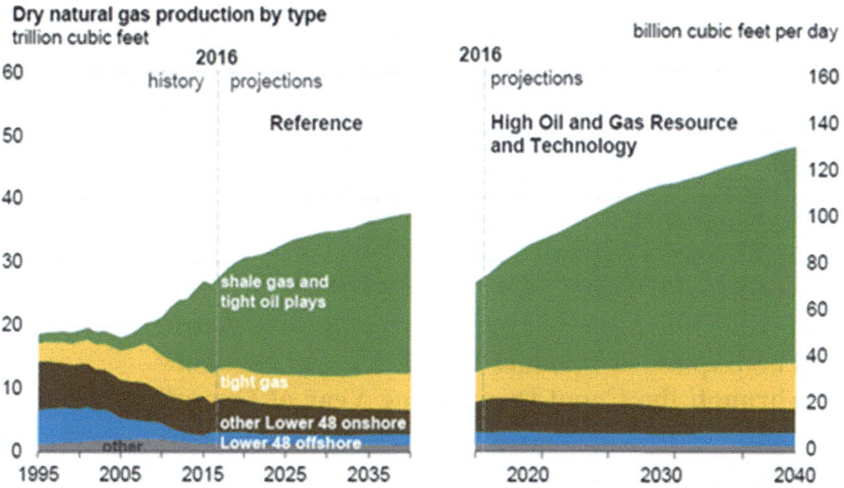

Fig. 3 Evolution over time of unconventional gas extractions in the United States. *Source* Ref. [23]

The evolution of the Asian economies and the Asia Pacific region will be key to the evolution of LNG prices [20], because these two regions are not only the biggest consumers—between the two they consume more than 75% today [20] and their demand has grown in a similar proportion in recent years—but will also be the regions where more than 50% of world production will be concentrated in the future.

2.3 *Maritime Routes for LNG Traffic*

Historically, LNG has been a fuel with limited trade within the same maritime area in which it is produced [24]. LNG producers of the Atlantic area, including Nigeria, Algeria, Trinidad and Tobago, Equatorial Guinea, Norway, Libya, and Egypt supplied approximately 76% of their production to LNG importers in the same area, most of them European countries, in 2010 and 2011.

The difference in the price of gas between Asia and Europe was small, not covering the high cost of shipping, before the 2010 nuclear accident in Japan. The price divergence between the areas increased since mid-2010 due to the strong demand in Asia, especially Japan. Therefore, the exportation of LNG between areas is profitable depending on the transport costs [15]. All these aspects have led to a notable increase in exchanges via the sea (Fig. 4).

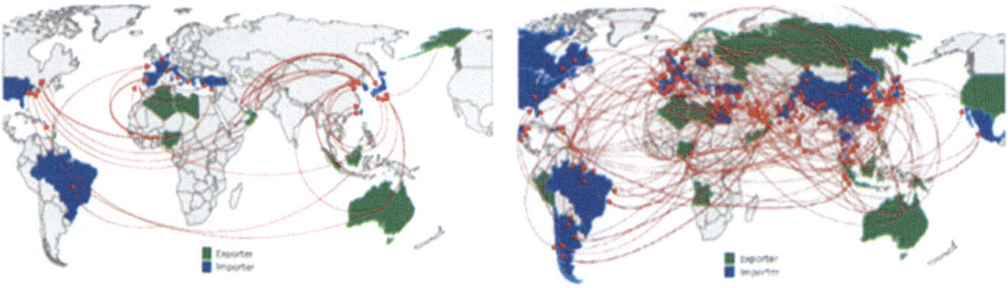

Fig. 4 Evolution of LNG exchanges by sea in 2000 (left) and 2015 (right). *Source* Ref. [25]

3 The Expanded Panama Canal: Transit Statistics Through the Canal During One Year of Operation

The transit of LNG through the Canal is a recent factor [4] as a consequence of the alteration of international trade patterns [10]. Commercial exchanges of LNG between the Atlantic and Pacific sea areas are possible and are conditioned by the costs of maritime transport. All the factors that modify production and consumption and the expansion of the Panama Canal [9] are modifying the global LNG transport market [7] by opening new market routes [26].

Gas vessels with capacities between 170,000 and 180,000 m^3 (known as NeoPanamax) have become the standard for newly constructed LNG with the expansion of the Panama Canal commissioned in 2016. The 87% of the 23 ships ordered in the shipyards in 2015 are registered as the new Panamax class [20]. The new NeoPanamax category can cross the expanded Canal and at the same time offer greater flexibility, especially in comparison with the Q-Class, regarding access to the main ports of discharge, particularly in Asia [10].

The extended Canal has been used by LNGC since its commissioning. In this way, it received a total transit of 167 vessels (Table 1) distributed according to a greater frequency in the months corresponding to winter in the northern hemisphere in the first year of operation.

Table 1 Monthly breakdown of LNGC transits in the fiscal year of 2017, indicating the type of containment of the ship's cargo

Type of containment	2016			2017									Total
	Oct	Nov	Dec	Jan	Feb	Mar	Apr	May	Jun	Jul	Aug	Sep	
Membrane	4	12	18	17	15	13	12	10	[a]				167
Moss	1	3	3	1	2	0	0	0					
Total	5	15	21	18	17	13	12	10	13.88	13.88	13.88	13.88	

10% of the transits corresponded to LNGC of the Kvaerner–Moss type

Source Self-elaboration based on ACP data

[a]Estimated values are used based on the data obtained in the first months of the fiscal year since June 2017

The LNG transport flows directed to Asian consumers are greatly influenced by the Canal rates due to geographical position, next to the Canal, of the exit ports of the exports of the United States–Gulf of Mexico and Trinidad and Tobago [8]. This effect leads to alternative exports from these points to Europe to Asia. The study of this effect, with the export estimates of these countries in the 2020 and 2035 horizons [15], is shown in Table 2, collecting data relative to the minimum and maximum LNG quantities to be transited by the Panama Canal with indication of the number of standard ships (175,000 m^3) in transit.

The maximum export to Asia corresponds to favorable Canal transit rates, which make transport costs very low and therefore the final price. On the other hand, the minimum value corresponds to the transport, under high toll rates in the Panama Canal, of the amounts of LNG agreed in long-term contracts. At that time, the majority of the exportable quantities of the United States and Trinidad and Tobago will be directed towards the European market (maximum value; Table 2). The interval that mediates between these extreme values corresponds to intermediate value fares in Canal transit.

Table 2 Forecast of United States and Trinidad and Tobago exports in the 2020 and 2035 horizons in billions of m^3, indicating the maximum and minimum values according to the different export opportunities that arise from the situation of the markets

Quantities × MM–m^3 (10^9 m^3)	**EUROPE (1)**		**ASIA (1)** Japan/South Korea		**ASIA (1)** China S-E/Asia India		Total capacity of exportation of the country
	Min.	Max.	Min.	Max.	Min.	Max.	
Horizon 2020							
USA–Gulf of Mexico[b]	0	12.7	13.6	58.40	4.6	4.6	63
Trinidad-Tobago[b]	8.43	17.28	0	21.62	0	0	125.6
Total GNL PC[c]	0	0	13.6	80	4.6	4.6	
Min/max transit[d]	104/483						
Horizon 2035							
USA–Gulf of Mexico[b]	25.09	46.62	12.69	58.40	4.6	4.6	74
Trinidad-Tobago[b]	10.45	15.70	0	25.09	0	0	125.6
Total GNL PC[c]	0	0	12.69	83.49	4.6	4.6	
Min/max transit[d]	99/503						

The amounts that can transit through the Canal are indicated in bold

Source Self-elaboration based on [9]

[a]Import geographic zones

[b]Exporting countries

[c]Maximum and minimum quantities of LNG that may transit through the Panama Canal

[d]Minimum and maximum number of LNG transits loaded, considering a conventional vessel of 175,000 m^3

Table 3 Forecast of United States and Trinidad and Tobago exports to Japan and Korea in the 2020 and 2035 horizons expressed in billions of m^3

Quantities × Billions m^3	ASIA[a] Japan and South Korea
Horizon 2020	
USA–Gulf of Mexico[b]	58.40
Trinidad–Tobago[b]	20.195
TOTAL GNL PC[c]	78.95
Loaded transits[d]	451
Total transits	902
Horizon 2035	
USA–Gulf of Mexico[b]	58.40
Trinidad–Tobago[b]	23.45
Total GNL PC[c]	81.85
Loaded transits[d]	468
Total transits	932

The indicated transit values have been estimated based on the Canal transit rate for the round trip proposed for October 2017 (ACP, 2017). The average value of this tariff was 2026 $/m^3$

Source Self-elaboration based on [8]

[a]Import geographic zones

[b]Exporting countries

[c]Estimated quantity of LNG that may transit through the Panama Canal

[d]Number of LNG transits loaded, considering a conventional vessel of 175,000 m^3

Therefore, it is expected that a maximum of 503 loaded vessels will transit annually through the Panama Canal (standard size 175,000 m^3). The minimum value will appear with favorable conditions to export to Europe, representing 99 transits charged per year (Table 2).

Considering the value of the Canal tariff for October 2017 [27], it is expected that a total of 451 loaded vessels will transit annually through the Panama Canal in the year 2020 (standard size 175,000 m^3), and it reaches 468 loaded vessels in 2035 (Table 3).

On the other hand, Table 4 collects data related to past years together with the estimated data (2020, 2035), representing the whole evolution of LNG transit between 2014 and 2035.

Table 4 Evolution of LNG traffic through the Panama Canal on the basis of the ACP data (2014–2017) and the traffic forecast made in Table 2

Year	2014	2015	2016	2017	2018[b]	2020[c]	2035[c]
Number of loaded transits	–	2	9[a]	94[a]	126[a]	451	468
Load × 10^9 (MMT)	–	–	0.58	–	8.9	37.1	38.45
Volume × 10^9 (MMM^3)	–	–	1.17	14.6	18.9	78.95	81.85
% traffic on total expanded Canal	–	–	6	9.4	10[d]	35	36.6
North transits	–	–	–	87[e]	106	190	197
South transits	–	–	–	101[e]	146	261	271

The intervals corresponding to future years are established based on the variable rates over time applied by the ACP

Source Own elaboration based on [8, 27]

[a]Values corresponding to transits of loaded ships. It is estimated based on 50% of the total transits indicated by the ACP

[b]Data obtained based on the extrapolation of the ACP data corresponding to the maximum transit value of 2017

[c]Only loaded vessels are considered, according to the values contained in Table 3

[d]Estimated value based on the year 2016

[e]Data of the ACP

4 Conclusions

There will be a significant flow of LNG through the Panama Canal in the medium and long term due to the adaptation of maritime trade patterns to the evolution of the global world energy market and, in particular, to the gas market.

The LNG transport flows directed to Asian consumers are greatly influenced by tariffs due to the geographical position close to the Canal of the ports in existence of the two major exporters, such as the United States–Gulf of Mexico and Trinidad and Tobago. This temporary effect leads to alternative exports from these spots to Europe or Asia, determining extreme values (maximum and minimum) of transits through the Panama Canal. Undoubtedly, this will be one of the future conditions for the exploitation of the Canal.

Estimation of minimum and maximum LNG loads corresponding to the horizon years 2020 and 2035 have been used based on the application of global gas models for the study of this effect. Estimations show that these extreme values of traffic of LNGC vessels per year will be between 99 and 503 loaded vessels in the long term, all conditioned by transport costs, with the influence on the Canal transit fees.

Consideration of the present fees in the Canal leads to values of vessels loaded in transit through the Canal close to the maximum value considered. These values and the fact that the ships can do ballast return, can determine higher Canal occupancy values than those estimated in its planning. Therefore, this traffic will undoubtedly be one of the factors to take into account when analyzing the possibility of a new expansion of the Panama Canal.

References

1. Huebner, G.G.: Economic aspects of the Panama Canal. Am. Econ. Rev. **5**(4), 816–829 (1915)
2. Sabonge, R.: La ampliación del Canal de Panamá. Impulsor de cambios en el comercio internacional. Naciones Unidas—Comisión Económica para América Latina y el Caribe (CEPAL) (2014)
3. Jaén, O., Alvarado, C., et al.: The Panama Canal. Ed. Balboa, Panama (1999)
4. Pagano, A., Wang, G., Sánchez, O., Ungo, R., Tapiero, E.: The impact of the Panama Canal expansion on Panama's maritime cluster. Marit. Policy Manage. **43**(2), 164–178 (2016)
5. Lindstad, H., Jullumstro, E., Sandaas, I.: Reductions in cost and greenhouse gas emissions with new bulk ship designs enabled by the Panama Canal expansion. Energy Policy **59**, 341–349 (2013)
6. Ungo, R., Sabonge, R.: A competitive analysis of Panama Canal routes. Marit. Policy Manage. **39**(6), 555–570 (2012)
7. Martínez, G., Vanesa, K.: Análisis de la Capacidad Operativa del Canal Actual y Ampliado, en Función de la Cantidad de Tránsitos. Repositorio UVA. (2016). https://uvadoc.uva.es/bitstream/10324/7078/1/TFM-P-116.pdf. Accessed July 2016
8. Moryadee, S., Gabriel, S., Rehulka, F.: The influence of the Panama Canal on global gas trade. J. Nat. Gas Sci. Eng. **20**(2014), 161–174 (2014)
9. Carral, L., Tarrío-Saavedra, J., Naya, S., Bogle, J., Sabonge, R.: Effect of inaugurating the third set of locks in the Panama Canal on vessel size, manoeuvring and lockage time. J. Navig. 1–19 (2017a). https://doi.org/10.1017/s0373463317000285
10. Carral, L., Tarrio-Saavedra, J., Castro-Santos, L., Lamas-Galdo, I., Sabonge, R.: Effects of the expanded panama canal on vessel size and seaborne transport. Aceptado en PROMET – Traffic & Transportation (2017b)
11. Carral, L., Tarrio-Saavedra, J., Alvarez-Feal, C., Naya, S., Sabonge, R.: Statistical methodology to determine the transit time required for neopanamax vessels in the Panama Canal. Aceptado en Transport Policy (2017c)
12. Rios-Mercado, R., Borraz-Sánchez, C.: Optimization problems in natural gas transportation systems: a state-of-the-art review. Appl. Energy **147**, 536–555 (2015)
13. IEA. World Energy Outlook 2016. http://www.iea.org/newsroom/news/2016/november/world-energy-outlook-2016.html. Last Accessed June 2107
14. Wood, D.A.: A review and outlook for global LNG trade. J. Nat. Gas Sci. Eng. **9**, 16–27 (2012)
15. Moryadee, S., Gabriel, S., Avetisyan, H.: Investigating the potential effects of U.S. LNG exports on global natural gas markets. Energy Strategy **2014**, 273–288 (2014)
16. Leather, D.T.B., Bahadori, A., Nwaoha, C., Wood, D.A.: A review of Australia's natural gas resources and their exploitation. J. Nat. Gas. Sci. Eng. 10, 66–88 (2013)
17. Aguilera, R.F., Inchauspe, J., Ripple, R.D.: The Asia Pacific natural gas market: large enough for all? Energy Policy **65**, 1–6 (2014)
18. Aguilera, R.F., Ripple, R.D.: Modeling primary energy substitution in the Asia, Pacific. Appl. Energy **111**, 219–224 (2013)
19. Alaskan Business Monthly: Expanded Panama Canal will reroute LNG industry (2012). Available from http://www.akbizmag.com/Alaska-Business-Monthly/November-2012/Expanded-Panama-Canal-Could-Reroute-LNG-Industry/
20. IGU (International Gas Union): World LNG Report 2016 LNG 18 Conference & Exhibition Edition (2016)
21. Locutura-Ruperez, E.: Geoestrategia del Gas Natural Licuado (LNG) In Energía y geoestrategia 2016. Instituto Español de Estudios Estartégicos (2016). www.ieee.es/Galerias/fichero/cuadernos/Energia_y_Geoestragia_2016.pdf
22. Palstev, S.: Scenarios for Russia's natural gas exports to 2050. Energy Econ. **42**(2014), 262–270 (2014)

23. EIA: Expanded Panama reduces time for shipments Canal of US LNG to Asian markets (2016). Available from https://www.eia.gov/todayinenergy/detail.php?id=26892. Accessed Mar 2017
24. Mokhatab, S., Economides, M.J., Wood, D.: Natural gas and LNG trade—a global perspective. Hydrocarb. Process. **85**, 39 (2006)
25. Poten & Partners: LNG Hipping Market Discussion—Madrid LNG & Shipping Forum 2016. May 2016
26. Sirvent-Zaragoza, G.: Visión geoestratégica de las rutas marítimas de la energía. In Energía y geoestrategia 2016. Instituto Español de Estudios Estartégicos (2016). www.ieee.es/Galerias/fichero/cuadernos/Energia_y_Geoestragia_2016.pdf
27. Autoridad del Canal de Panamá (ACP): Tarifas propuestas (2017). http://www.pancanal.com/peajes/2016.html. Accessed June 2017
28. Taheri, Z., Shabani, M.R., Nazari, K., Mehdizaheh, A.: Natural gas transportation and storage by hydrate technology: Iran case study. J. Nat. Gas Sci. Eng. **21**, 846–849 (2014)
29. GII LNG: The International Group of Liquefied Natural Gas Importers: the LNG Industry (2013). Available at: http://www.giiLNG.org/publications/lng-industry-2013. Accessed 31 July 2015

Implementation of a Yard for a Container Terminal in the Commercial Port of Esmeraldas

Rafael A. Plaza Perdomo, Roberto González González and David A. Plaza Mendoza

Abstract The Commercial Port of Esmeraldas has increased the traffic of containers with goods of the northern region of Ecuador but lacks infrastructure and specialized equipment to handle containers. To improve this service, it is proposed to determine the feasibility of the courtyard's space distribution for a terminal of containers to be located on the back side of the main breakwater of the port, which is 400 m in length. The methodology used consists in optimizing the sizing of the courtyard with TEU cells for containers, using different port equipment systems and determining the economic impact generated by each system, based on the forecast of annual container transiting in the logistics corridors within and outside the area of influence of the port. It was determined that the system of playground gantry crane equipment is the most suitable, allowing handling 400,000 TEUs annually, in an area of 18.50 ha, of which 11.50 ha are exclusive for stacking and equipment movement with the lowest cost of movement per TEU. With the current port rates, it is feasible to finance the construction of the container terminal and the acquisition of the port equipment, whose value is USD$87,320,000. The optimal configuration and arrangement of the terminal is for 3312 TEU cells, demarcated for the stacking direction parallel to the dock. The analysis tool of risk priority for decision making was applied, verifying that a gantry crane equipment system presents less risk.

R. A. Plaza Perdomo (✉)
Central University of Ecuador, Research and Postgraduate Institute,
University Citadel, Quito, Ecuador
e-mail: rafaelplaza55@hotmail.com
URL: http://www.uce.edu.ec/web/ingenieria-ciencias-fisicas-y-matematica/posgrado

R. G. González
School of Economics, State University of Guayaquil, University Citadel,
Guayaquil, Ecuador
URL: http://www.ug.edu.ec

D. A. Plaza Mendoza
Faculty of Maritime Engineering, Biological, Oceanic and Natural Resources,
Higher Polytechnic School of the Litoral, University Campus, Guayaquil, Ecuador
URL: http://www.espol.edu.ec

A. Vega Sáenz et al. (eds.), *Proceedings of the 25th Pan-American Conference of Naval Engineering—COPINAVAL*, https://doi.org/10.1007/978-3-319-89812-4_24

Keywords Commercial port of esmeraldas · Containers traffic Sizing of stacking courtyards

1 Introduction

In 1979 the Commercial Port of Esmeraldas was inaugurated, having a limited connection with the rest of the country, especially with the area of influence and outside it, which includes the provinces of Santo Domingo de los Tsáchilas, Pichincha, Cotopaxi, Tungurahua, Napo, Francisco de Orellana (route E-20, E-30, E-35), Carchi, Imbabura, and Sucumbíos (route E-15, E-10).

As of 2007, with the government investment in road infrastructure, this port had a significant volume of cargo and container traffic, moving containers and different cargo clusters with a volume close to 100,000 TEU per year [1] from and towards the aforementioned provinces.

The Competitiveness Report of the World Economic Forum, on the website http://reports.weforun.org/global-competitiveness-report-2015–2016/, determined Ecuador as the third country of the South American Region that invested in infrastructure works, building the best communication channels in the region.

The city of Esmeraldas has been transformed into a platform or natural logistic node of the region of the country, as shown in Fig. 1, counting on the commercial port, oil and fishing, the airport, and the logistical support area, supported with a fully modernized customs.

The Commercial Port of Esmeraldas is located to the north of the province of Esmeraldas, left margin of the mouth of the Esmeraldas River to the Pacific Ocean,

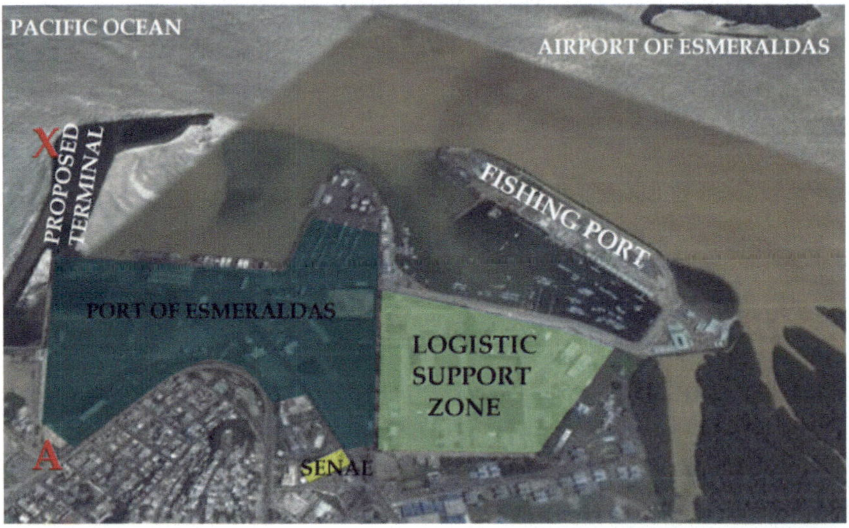

Fig. 1 Esmeraldas logistics platform

with a geographical location between the meridians 79° 38′ west and 79° 39′ west and the parallels 00° 59′ north and 01° 02′ north, and is the only port in Ecuador in the line direction that vessels must follow from the sea buoy to the dock of the port. The depths of the seabed range from 50 to 200 m, located at the entrance of the dock of the port [2]. This natural nautical feature proves the Port of Esmeraldas as a deepwater port.

Currently a container yard has been improvised in this port with an area of four hectares, using a system of equipment to handle ship-port containers, cranes' own ships for loading and unloading, truck-trailers for travel, and reacher-type reacher stackers to stack in the storage area, which causes operational delays.

International maritime transport on overseas routes is currently carried out by container ships of more than 18,000 TEUs, which are displacing ships with smaller capacities of equatorial navigation, forcing the latter to travel in a pendular manner [3].

Therefore, the Commercial Port of Esmeraldas must prepare technically and operationally to attend to these vessels that will have routes from north to south, departing and arriving at a regional loading port and handling cargo from the different ports of connection or feeders, as shown in Fig. 2.

The mentioned characteristics of the Port of Esmeraldas and of maritime transport, motivated for the project, have as their objective to determine the feasibility of a spatial redistribution of the yard of a container terminal in the port in order to improve its service and competitiveness at the national and international levels, accounting for it with the theoretical sustenance of handling an international terminal with the configuration and firm disposition of the terminal of containers, and with the evaluation of risks of the service.

The methodology used allowed optimizing the sizing of the yard with TEU cells for containers, using different systems of port equipment, and calculating the economic impact that each system generates, starting from the forecast of annual traffic of containers that transit in the logistics corridors inside and outside the zone of influence of the port [4, 5].

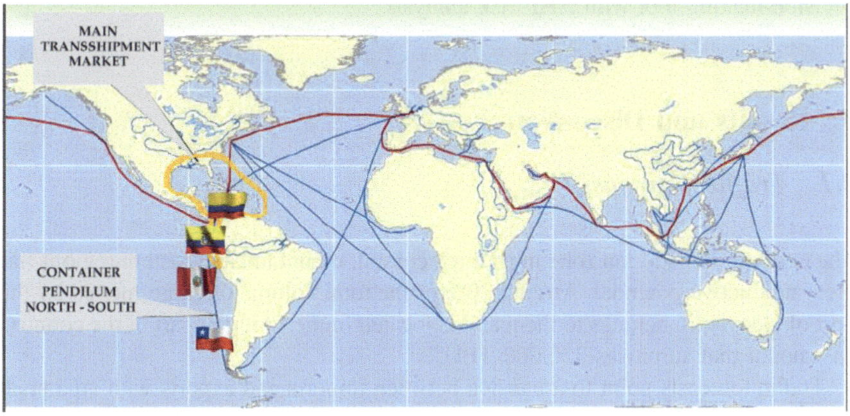

Fig. 2 Main international shipping routes

It turned out that the system of cranes–patio gantry equipment is the most convenient, being able to handle 400,000 TEU per year in a total extension of 18.50 ha, with the lowest cost of movement per TEU; and for the current charge made in this port by TEU, it's feasible to finance the construction of the container terminal and the acquisition of port equipment, which has a total value of USD $87,320,000.

The configuration and optimal layout of the container terminal yard in the Commercial Port of Esmeraldas is for 3312 TEU cells, demarcated for the stacking of containers parallel to the dock. Its location is offshore from the breakwater wall of the breakwater main of the port.

When applying the decision-making tool with AHP (analytic hierarchy process) risk analysis [6], it was verified that the gantry crane system chosen as optimal presents fewer risks, obtaining a high ranking with the most descriptive criteria of a port, thus guaranteeing the suitability of the project.

2 Materials and Methods

The methodology used for the implementation of the yard for a container terminal in the Commercial Port of Esmeraldas, had as a basis the forecast traffic of the containers to be mobilized annually by the aforementioned port, from and to the transverse internal logistics corridor of the north of the country, connected with the overseas external brokers of Asia and Europe and with the brokers of Colombia and the United States–Mexico (North American Free Trade Agreement–NAFTA); then it was optimized for several systems of port equipment used in the container terminals, the cells to be required in stacking containers inside a paved courtyard with their respective streets for the circulation of the port machinery in such a way that the cost/benefit was generated by the operation of these systems, prioritizing and checking the most convenient system, under the application of the decision-making tool with AHP risk analysis.

3 Results and Discussion

3.1 Traffic Forecast

The northern Ecuadorian zone in its three coastal, mountain, and eastern regions has great productive potential. Around 20% of the total volume of cargo moved by the port of Guayaquil belongs to them and is located in the southern part of the country, an amount that surpasses 320,000 TEU.

Traffic forecasts point to a balance between imports and exports with an annual growth rate of 5%. The containers moved by the terminal in the 20 years following the year 2020, were calculated with projections that have a compound annual growth rate of 5%.

The movement in port of the containers in their different loading conditions is greater with respect to their quantity. In the container terminal in 2020, 400,000 TEUs will be handled, of which it is expected that 50% will be transshipped in the region's traffic, 90% of them loaded and 10% empty.

The handling of the rest of the containers is the same, both for import and export. Forty percent of import and export containers that are loaded will be LCL, because at that date a very dynamic development is expected of ZALSA C.A.

The strategic geographical position of the Commercial Port of Esmeraldas, which, having the services of a container terminal, has the possibility of being a transshipment port for containers that are moving in the region, especially in Colombia, a country that only has the Port of Buenaventura on the Pacific Ocean. It will also allow the incorporation of the maritime connectivity system of cabotage or of short routes known as Short Sea shipping, and that it is an alternative to mobilize the large volume of merchandise of binational trade with Colombia that is currently transported with difficulty by land. As for the connectivity with the port of Callao in Peru, it is feasible for the commercial exchange registered as part of the Andean Community.

Table 1 shows the annual movement of containers in their FCL, LCL, and empty conditions.

3.2 Determining the Demand for Storage

The figures for the future movement of container traffic expressed in TEU and their respective movements in the terminal require a relationship with the storage space, for which other factors must be taken into account. The first of these is the safety factor, and in this case values of that factor are recorded in a similar terminal that operates in the region, where inventories of stored containers have been carried out, grouping them into ranges, and determining the frequency distributions of the different classes, as well as their cumulative relative distribution.

Graphically you can determine the percentage of time in which a number of containers would be stacked in the terminal and choose three safety factors expressed in proportion: accommodate 95% of the maximum number of containers, a safety factor of 1.4; accommodate a maximum of 100% of the number of containers, a safety factor of 1.6; and accommodate 90% of the maximum number of containers, a safety factor of 1.3. It is also necessary to consider the average dwell time of the containers in the terminal, which is 5 days, as well as the separation factor between containers, that is 1.25. Table 2 shows the number of TEU cells needed daily.

Table 1 Annual and average daily movement of containers within the terminal and entry/exit with the container station and empty park inside or outside the terminal (year: 2020)

Movements		EC and empty park inside the terminal	EC outside the terminal; empty park inside the terminal	EC and empty park outside the terminal
Transfer at dock				
Import	FCL	57,000	57,000	57,000
	LCL	38,000	38,000	38,000
	Empty	5000	5000	5000
Export	FCL	57,000	57,000	57,000
	LCL	38,000	38,000	38,000
	Empty	5000	5000	5000
Transshipment		400,000	400,000	400,000
Mobilization via pier		10,000	10,000	10,000
Subtotal transfer at the dock		610,000	610,000	610,000
From/to the EC				
Import LCL to EC		38,000	38,000	0
Empty EC to your park		28,000	28,000	0
Export LCL from EC to the stacking area		38,000	38,000	0
Empty from your park to EC		28,000	28,000	0
Subtotal of/to the container station		132,000	132,000	0
Enter/exit				
Enter	FCL import	57,000	57,000	57,000
	LCL import	0	0	38,000
	Empty import	5000	5000	5000
Exit	FCL export	57,000	57,000	57,000
	LCL export	0	0	38,000
	Empty export	0	0	5000
Input FCL import as empty[a]		0	0	0
Empty park exit[b]		0	0	0

(continued)

Table 1 (continued)

Movements	EC and empty park inside the terminal	EC outside the terminal; empty park inside the terminal	EC and empty park outside the terminal
Subtotal input/output terminal	119,000	119,000	200,000
Total annual movement	861,000	861,000	810,000
Average daily movement	2359	2359	2219

[a]These are import containers that return empty (recirculation)
[b]These are empty containers left from your park for users (loaders) to be filled

Table 2 Number of TEU cells needed daily (year: 2020)

State of the containers	Number of TEU cells needed daily			
	Average	95%	100%	90%
FCL import	975	1369	1563	1269
LCL import	650	913	1038	844
Empty imports	88	125	138	113
FCL export	975	1369	1563	1269
LCL export	650	913	1038	844
Empty export	88	125	138	113
Return	481	675	769	625
Transshipment	3425	4794	5481	4450
Total	7332	10,283	11,728	9527

3.3 *Superficie de almacenamiento y espacios necesarios*

Table 3 shows the required capacity of the container yard expressed in TEU surfaces, where the stacking height was considered, the different port equipment systems, and the results of the most conservative safety factor of 90% of Table 2.

Table 4 contains the necessary capacity of the container yard for the port equipment systems, considering the TEU average surface area for maneuverability, an esplanade 350 m long and 50 m wide by berth, an area of 2.25 ha for the container station, and an area of 3 ha for facilities and support services and green areas.

3.4 *Economic Analysis of the Sizing of the Terminal*

This analysis is the calculation of the cost per container movement in the selected equipment systems as shown in Table 5.

Table 3 Required capacity of the container yard expressed in TEU surfaces (year: 2020)

Container status	Required daily capacity (90%)	Tractor and trailer system		Trolley–gantry system		Crane system–patio porch		Front loader system		Combined system	
		Medium height	Area	Medium height	Area	Medium height	Area	Medium height	Area	Medium height	Area
Import. FCL	1269	1	1269	1.5	846	2.5	508	1.5	846	1.5	846
Import. LCL	844	1	844	2.0	422	2.5	338	1.5	563	2.0	422
Import. Empty	113	1	113	1.5	75	2.5	45	1.5	75	1.5	75
Export. FCL	1269	1	1269	2.5	508	3.0	423	2.5	508	3.0	423
Export. LCL	844	1	844	2.5	338	3.0	281	2.5	338	3.0	281
Export. Empty	113	1	113	2.5	45	3.0	38	2.5	45	3.0	38
Return	625	1	625	2.0	313	3.5	179	3.5	179	3.5	179
Transshipment	4450	1	4450	2.5	1780	3.0	1483	2.5	1780	3.0	1483
Total	9527		9527		4327		3295		4334		3747
Average stacking height		1		2.20		2.89		2.20		2.54	

Table 4 Area of land required (year: 2020)

Container handling system	Average area per surface TEU (m^2)	Average area per surface TEU (m^2)	Total zone of the stacking area (ha)	Total support area (ha)	Total area of the terminal (ha)	Width of the land (m)	Pier length that would be used (m)	No. of berths
		A	B	C	D	E	F	G
Tractor and trailer	50	9527	47.6	8.8	56.4	806	700	2
Gantry forklifts	45	4327	19.5	7	26.5	757	350	1
Patio–gantry cranes	35	3295	11.5	7	18.5	529	350	1
Front loader	80	4334	34.7	8.8	43.5	621	700	2
Combined	40	3747	15	7	22	629	350	1

Table 5 Cost per container movement (USD)

Annual	Systems			
	Gantry forklifts	Forklifts–gantry with tractors and trailers	Cranes–gantry on tires	Combined
Cost of equipment per system (millions of USD)	8.17	8.73	9.62	10.09
Total movements of TEU	861,000	861,000	861,000	861,000
Cost of equipment per movement of TEU	9.49	10.14	11.17	11.72
Cost of infrastructure per system (millions of USD)	6.86	6.86	5.17	5.91
Total movements of TEU	861,000	861,000	861,000	861,000
Cost of infrastructure per movement of TEU	7.97	7.97	6	6.86
Total cost per movement	17.46	18.11	17.17	18.58

Given that the system of container handling using patio cranes on tires is the most economical, with a total cost of USD$17.17 per movement of TEU, you have the first option, so that the terminal is sized according to the data that were obtained in this study.

There are 200,000 import/export TEUs that normally require three movements, which at a cost of USD$17.17 per movement, gives a handling cost of USD $10,302,000. There are 200,000 TEU transshipments that normally require two movements, which with a cost of USD$17.17 per movement, gives a handling cost of USD$6,868,000.

For administrative expenses, 20% of the income is considered, that is, USD $5,600,000; therefore the cost of handling the 400,000 TEU for the year 2018 in the terminal to be built is USD$22,770,000. The rates that are currently charged in the Commercial Port of Esmeraldas to move a container is around USD$80 per TEU of import/export and USD$60 for TEU of transshipment; they are then fully justified, because it will generate an income of USD$28,000,000 and a profit of USD $5,230,000.

3.5 Determination of the Configuration and Arrangement of the Terminal

In the sizing of the yard for the container terminal of the Commercial Port of Esmeraldas, it was determined that the system of patio–gantry cranes is the most convenient, allowing handling 400,000 TEU per year in a total area of 18.50 ha.

The spatial distribution of the container terminal 350 m wide (length of the pier) and 529 m depth is: 11.50 ha exclusive for the stacking area with the patio–gantry cranes on tires and for the circulation of equipment of support (tractors and trailers), 2.25 ha for the EC container station, 3.00 ha for facilities, support services, and green areas (power plant, fuel station, access doors, offices, weighing of vehicles, washing machine, workshop maintenance, lighting towers, and others), and 1.75 ha for the dock with gantry cranes.

In Plan 1, there is the configuration and disposal of the container terminal to handle 400,000 TEU per year, which will be located within the Commercial Port of Esmeraldas, taking advantage of the main breakwater wall of approximately 400 m in length, suitable for the dock capable of operating container ships of around 8000 TEU capacity. It will also use the same port dock that will have to be dredged from the current 11.6–14 m, building the container yard from the waterfront of the dock inwards.

As regards the superstructure for handling containers, the system of patio–gantry cranes on tires will be used, with 3312 TEU cells, demarcated for the stacking of containers parallel to the dock, allowing 24 cranes to work simultaneously with the gantry on pneumatic for 24 stacking areas composed of 23 sections and 6 rows.

The breakwater wall of the main breakwater (shown in Fig. 1) is on a reference line that starts at point A, on the northern edge of Kennedy Avenue of the urbanization "Las Palmas," located at coordinates 9985.21 north and 10,216.35 east, and ends at point X, open sea, located at coordinates 10,864.98 north and east 11,096.09 [7] .

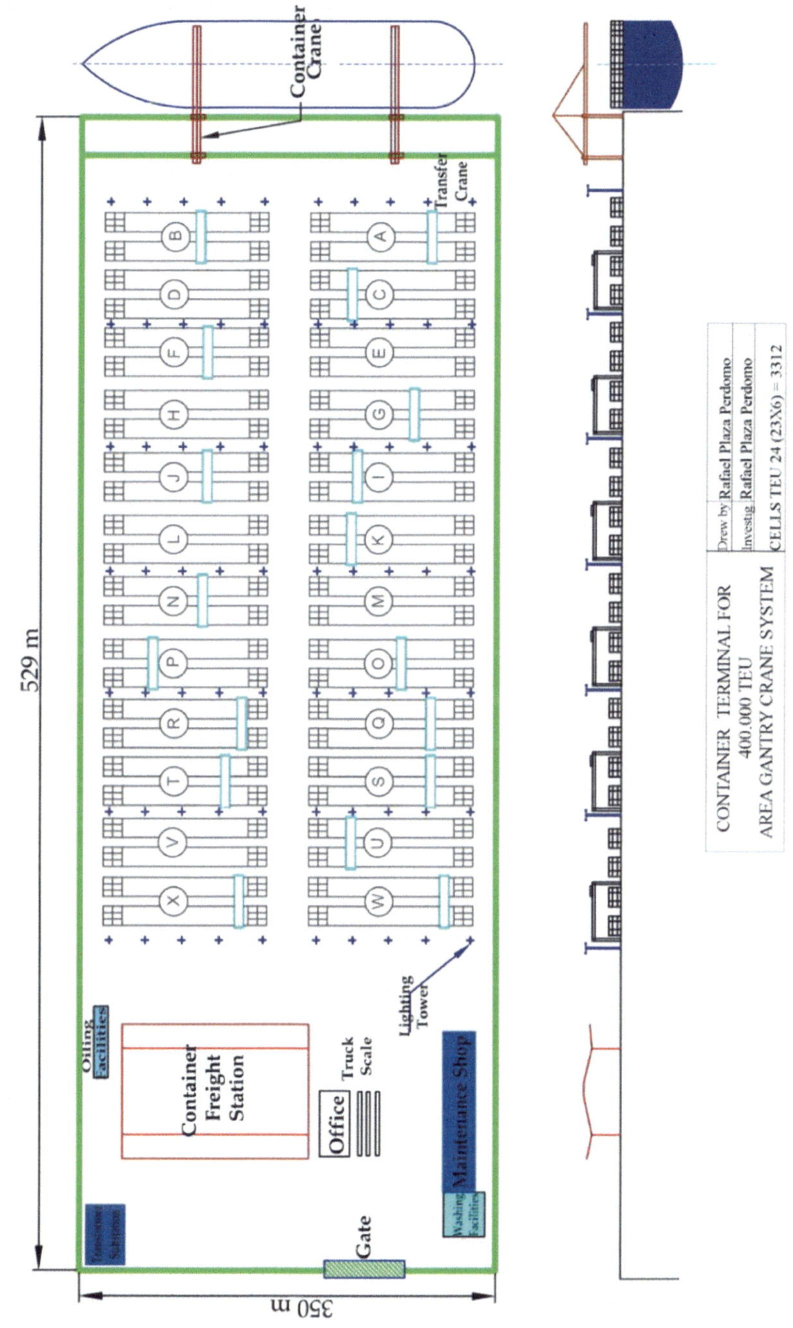

Plan 1 Configuration and layout of the container terminal of the Commercial Port of Esmeraldas

Table 6 Final ranking

1	Combined system	2	Trolley–gantry system with tractor-trailer
3	Trolley–gantry system (direct)	4	System of cranes–portico of patio on tires

3.6 Feasibility Analysis of Alternatives

Applying the decision-making tool with risk analysis AHP, runs were made with matrices, synthesis of weights, normalization, calculation of vector priority for the criteria of use of space, factors of the development of the terminal, cost of the equipment, maintenance factors, and resources of human talent, versus comparisons between the different port handling systems of containers studied.

In Table 6 we have the weighted result of the analysis of higher to lower risk of the use of container handling equipment systems, confirming that the container yard must have the configuration and arrangement capable of using patio-on-tire gantry cranes to occupy less space and have operational ease in the stacking of containers in the specialized terminal.

4 Conclusions

The Commercial Port of Esmeraldas with its technical-nautical features and location is the natural port of the entire northern region of Ecuador. Once the container terminal is built, the possibility of growth is high, given the good terrestrial and maritime connectivity, with excellent roads in the hinterland and deep sea accessibility from the foreland.

The container terminal must be built on the outside of the main breakwater wall, and the dock for container ships on the inner part of this wall that faces the port dock.

In the sizing of the yard for the container terminal of the Commercial Port of Esmeraldas, it was determined that the system of equipment crane–patio gantry is the most convenient, allowing it to handle 400,000 TEU per year in a total area of 18.50 ha, with 3312 TEU cells, demarcated for the stacking of containers parallel to the dock.

References

1. Subsecretaría de Puertos y Transporte Marítimo y Fluvial: Estadísticas Portuarias y de Transporte Acuático, Guayaquil (2010, 2011, 2012, 2013, 2014)
2. Dirección General de Intereses Marítimos-Inocar: Carta Náutica del Puerto de Esmeraldas, 28 July 2012

3. World Shipping Council: Industria-Rutas de Comercio. http://www.worldshipping.org/about-the-industry/global-trade/trade-routes. último acceso 2 June 2015
4. Secretaría-Unctad: Planificación para el Desarrollo de una Terminal de Contenedores. Seminario de las Naciones Unidas, Guayaquil, 87p (1991)
5. The Japan Port & Harbor Association: Technical Forum on Construction Technology of Container Terminal. Publication of Technology Transfer Institute (TTI), Tokyo, 307p (1978)
6. Universidad Central del Ecuador: Maestría en Gestión y Logística del Transporte Multimodal. Módulo de herramientas de ayuda para tomas de decisiones, Quito (2014)
7. Gobierno del Ecuador: Registro oficial 34, Decreto Supremo 109, 26 Feb 1976

Part III
Marine Sustainability

Environmental Impact of a Garbage Dump in an Estuary

Silvina Medus, Daniela Escudero and Olga Cifuentes

Abstract Former garbage dumps and clandestine garbage dumps are an environmental problem. The main goal of this work is to present how the lack of an adequate management allowed the location of a dump on the floodplain of the Bahía Blanca estuary (Province of Buenos Aires, Argentine Republic), putting it at risk. For this, the area, the chronological evolution of the former landfill from its beginnings to its current situation as a coastal seafront, and the parallel with the evolution of environmental legislation is described. An evaluation of the results detected in estuary water is done at the stations nearby the former landfill in order to confirm whether the detected dissolved metals originate in their leachates. As a support tool, a geographic information system is used to show the spatial and temporal evolution of the former landfill, which has an area of more than 50 ha shaped like a peninsula, surrounded by a sector that appears at low tide, with muddy consistency and at high tide is covered with water. In addition, it allows us to visualize clandestine overturns even though the landfill was closed in 1992. The conclusion is that the absence of appropriate legislation, planning, and compliance of the development plans, as well as the lack of continuity of all of these and the need to adopt immediate solutions, led to the location of the former landfill that compromises the sustainability of the estuary.

Keywords Estuary · Sustainability · Environmental liability

1 Introduction

The detection of dissolved metals in water of the Bahía Blanca estuary (Province of Buenos Aires, Argentina) in monitoring stations close to the former Belisario Roldán landfill, motivated research of the management of the area and the municipal solid waste (MSW) overturned on the floodplain for more than four

S. Medus (✉) · D. Escudero · O. Cifuentes
Universidad Tecnológica Nacional FRBB, 8000 Bahía Blanca, Argentina
e-mail: silvina_medus@yahoo.com.ar

A. Vega Sáenz et al. (eds.), *Proceedings of the 25th Pan-American Conference of Naval Engineering—COPINAVAL*, https://doi.org/10.1007/978-3-319-89812-4_25

decades. Also, environmental regulations, from its beginnings to its current situation as a recreational seafront were considered.

The objective is to demonstrate that the lack of adequate management allowed the location of a dump on the floodplain of the estuary, which does not warranty the preservation of the water resources.

To achieve this, the following topics are described: the study area, the chronological evolution of the coastal area in parallel with the environmental legislation, and the evaluation of monitoring results carried out in estuary stations near the former landfill.

This work is done within the frame of the research project "Risks of an environmental liability on Bahía Blanca estuary" from Universidad Tecnológica Nacional Facultad Regional Bahía Blanca (UTN-FRBB), which considers the problem of the former landfill from the aspects: legal, territorial management, and environmental impact.

2 Methodology

The Bahía Blanca estuary, has an approximate length of 80 km in the NW–SE direction. It is located southwest of the Province of Buenos Aires (Argentina). Next to it coexist the cities of General Cerri, Ingeniero White, and Bahía Blanca, a natural reserve, a municipal beach, a petrochemical complex, an industrial park, and an industrial port area integrated by Puertos Cuatreros, Galván, and Ingeniero White. Due to these varied activities the estuary has received contributions of exogenous substances to the system, coming from sewage, industrial discharges, leachate from landfills, agrochemicals, and others from the port activity. The study area of the former landfill on the estuary has a size of more than 50 ha in the shape of a peninsula, surrounded by a sector that at low tide presents itself with muddy consistency and at high tide is covered with water (Fig. 1).

First the search of satellite images of the former landfill was done, using images from Google Earth (Satellite Digital Globe), which for the study area appeared only from the year 2006, and an image of the Ikonos satellite, of the year 2001.

Also, research on the existence of aerial photos was done for the period before 2000. The images are collected from flights dated 1967, 1983, 1990, and 1996.

The photos and images not geographically located, are georeferenced to digitize the areas filled with garbage. These are digitized in the corresponding periods elapsed between the dates of images or photos. In each resulting map, the changes of the coastal morphology, introduced both anthropogenically, and by the natural response of the system are visualized. At the same time, a chronology of the applicable environmental regulations was made, linked to a brief historical review on the management of the MSW in the city of Bahía Blanca, which includes the location and subsequent abandonment of the former landfill until its current use as a seafront. This was done in order to explain the consequences generated by the lack of appropriate management.

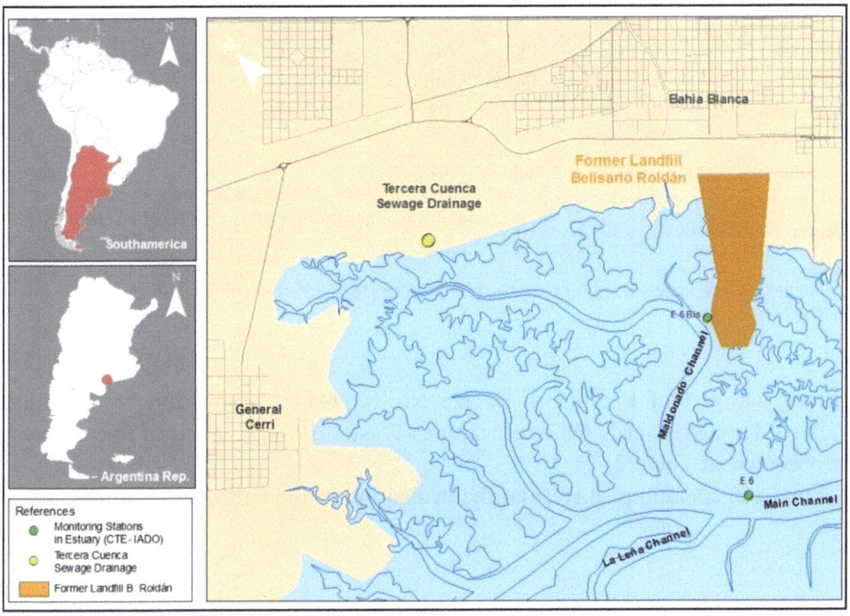

Fig. 1 Bahía Blanca estuary (Buenos Aires, Argentina)

Table 1 Monitoring stations close to the former landfill Belisario Roldán

Station	Name	Latitude	Longitude
E6	Proximity to Maldonado Channel flow	38° 45′ 54.1″S	62° 20′ 25.2″O
E6-BIS	Maldonado Channel	38° 44′ 50.86″S	62° 19′ 31.54″O

Field visits were made to confirm in situ the image interpretation, as well as to incorporate details and information that does not appear in the digital model.

On the other hand, the monitoring results of the period 2006–2012 done in estuary water in stations close to the environmental passive [1, 2] were compiled and evaluated. The locations of the monitoring stations on the estuary close to the landfill are shown in Table 1 and in Fig. 1.

A bibliographic review was carried out in order to have an updated theoretical framework about products containing metals (Cd, Pb, Cu, Zn, Cr, Ni, and Hg) that have been found dissolved in water. This is to establish if their origin is in the leachates of waste dumps from the former landfill.

The concentrations of metals dissolved in water detected in the monitoring results were compared with the reference levels of the National Oceanic and Atmospheric Administration (NOAA) [3], both for chronic exposure and for acute exposure. The latter is defined as the average concentration for 1 h of exposure and chronic exposure as the average for 96 h (4 days). Likewise, the values were compared with the water quality guide levels for the protection of aquatic life in

Table 2 Reference levels established by NOAA and by National Decree No. 831/83

Parameter	Cd (μg/l)	Pb (μg/l)	Cu (μg/l)	Zn (μg/l)	Cr[a] (μg/l)	Ni (μg/l)	Hg (μg/l)
Acute exposure (NOAA)	40	210	4.8	90	Not established	74	1.8
Chronic Exposition (NOAA)	8.8	8.1	3.1	81	Not established	8.2	0.94
Decree No. 831/93	5	10	4	0.2	Not established	7.1	0.1

[a]There are no reference values for total Cr. NOAA and Decree No. 831/93 established guide values for Cr^{+6}

surface seawater of Regulatory Decree No. 831/93, corresponding to Argentine National Law No. 24,051/93 on Hazardous Waste. All these levels of reference can be observed in Table 2.

Final considerations arise from the evaluation of the collected information.

3 Chronological Evolution of the Former Landfill and Its Parallelism with the Evolution of Environmental Legislation

This case study presents, through the GIS (geographic information system) tool [4], the progress of an endeavor born when there was still no environmental legislation. The initiative generated over time an environmental liability difficult to remedy, even if nowadays there was an intention to give legal compliance to its mitigation.

Starting in 1950, the urban area of Bahía Blanca began to expand due to population growth, appearing as a demand for new areas for waste disposal. This situation gave rise to an open-air dump called Belisario Roldán, located in the vicinity of a municipal beach. The landfill advanced over the floodplain of the estuary by dumping waste. The lack of knowledge and absence of legislation on the proper treatment of the MSW and its environmental impacts reinforced the idea of using them to gain land to the sea, without considering that an environmental liability was being generated that would leach, threatening the sustainability of the estuary and its use for future generations.

Figure 2 shows the map that results from digitizing the landfill with MSW from the first years until 1983. The landfill advanced on the coastline, outlining the area of a new peninsula on the floodplain. The map is elaborated unifying the photos corresponding to the years 1967 and 1983. In both images the surface of the fill, which initially can be observed in the 1967 image, remained constant in the 1983 image. As during those years the landfill was active, it is deduced that during that period of time, the MSW increased the sector in height.

Fig. 2 Evolution of the former landfill until 1983

In the 1970s, the first international environmental regulations emerged with the Stockholm Convention that set a precedent for the birth of environmental awareness, and national regulations with the first laws for the prevention of marine pollution by waste dumping. Following this trend in the 1980s, controls started on the water quality of the Maldonado Municipal Beach (Balneario Maldonado), due to its proximity to the Belisario Roldán landfill, because its pools were fed with water from the estuary and the possibility of relocation of the MSW began to be evaluated.

In Fig. 3, the situation of the landfill in 1983 is observed and its evolution between 1983 and 1990 is incorporated.

In 1992, at the Rio Earth Summit, a global growth and development model was proposed, articulating ecological, social, and economic aspects. In 1993, Argentina enacted National Law No. 24,051 on Hazardous Waste, motivated by this global trend. At the local level, in 1992, the Belisario Roldán site was closed for the disposal of the MSW, designating it only for inert waste deposit (mainly rubble) in order to fill the place where the creation of the urban park Almirante Brown on the seafront was projected. The MSW was relocated into a Sanitary Landfill nearby Route 229, southeast of Bahía Blanca.

In Fig. 4, the spatial evolution of the area between 1991 and 1996 is incorporated into the map. It can be observed that the outer limit of a main enclosure was completed with waste. In addition, Fig. 5 shows the evident growth of the area, covering 70% of the interior of this site built in the previous period. The modification of the coast shape in the south sector of the peninsula generated by the landfills also appears in this map. This variation is an environmental impact not previously contemplated, which was due to the accumulation of sediments

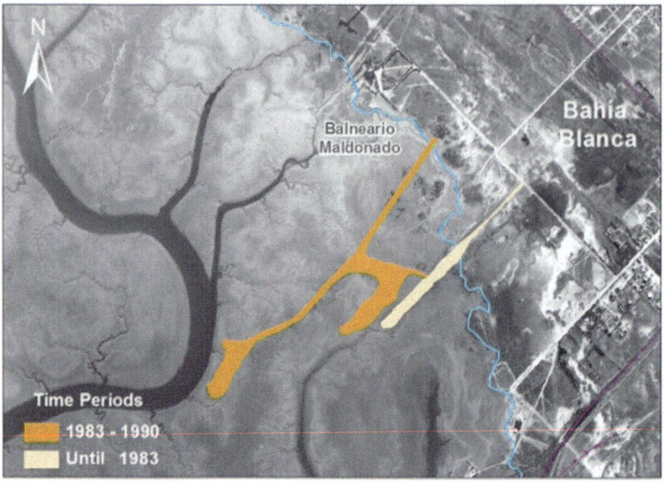

Fig. 3 Evolution of the former landfill until 1990

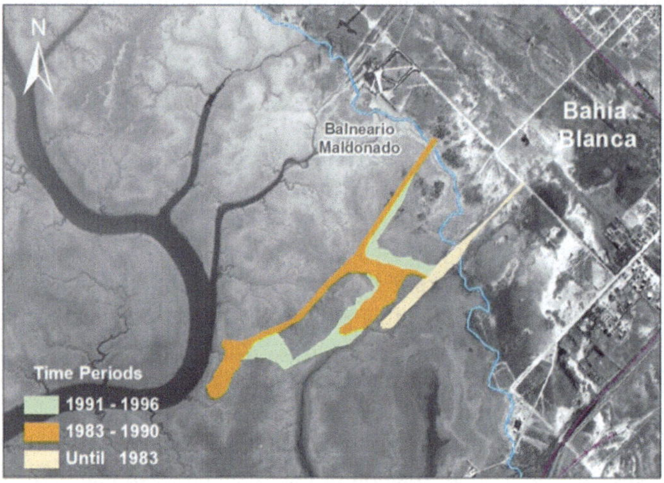

Fig. 4 Evolution of the former landfill until 1996

produced by the waves and tidal currents when interposing the new construction that disrupts the normal littoral dynamics.

In the map of Fig. 6 it is observed that between 2002 and 2006 the main enclosure was completed. In addition, to avoid the erosive effect of the strong southeast winds on the end of the peninsula, and to dissipate the energy of the waves and avoid the dragging of the waste, a new enclosure was defined with large debris and remains of concrete constructions.

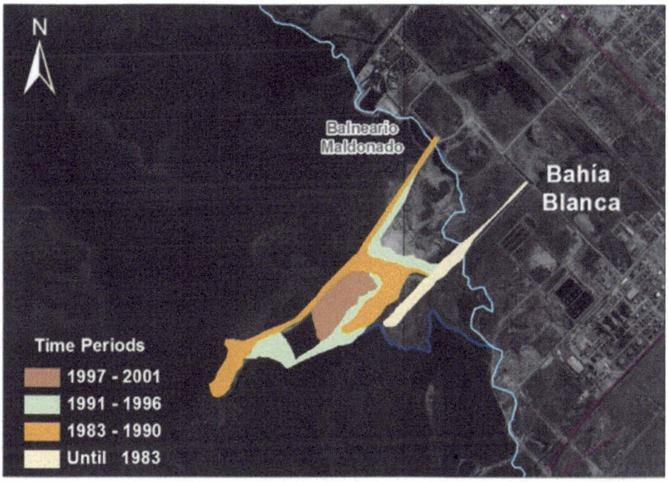

Fig. 5 Evolution of the former landfill until 2001

Fig. 6 Evolution of the former landfill until 2006

Only in the year 2006, in the Province of Buenos Aires, was Law No. 13,592 enacted for the integral management of the MSW. Figure 7 shows how, despite this law and the closing of this landfill, its area continued growing for the period 2007–2009, at the northern and southern ends of the peninsula, without specific controls to prevent the clandestine dumping.

In 2012, the Law of the Province of Buenos Aires No. 14,343 proposed the identification of environmental liabilities, and the duty to recompose contaminated

Fig. 7 Evolution of the former landfill from its beginning until 2016

sites or areas risking population health, to mitigate negative impacts on the environment. In Fig. 7, the advance sequence of the fills in the former Belisario Roldán landfill is summarized from 1967 to 2016, showing that despite its closure in 1992, it continued to receive clandestine overturning, for more than 25 years.

As a complement, in the field visits to the study area, the confirmation of the observed evolutions was verified and georeferenced points with terrain height values were taken. The site was surveyed, documenting the presence of new MSW mixed with debris in different sectors of the former landfill. It is necessary to point out that in 2011, in order to promote the approach of the population to the sea and the integration of urban uses in the coastal zone, the Municipality of Bahía Blanca executed the Wetland Maritime Walk Project on the former landfill, without considering that the place continues to be a recipient of clandestine dumping.

Using GIS [4], the surfaces of the different areas of evolution and their respective volumes were estimated. With the altitude values taken in the field, a digital elevation model (DEM) was defined adding the coastlines, the internal lagoons, and some approximate heights at inaccessible spots. This model tries to estimate the minimum volume of accumulated waste. By intercepting this model with the areas of evolution, volumes were obtained by period. The obtained results are shown in Table 3.

Table 3 Areas and volumes of the former landfill

Period	Area DEM (m^2)	Volume DEM (m^3)
Until 1983	31,136.17	47,252.81
1983–1990	157,651.48	413,379.28
1991–1996	102,584.77	316,347.48
1997–2001	72,741.01	207,697.65
2002–2006	63,661.82	193,776.15
2007–2009	20,117.39	64,812.53
2010–2015	52,846.99	92,214.35
Total	500,739.62	1,335,480.25

4 Estuary Water Monitoring Results from Stations Near the Former Landfill

Table 4 shows the maximum concentrations detected, highlighting those that exceeded any of the guide levels. This table is the result of processing and evaluating the analysis from the stations E6 and E6-BIS [1, 2], and of the comparison of those results with the reference values proposed by the NOAA [3] and the Decree No. 831/93.

From this table, it is clear that all metals analyzed (Cd, Pb, Cu, Zn, Cr, Ni, and Hg) have been present at some point in the estuary water near the former landfill. Particularly, at station E6, Pb and Hg concentrations have once exceeded the NOAA Chronic Exposure reference levels and Decree No. 831, and those of Zn, the Acute Exposure Level of NOAA and the Decree. For the E6-BIS station, the most significant concentrations are those of Cu (still under review) exceeding all guide

Table 4 Maximum concentrations of metals dissolved in estuary water (2006–2012)

Station	Cd (μg/l)	Pb (μg/l)	Cu (μg/l)	Zn (μg/l)	Cr (μg/l)	Ni (μg/l)	Hg (μg/l)
E6	0.40[b]	19.63[d] 10.14[d] 9.57[c]	1.80[d]	6.40[d] 4.93[d] 150.44[c]	2.26[a] 2.21[d]	2.39[b]	0.12[d] 0.41[d] 0.35[d] 1.06[c]
E6 BIS	0.014[e]	4.30[e]	1227[e,f]	[g]	[h]	18.80[e]	0.83[e]

[a]IADO, 2008
[b]IADO, 2009
[c]PIM Adenda, 2009/2010
[d]IADO, 2011
[e]IADO, 2012
[f]The value, although published, is not incorporated into the database because it is in the process of determining if it is an anomalous value (PIM 2012, Sub Program Ría de Bahía Blanca, page 14)
[g]All the values detected in the monitoring period exceed the guide level of Decree No. 831, without specifying them (PIM 2012, Sub Program Ría de Bahía Blanca, page 16)
[h]Detectable values are indicated, without specifying them (PIM 2012, Sub Program Ría de Bahía Blanca, page 18)

reference levels, the Ni that exceeds the value of the Chronic Exposure of NOAA and the Decree, and the Hg only higher than that of the decree.

Although the results evaluated are not statistically representative nor reflect alarm signals, according to the IADO reports [1], the presence of metals dissolved in water is an indicator of recent depositing, because this phase of metallic compounds is extremely ephemeral and is rapidly displaced to other system compartments (e.g., to particulate matter in suspension, sediments, organisms). Therefore, the detection of concentrations of these metals dissolved in water at stations E6 and E6-BIS indicates that they are entering the system and, therefore, that there is a nearby source that is introducing them.

5 Final Considerations

The results of water monitoring in the Bahía Blanca estuary (Province of Buenos Aires, Argentine Republic), close to the former Belisario Roldán landfill, show the presence of concentrations of some metals that could be contributed by leaching through more than four decades on the floodplain. The waste dumped without adequate waterproofing, its degradation, its leaching, and its constant flooding with seawater, would have modified the water quality of the estuary in the area of its location.

The lack of an adequate management allowed the location of the dump on the water of the estuary putting at risk its preservation, threatening its sustainability and its use for future generations.

In Argentina, the management of the MSW is the responsibility of the municipalities. The provinces, as the owners of natural resources, are responsible to care for and preserve them, avoiding potential negative impacts on the environment and the health of the population, which may occur due to inadequate management of the MSW. This shared management requires planning that guarantees health and provides protection and preservation of the environment, which in this case engages areas of recreation, flora, and fauna.

The lack of regulations and the absence of knowledge about the risks that an open dump could generate on the floodplain of the estuary, enabled the location and progress of a landfill whose leaching became a diffuse discharge difficult to control and its remediation a challenge. Although different alternatives have been evaluated to recompose it, the truth is that using removal and transferring techniques, as has been implemented in other areas, would increase the environmental impact on the estuary. Other alternatives for recomposing contaminated sites as established by national legislation are economically impracticable; on the other hand, phytoremediation would be a palliative to mitigate the impact.

Finally, the use of new technologies allows quantifying areas and volumes of environmental liabilities. In this case, the evolution of the former landfill could be confirmed through the GIS. This tool allows observing when an event began, who was responsible for its start, how it evolved in the different stages, and even identifying if its progress was interrupted at some point.

References

1. IADO, Instituto Argentino de Oceanografía: Programa de Monitoreo de Cuerpos Receptores. Ría de Bahía Blanca. Informes Finales 2006 a 2011. http://www.bahiablanca.gob.ar/equipos-de-trabajo/gestion-ambiental/comite-tecnico-ejecutivo/informes-medioambientales/
2. CTE, Municipalidad de Bahía Blanca, 2003–2012: Programa Integral de Monitoreo-Polo Petroquímico y Área Portuaria Bahía Blanca. Informes anuales 2003 a 2012. http://www.bahiablanca.gob.ar/equipos-de-trabajo/gestion-ambiental/comite-tecnico-ejecutivo/informes-medioambientales/
3. Buchman, M.F.: National oceanic and atmospheric administration, NOAA screening quick reference tables, NOAA OR&R Report 08-1, Seatle WA, Office of Response and Restoriation Division, 34 pages (2008)
4. ESRI ®, ArcGIS 9.2™. Copyright 1999-2008. ESRI Inc. All Rights Reserved. www.esri.com

Advantages of Solar Energy Uses in Brazilian Public Port Terminals

Newton Narciso Pereira, Joyce A. Oliveira de Sousa and Caio M. da Silva Anastácio

Abstract Sunlight is a great source of available energy on the planet. South America is a region near the line of the Equator, and Brazil is a country with a large amount of solar radiation. In this paper, only 4 of the 37 TPPs of the Brazilian coast that transport million tons of cargo per year are considered. The TPPs are distributed around the coast and have many roof warehouses and covered carport areas available or idle. The times of insolation and solar irradiance data were collected and correlated with these PPs to determine a solar energy generation using photovoltaic system modules available on the Brazilian market for each of these ports. Then we estimate the total potential energy lost in selected ports around 96.81 mW per day, due to the lack of SPVm in TPPs installed in the TAs. Thus, we suggest that Brazilian port authorities should develop programs to encourage an application for areas to generate solar energy.

Keywords Solar energy · Photovoltaic panels · Brazilian ports
Green energy · Sustainability

1 Introduction

The use of solar energy in electricity production dates back at least three decades [4]. This energy is produced when solar radiation enters a photovoltaic panel and then the electrons of this cell are set in motion [3].

In a moderate scenario, according to the International Energy Agency (IEA), solar energy power could account for around 11% of the global electricity supply by 2050, close to 5000 tWh. The area covered by the facilities would be 8000 km^2

N. N. Pereira (✉) · J. A. O. de Sousa · C. M. da Silva Anastácio
Department of Production Engineering, School of Industrial,
Metallurgical Engineering at Volta Redonda—EEIMVR, Federal Fluminense
University, Av. dos Trabalhadores 420 – Sala C77, Vila Sta. Cecília,
Volta Redonda, RJ 27255-125, Brazil
e-mail: newtonpereira@id.uff.br

A. Vega Sáenz et al. (eds.), *Proceedings of the 25th Pan-American Conference of Naval Engineering—COPINAVAL*, https://doi.org/10.1007/978-3-319-89812-4_26

(300 W/m^2 using cells of 25% efficiency), equivalent to a square of 90 km/side. It should be noted that the efficiencies of the cells available on the market have not yet reached such an efficiency level, where they are on the order of 11–21% [2].

In view of the opportunities generated by the use of solar energy in various segments, it has also been incorporated into the port terminals. For example, in Singapore, Jurong Harbor announced the success of a 9.5 mW solar power plant with the installation of photovoltaic system modules (SPVm) on the roofs of the warehouses. Other projects such as this have already been successfully completed, such as the Port of Rotterdam where an area of 7500 m^2 was covered with 3100 SPVms installed to generate 750 kWh/year. Meanwhile, the Port of Long Beach is doing its part to contribute to a sustainable environment with the installation of 904,75 kW with SPVms from Mitsubishi Electric. This system in turn includes 3290 SPVms and has the potential to generate approximately 1547 mWh/year. In addition to warehouse roofs, areas are also being used for parking cars such as open garages that are receiving coverage with SPVms for power generation.

Considering that the largest solar irradiances in Brazil are in areas of low economic development (northeast region), the port terminals could use this alternative of available energy free of charge.

Thus, the objective of this paper is to demonstrate the amount of energy lost daily in Brazilian public ports for not using SPVms to capture solar radiation. The ports of the four regions (north, south, southeast, and northeast) are analyzed.

2 Methodology

2.1 Brazilian Port Terminals

Brazil has the largest economy in South America with 27 states and a coastline of 8511 km. The port terminals in Brazil are divided into three categories: public port terminals (TPP), mixed between TPP and private use terminals (TUP), and TUP with private sector use only.

In total, there are 37 TPPs and more than 150 TUPs in operation in the country. TPPs are divided into four macro regions (north, northeast, southeast, and south) of the country and, for each region, have particular cargo characteristics such as solid, liquid, general cargo, tourism, and containers.

2.2 Identification of Port Terminal Areas

We analyzed all the National Plans for Port Logistics (PNLP) and Port Master Plans (PMP) developed by the Special Secretariat of Ports (SEP) that describe the main characteristics of each TPP selected in this research. We also analyzed the

Development and Zoning Plans for Ports (PDZ) that show the areas to be explored in these TPPs. Based on these reports, it is possible to analyze all the public areas distributed to be explored to transport cargo in TPPs. Then, four representative TPPs were chosen for the analysis of the electric energy potential produced by solar panels: Rio de Janeiro Port, Itaqui Port, Vila do Conde Port, and Paranaguá Port. These TPPs were selected to evaluate the solar energy potential of each macro region.

Once these areas were determined, the polygonal areas that define the organized port areas were analyzed for each TPP. These polygonal areas were mapped by the SEP and their geographical coordinates were used during the mapping. Thus, we used the Google Earth® (GE) platform to identify and map all these available areas in polygonal areas of the TPPs chosen for analysis. First, we identified all roofs, considering warehouses installed in the TPP area. For each identified roof, the total available area was estimated.

We did a sweep on all possible roofs based on the GE platform. Some images have had problems due to shadows, trees, clouds, distortions, and poor image quality that caused difficulties for estimation. Therefore, we tried to estimate based on what we could evaluate. The areas that presented difficult visibility in the identification were not considered in this evaluation.

In these cases, the GE information with PNLP, PMP, and PDZ was correlated to delimit the area as correctly as possible. For simplified estimation, all areas of the roof collected were considered as a rectangle considering the tools available in GE for capturing the area.

Finally, we created a code to identify each area considering terminal areas during the process of quantification of available areas.

2.3 Estimation of Solar Energy

To estimate the solar energy generation in TPPs, we collected SPVm data available on the Brazilian market. These data were collected from the National Institute of Metrology, Quality and Technology (INMETRO), with the responsibility of evaluating all equipment before being introduced to the market. Several tests were performed to determine the efficiency of the system, processes, and equipment.

Considering the rigor of the tests, we identified the SPVm presented by IMETRO with power above 250 Wp and efficiency above 15%. For this, we analyzed 804 PV models available on the market, however, only 7 of these models were selected for this evaluation.

A SPVm model was selected considering the mean values characteristic of these SPVms as power, efficiency, and area of the module.

The efficiency of SPVm depends on the incidence of sunlight on it. As there are many port terminals scattered on the Brazilian coast, solar insolation and irradiance were identified for each port terminal considering its main region. We then produced a map that was carried out by the database available from the Reference

Center for Solar and Wind Energy Sérgio de Salvo Brito (CRESESB), which in turn through the input of geographic coordinates and the distance from each access terminal allowed us to obtain the parameters of the period of solar insolation in hours and area of irradiation.

To validate the hours of insolation, we collected information and INMET data on the insolation of 2014. Data were collected from four different regions where the port terminals were installed and compared to the values offered by CRESESB and the Energy Research Center Electrical (CEPEL). Thus the time of insolation was determined considering the average time for each region.

To determine the total solar energy that can be produced, we first made an estimate of the installed power capacity considering the nominal power of the panel by the number of modules. However, this determined value does not imply losses that occur through the module. Then the estimated solar capacity installed in each port was calculated by expression (1) [1]:

$$\text{Installed Capacity (Wp)} = \text{Number of SPvms} \times \text{Nominal Potency(Wp)} \tag{1}$$

In sequence we calculate losses that vary from place to place for SPVm installation. The hours of insolation and irradiation (kWh/m^2. day) were considered and losses under normal conditions may exceed 30% of the nominal power supplied by each SPVm.

Considering that the energy produced depends on the internal and external factors of the modules, each PV is constructed in different ways and with different materials which influence the variation of the same generation that these modules have the same efficiency. The calculation that defines the energy produced (EP) is represented by expression (2):

$$\text{EP(Wh)} = \text{Module Efficiency (\%)} \times \text{Average Daily Irradiation} \times \text{Module Area} \tag{2}$$

We consider that the energy produced by PV modules is not available 24 h a day. The solar heat was based on data collected by CRESESB and CEPEL. This allowed us to make the average estimate of solar insolation and irradiation in areas where the ports are as shown in expression (3).

$$\text{Average Module Power(Wp)} = \text{EP/(Average Insolation Hours)} \tag{3}$$

We considered that the PV module has natural operating losses. We then deduced that these energy losses and the actual value of operation for SPVm at these port terminals were shown in expression (4).

$$\text{Installed Average Capacity(Wp)} = \text{Number of PVm} \times \text{Average Module Power} \tag{4}$$

Then we estimated for each port terminal the total energy produced by the PV modules. This allowed us to identify the generation that each region can provide with SPVm.

For the estimation of the number of SPVm that will be needed in the useful areas of the ports, we divided the total area available by the area of each PV module (1.63 m^2).

3 Results

In Brazil, there are 37 public ports located in more than 8500 km of the coast. In this research, only 4 TPP were considered, each representative of each region (north, south, northeast, and southeast). In these public ports, there are several terminals that serve many types of cargo, considering export and import operations. Port terminals mainly handle bulk cargoes and use silos and warehouses to store grain, soy, corn, sugar, wheat, and others. Warehouses in port areas have large roofs with high potential to receive SPVms for electric power generation. Brazil is a major exporter of bulk; these warehouses have large static storage capacities.

3.1 Mapping of Port Areas

The coordinates of TPPs in Brazil were collected and inserted in Google Earth. The purpose of this evaluation was to identify areas of public ports as warehouse roofs available to install SPVms. For each port terminal area, we estimated and identified using the GE as shown in Fig. 1a. Only the areas within the polygonal areas were considered in this evaluation as shown in Fig. 1b.

Figure 1a shows some parts of the roof areas of the warehouses of the mapped Rio de Janeiro port. Many of the warehouses are intended for the storage of various

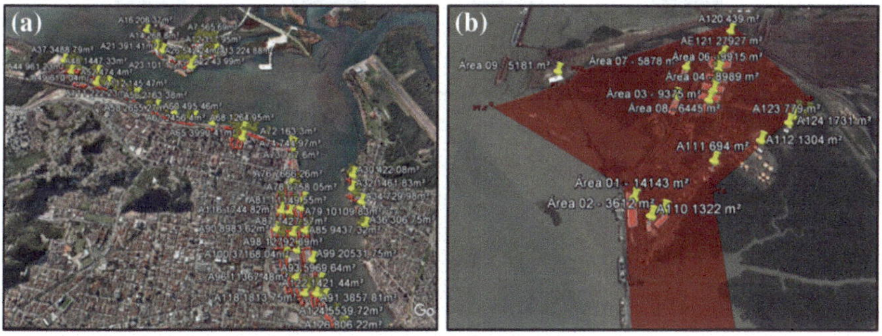

Fig. 1 Mapped areas in the port terminals inside the port polygon

loads with areas ranging from ±10,000 to ±27,000 m^2 of coverage available for installing SPVs. Figure 1b shows that we considered only the polygonal (red) area. This example is the Port of Itaqui installed in the north region, where roof and covered carport (CC) areas were identified and mapped.

One limitation of this procedure for identifying areas was the image quality of the GE viewer. Therefore, we believe that some small areas cannot be considered in these analyses using this procedure. However, the main port areas have been identified and PV power generation can be estimated. Figure 2 shows the total of areas available to receive the installation of SPVms considering the roofs of the warehouse and the CC.

Figure 2 shows that the Ports of Rio de Janeiro (southeast) and Paranaguá (south) present a large amount of free areas on roofs to receive the SPVms. Following this trend, in the Port of Rio de Janeiro (also located in the southeast region) only one warehouse has a coverage area of 290,000 m^2, where there is no other building that can influence in terms of solar insolation. In this port terminal, many cars deal with an area of ±4000 m^2 that can be adapted to the installation of SPVms.

The northern port of Vila do Conde (north), has some warehouses with areas ranging from 2000 to 4000 m^2 capable of installing SPVs. The Port of Itaqui

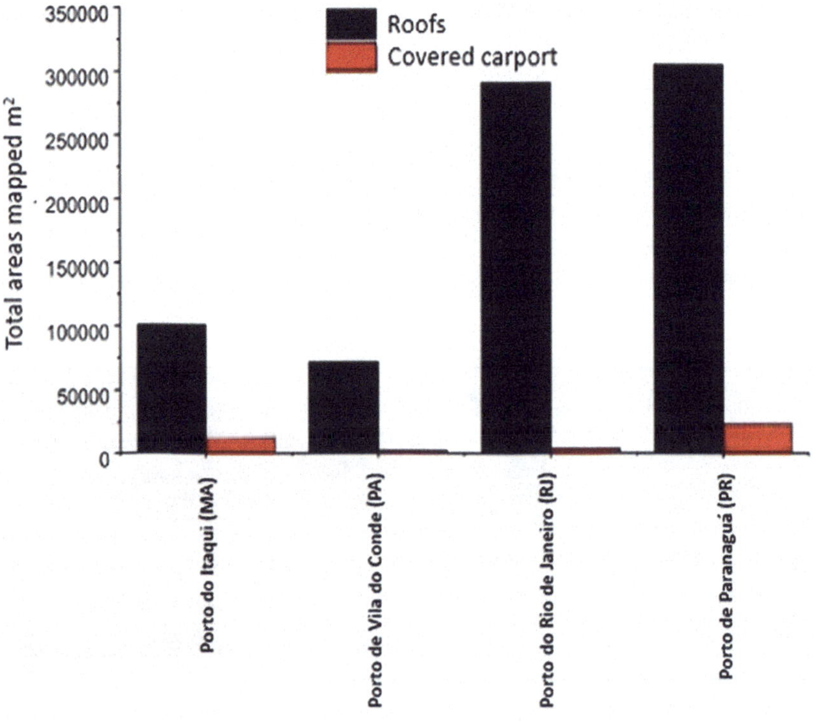

Fig. 2 Available areas to install PV to generate energy in selected public ports

(northeast) also has a large number of warehouses with a range between ±6500 and ±8500 m^2.

In the Port of Itaqui, the Tegram Terminal dedicated to the export of grains is composed of four warehouses with ±9000 m^2 of roofs each and more than two warehouses of smaller capacity, but with areas ranging from ±5000 and ±6000 m^2.

This shows that the northern TPPs do not present a large number of roofs with available areas when compared to the southern ports.

However, the Port of Itaqui presents an excellent area of polygonal line for trucks that wait to unload grains in the terminal. Only this area has ±27,000 m^2 free to receive a SPVms structure, but must be adapted to avoid its influence on the operation of the trucks. Within this port terminal, there is an area that has been used as a CC for the regulars of the port. We consider this area available to receive SPVms. The same procedure has been done for all ports to identify more areas to install SPVms. A total of 2546 mapped areas are shown in Fig. 3.

As expected, Fig. 3 shows that the Port of Rio de Janeiro presents more areas available to receive SPVms. This fact is associated with the number of warehouses installed in relation to other ports in Brazil. Thus we can note all port terminals that have a large number of idle areas available to install SPVms operate with dry bulk

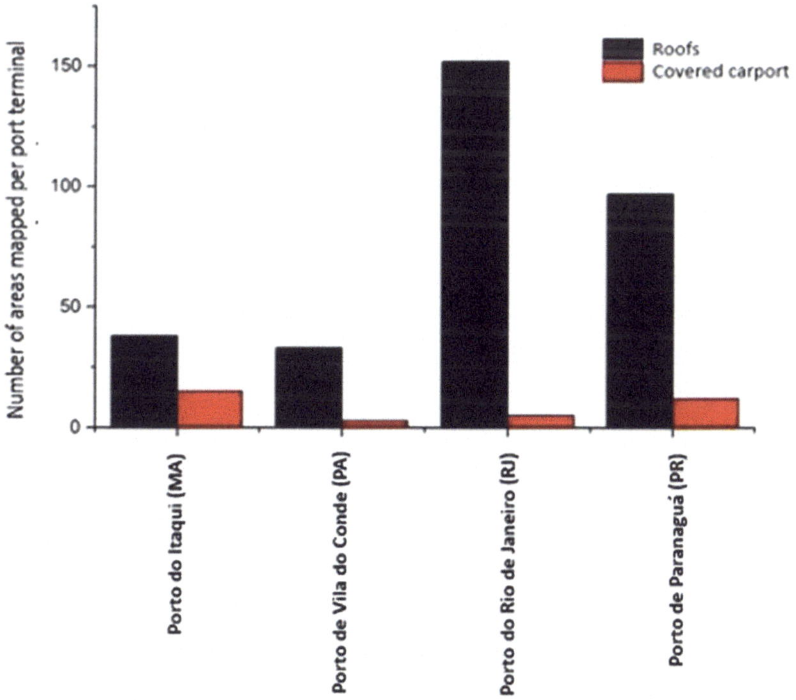

Fig. 3 Number of areas of port terminals mapped with potential to receive PV modules

cargo because they are exported or imported with high concentration in the south and southeast. Some TPPs have the potential to use CC, but with a small participation (6%) when compared to the roofs' warehouse areas. If these areas were not used, this would not affect the potential of electric power generation. Among the TPPs installed in the northern region only the Port of Itaqui with the Tegram terminal has a plan to expand its warehouse infrastructure and will have more roof areas in the future.

3.2 *Estimation of Solar Energy*

The estimated solar energy generated in these TPPs depends on the areas available to install the SPVs. Another important aspect is the solar panel technology (SP) available to be installed in the TPPs. Our estimation considered the data of the solar system certified by INMETRO. There are about 804 models of photovoltaic panels available for the market, however, only seven of these models were used for this estimate. Our criteria of choice were the potential, efficiency, and area for each SPVm. The potential should be higher than 250 Wp, because with these characteristics the area of each SPVm is around 1.63 m^2 and energy efficiency has a range between 15 and 16%.

The commercial models have a narrow spectrum of efficiency; that is, they generally vary from 1 to 2% of the difference. Table 1 indicates the main technical parameters of each SPVm used for TPP sizing.

We can note that SPVms have similar characteristics in terms of power, area, and efficiency. In terms of efficiency, it is correlated with temperature. Although Brazil is a tropical country, all port locations are within the temperature range; that is, the upper temperature limit present in the Brazilian territory is not capable of affecting the efficiency of the SPVm.

Table 1 Characteristics of PV modules available on the Brazilian market

Brand (Portuguese)	Model	Capacity (Wp)	Area of panel (m^2)	Efficiency	NOCT (°C)
A do Brasil Informatica Ltda	3TS260Wp	260	1.63	16.0	45
AJ Imports Soluções Sustentáveis para Energia	S/M-270	252	1.64	15.4	47.5
Atria Engenharia Civil e Implementações Energéticas	XP460-260I +35	262	1.63	16.1	46
Green Qualy	Chrome 250	250	1.59	15.8	47.5
Alper Energia S.A.	ALP250 M-60	250	1.64	15.3	45
Alper Energia S.A.	ALP260 M-96	260	1.7	15.3	45
Alper Energia S.A.	ALP265 M-60	265	1.7	15.6	45

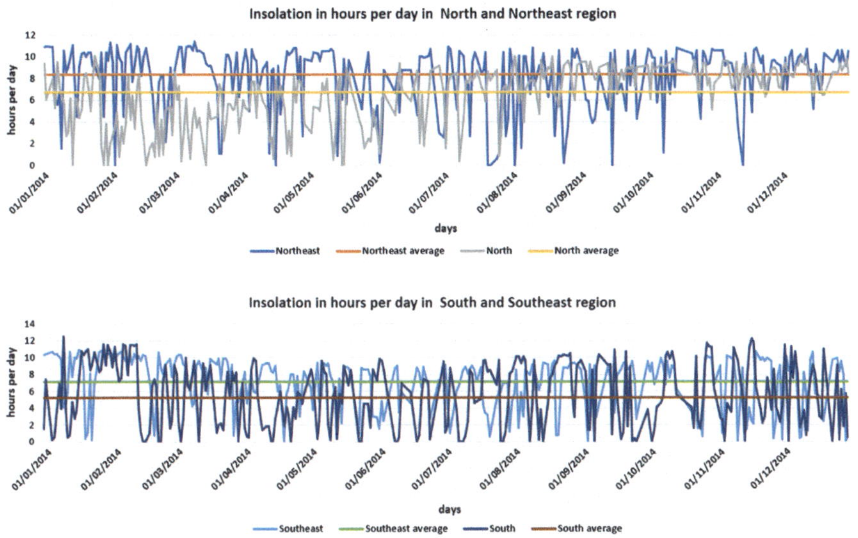

Fig. 4 Solar insolation during 2014 in the regions of TPPs

Thus the generation of solar energy also depends on daily solar insolation. Brazil is a country with continental dimensions and there are ports distributed along the coast from north to south. We evaluated the insolation data of the National Institute of Meteorology (INMET) considering the four Brazilian regions where the port terminals are installed. Figure 4 shows the results of the reading of sunshine sensors in the Brazilian territory in hours per day.

We selected the data from some stations that represent the location of the TPPs to estimate the average insolation per year. Figure 5 shows that in the north and northeast, mean sunshine is between seven and eight hours per day. In the south and southeast, average insolation is between five and seven hours.

Solar radiation depends on the power per unit area received from sunlight. Irradiation is a function of the distance from the sun. It is measured by kWh/m^2 day and for each TPP region, we considered these data. Figure 6 shows the distribution of sunshine and the incidence of solar radiation on these port terminals.

Figure 5 is a map that has been adapted from INMET, CRESESB, and CEPEL. We used this information as a basis for estimating the energy generated through the use of SPVms for each TPP. First, we estimated the total energy production for each TPP, one for each region shown in Fig. 6.

Figure 6 shows the results considering the port terminals installed in each region. The ports of Paranaguá (south) and Rio de Janeiro (southeast) present high potential to generate energy from SPVms (± 39 and ± 35.44 mW), respectively. The northern and northeastern regions have less potential, due to the characteristics of the ports and the areas available to install SPVms. However, the northern region presents high insolation and solar irradiation when compared with other Brazilian regions.

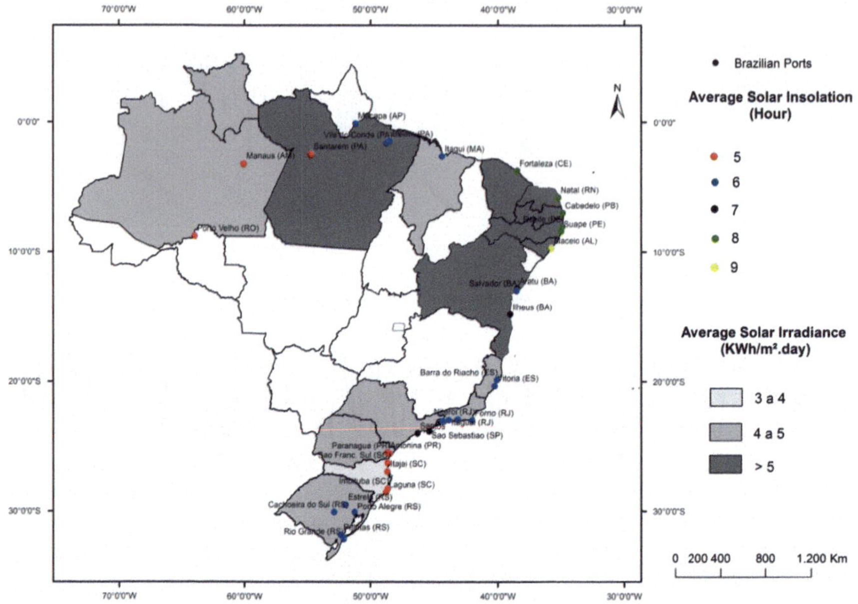

Fig. 5 Solar insolation (hours) and irradiation (kWh/m^2 day) in the areas of the port terminals

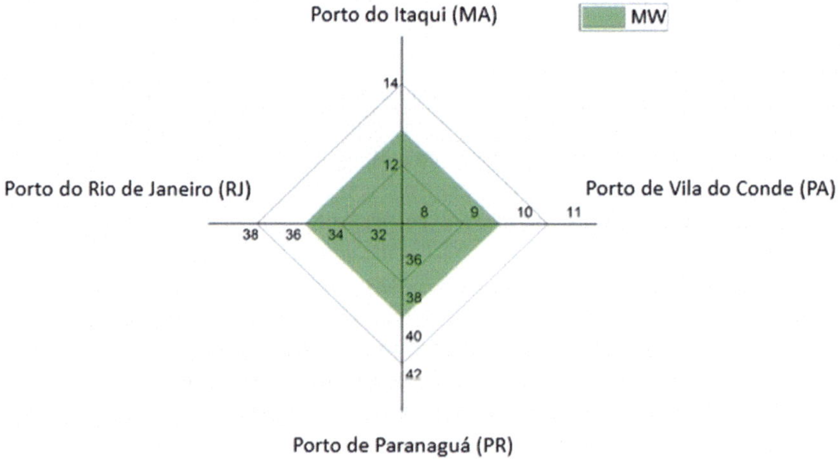

Fig. 6 Total electrical energy potential generated by TPPs installed on roofs

For these estimates, we considered only roof areas available. This may be influenced by some area losses during the installation process, but showed the potential of using these idle areas for the installation of photovoltaic panels and solar energy production. Other areas such as CCs can contribute to improve the amount of energy generated in TPPs. Considering that the available areas are identified with the GE visualizer, the energy potential was estimated and is presented in Fig. 7.

These uncovered areas need to be interpreted with care because it is a hypothesis that CC areas may be candidates for capturing solar energy. These areas have been identified with the images available in the GE viewer. Given that our findings were based on areas identified by the GE platform and the port authorities were not consulted, we estimated the number of photovoltaic modules needed to produce solar energy in these TPPs. It considered a SPVm standard with 1.63 m^2. Thus Fig. 8 shows the number of SPVms distributed in each TPP considering roofs, warehouse, and CC areas.

In this estimation, we considered only the characteristics of the roof area with respect to the mean SPVm size presented in Table 1. Figure 8 shows that to generate solar energy at these port terminals, a large amount of SPVms will be required to cover these areas. We are aware that our search may have a limitation in terms of total area available and actual conditions to install these SPVs in these TPPs. Some areas cannot have these modules, due to the limitations of the structures and the

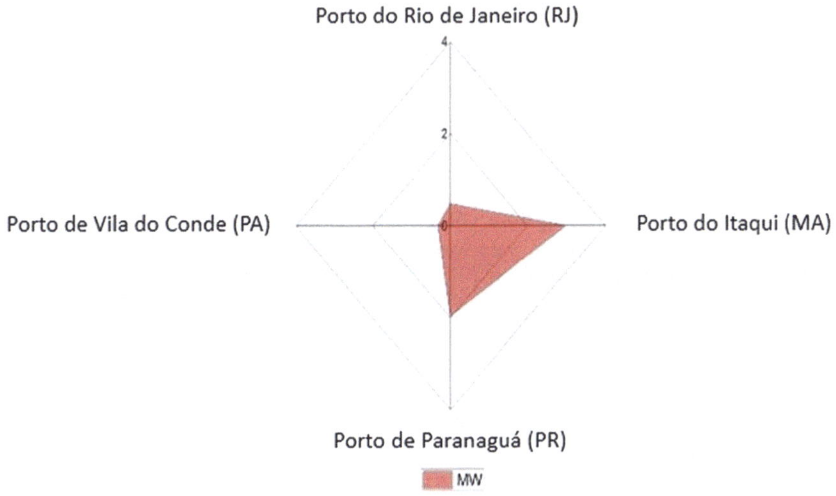

Fig. 7 Possible additional areas for installing SPVms in port terminals

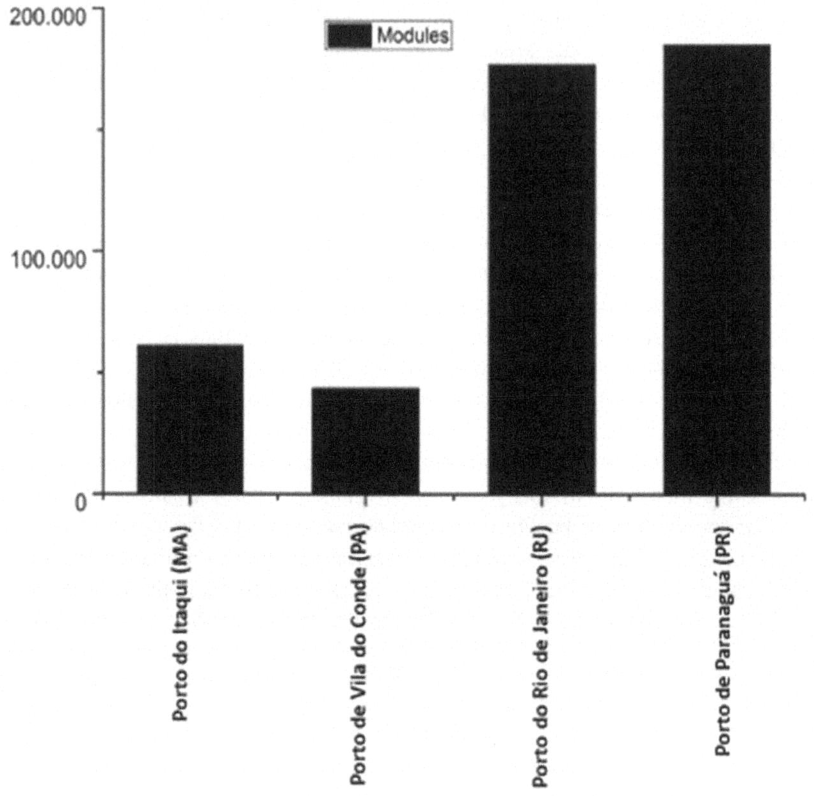

Fig. 8 Number of PV modules for each terminal

characteristics of the project. Other components and adaptations needed to recover solar energy at these port terminals were not considered in this investigation. We assume that all the electrical energy produced for these SPVms must be entered directly into the network because no system was considered to accumulate the energy generated by these SPVms.

4 Conclusion

The Brazilian public ports lose a large amount of solar energy per year due to the absence of photovoltaic systems to capture the radiation and convert it to electricity. The total energy lost in the four analyzed ports is around 96.93 MW per day. This value was estimated considering photovoltaic systems with efficiency around 16%. This value was an estimate considering the conditions for making our observations using the Google Earth platform. We found some limitations in terms of image quality, but this had a small influence on the end result. In total, 320 roofs were mapped in warehouses and 17 covered areas, which together offered 768,772 m^2 of roofs and 42,720 m^2 of covered areas available to receive the solar panels. TPPs located in the south and southeast regions have greater potential for solar energy generation due to the amount of warehouse areas available for the installation of SPVs. Only the port of Paranaguá, which is the largest port in the south region, can generate 39 mW with 97 roof areas and 12 identified covered areas. Therefore, the result of this research shows that Brazilian ports, port authorities, and the government should look at the renewable sources of energy available in the areas around the ports. To invest in solar energy, the government could create subsidies for the implementation of SPVs in TPPs with idle AT and CC areas. Another alternative is to rent these areas' ATs and CCs for companies to install SPVs and receive the return on the sale of energy from concessionaires, considering that the TPPs do not have dedicated programs for the use of renewable energy.

References

1. Energia Solar Fotovoltaica, Conceitos e Aplicações, Sistemas Isolados e Conectados à Rede. Villalva G, M, Gazoli J, R, Editora Érica 1° Edição 2012 São Paulo
2. Green, M.A., et al.: Solar cell efficiency tables (Version 45). In: Progress in Photovoltaics: Research and Applications, vol. 23, issue 1, pp. 1–9 (2015). LNCS Homepage, http://www.springer.com/lncs. Accessed 21 Nov 2016
3. Lewis, N.S., Daniel, G.N.: Powering the planet: chemical challenges in solar energy utilization. In: Proceedings of the National Academy of Sciences vol. 103, issue 43, pp. 15729–15735 (2006). Author, F., Author, S., Author, T.: Book Title. 2nd edn. Publisher, Location (1999)
4. Solangi, K.H., et al.: A review on global solar energy policy. Renew. Sustain. Energy Rev. **15**(4), 2149–2163 (2011)

Statistical Methods for Automatic Identification of Seabed

Javier Tarrío Saavedra, Noela Sánchez Carnero and Andrés Prieto

Abstract This work proposes the use of statistical methodologies based on unsupervised classification for the automatic identification of seabed types in coastal areas. For this purpose, acoustic data obtained by means of a simple beam echo sounder (at 200 kHz) coupled to a small ship have been used as discriminant features. Each of the resulting acoustic curves has been preprocessed through the application of time corrections (elongation of the echo with depth), power (attenuation of the wave with distance), and ping length (deformation of the echo due to distance), with the aim of eliminating its dependence with respect to depth. The experimental data have been obtained in a controlled environment in the region of Cabo de Palos (Murcia, Spain), studying three different types of bottom: sandy, sandy with sparse vegetation, and rock. The statistical techniques adapted and applied to this particular case belong to the cluster classification from time series. In fact, taking into account that in actual identification problems the existing fund classes are not known in advance, the problem of identification has been addressed

J. T. Saavedra (✉)
Grupo MODES, Departamento de Matemáticas, Escola Politécnica Superior, Universidade da Coruña, A Coruña, Spain
e-mail: javier.tarrio@udc.es

J. T. Saavedra · A. Prieto
ITMATI, Santiago de Compostela, A Coruña, Spain

J. T. Saavedra · A. Prieto
CITIC, A Coruña, Spain

N. Sánchez Carnero
Centro para el Estudio de Sistemas Marinos (CESIMAR), Centro Nacional Patagónico (CENPAT-CONICET), Puerto Madryn, Argentina

N. Sánchez Carnero
Grupo de Oceanografía Física, Universidad de Vigo, Vigo, Spain

A. Prieto
Grupo M2NICA, Departamento de Matemáticas, Facultade de Informática, Universidade da Coruña, A Coruña, Spain

A. Vega Sáenz et al. (eds.), *Proceedings of the 25th Pan-American Conference of Naval Engineering—COPINAVAL*, https://doi.org/10.1007/978-3-319-89812-4_27

through an unsupervised classification perspective based on the previous calculation of dissimilarity matrices and the application of hierarchical cluster classification methods. The results obtained, correctly identifying 93% of the total funds—with little confusion between their classes—support the use of automatic classification techniques in this area for the correct characterization of the seabed.

Keywords Seabed · Sonar · Cluster classification · Time series

1 Introduction

The column of water represents the first and nearly breakable barrier when we want to study the seabed. The simple fact of their presence drastically limits the use of the two main methodologies used in the terrestrial environment: direct methods of data collection, in this case through diving, restricted for safety reasons, or the use of dredges, always limited by the need for infrastructure and investment of time for data collection; and, on the other hand, satellite images, limited to very shallow areas where the attenuation of the electromagnetic signal in the water still allows data collection [1]. Taking into account the restrictions of the aforementioned techniques, underwater acoustics represents the best methodological alternative for the study of the seabed [2]. The acoustic signal is a longitudinal wave that, in a dense medium such as water, is transmitted at high speeds (around 1500 m/s) and long distances before being attenuated below the ambient noise. Consequently, its use and study allow us to work at great depths. In scientific echo sounders, this signal, generated by a transducer, propagates to the bottom in the form of a spherical wave. Upon reaching the bottom, the acoustic signal insonicizes it with its wavefront, first vertically and, immediately afterwards, obliquely, due to the sphericity of the wavefront. Part of this acoustic energy is reflected by the most superficial layer and another part slightly penetrates in the bottom (depending on the frequency and intensity of the wave), to be reflected in turn by the immediately deeper layers [3].

The acquisition of the acoustic data obtained by reflection in the seabed allows us to infer characteristics of the bottom, taking into account that the characteristics of the initial wave are known or can be modeled, in addition to the geometry of the propagation of the wave (both in its advance through the water column and with respect to the insonification of the bottom surface) and the attenuation along the water column. In this line, valuable work has been carried out on habitat classification [4], underwater vegetation mapping [5], underwater geomorphology studies [6], and cartography of benthic communities [7], among others.

Schematically the acoustic wave is generated in the transducer; it goes down to the bottom where it is reflected and ascends towards the surface. Once there it undergoes a new reflection in the water–air interface. This fact makes the wave return to propagate towards the bottom, where it again suffers reflection and ascends. From the moment in which the transducer emits the wave, it begins to acquire information of the reflected wave. During the advance of the wave through the water column, the reflected acoustic energy is very low, due to changes in the characteristics of the medium, or even to some element (such as fish) found in the trajectory. When it reaches the bottom there is a strong reflection due to the great difference of acoustic impedances between the water and the sediment or the bottom rock. This process is repeated with the second bounce. In this way, the transducer picks up the acoustic intensity that it receives over time in the form of a background noise, of a very low intensity level, until it receives the first echo, coming from the first ping bounce. This echo begins with a maximum corresponding to the reflection of the wave part that propagates vertically followed by a "tail" produced by the oblique backscattering of the spherical wave. The shape of this tail depends on the type of background in which it is reflected, and it can present different maximums due to the interference of the wave, the penetration into the bottom, and so on. That second echo is followed by another second echo from the second reflection, with a similar but attenuated form and smoothed after the two previous bounces on the bottom and water surface.

The data acquisition rate with echo sounders is very high due to the high- speed acoustic wave advance in the water. For example, working in shallow waters (e.g., 50 m deep) with pulses of medium length (1024 μs) and acquiring both the first and the second bounce of the echo sounder in the bottom, scientific echo sounders allow the acquisition of about 450 pings per second (although the usual is about 10 pings per second). This high sampling rate results in huge databases that require the application of statistical analysis for the extraction of relevant information.

At the beginning of the application of acoustic methods for the seabed study, in the 1980s [8], the total energy integrated within each of these curves (first and second echo) was used to perform an unsupervised classification of each of the sampled points [9]. With this methodological approach good results were obtained [10], although having so much level of simplification (the whole curve characterized by two unique variables) limits the classification obtained. In order to capture more information of acoustic curves, multivariate approaches emerged that extracted variables (features) based on different types of signal analysis: Fourier analysis, fractal analysis, energy integration, power distribution along the echo, and the like. This methodology is still the main approach for the background acoustic study, either through commercial software [11] or with free applications [12]. Although these approximations drastically expand the number of variables from the initial focus (from 2 in RoxAnn, to 166 QTC), by subjecting the signal to a characterization process using variables, not always totally independent (functionally or statistically), a part of the information relative to the shape of the echo curve that could respond to relevant features of the fund is always being discarded. Finally, classification alternatives have recently been proposed, both supervised and unsupervised, using functional data analysis (FDA) techniques and clustering of time series [13], a research line that follows this work. This type of statistical classification alternative has the advantage of automating the process of seabed identification, minimizing the need for prior knowledge about the physical problem, and also providing a quicker response without the need for trained personnel.

2 Data Collection

Using a simple beam echo sounder (EA400P Simrad) coupled to a 5 m length boat, acoustic curves corresponding to a frequency of 200 kHz have been obtained in a controlled area of Cabo de Palos (Murcia, Spain) on July 19, 2014. These curves correspond to three different kinds of funds, depending on the substrate: sandy bottom (SP1), sandy bottom with sparse vegetation (SP2), and rocky bottom (SP3). To eliminate the effect of depth, we have proceeded to apply time corrections (lengthening of the echo with the depth), power (attenuation of the wave with the distance), and length of ping (deformation of the echo), the latter through convolution using a core-type function that compensates for the relationship between pulse-length and depth [7]. All the transformations described above have been applied using the ECOSONS software [12]. As a result, for each ping we obtained a curve of acoustic intensity as a function of time. It is composed of two echoes or peaks, the first one more related to the roughness of the background, whereas the second one had its hardness [9]. Finally, 1383 acoustic curves were obtained, one for each ping, with acoustic intensity values integrated in 710 different times. On the one hand, the values of the Z variable or true seafloor classes were collected, and the sound intensity values measured in decibels (dB) were obtained. These values measured as a function of time formed the time series used in this work to estimate the different types of seabed.

It is important to point out that, of the 1383 acoustic curves obtained, 678, 236 of the SP1 class (25% of the class totals), 160 of the SP2 type, and 282 of the SP3 class were taken into account for the analysis. We proceeded in this way to achieve a balanced design in order that the indices which measure the performance of the classification were not biased in a spurious manner by the sample size of each class.

3 Unsupervised Classification from Time Series

This study provides an alternative solution for the seabed type identification, taking into account the substrate that defines it. Therefore, this is a problem that can be approached from the perspective of statistical classification. Classification techniques can be divided into groups: (a) supervised classification techniques, whereby the seabed class Z (qualitative random variable) is estimated from a statistical model. This model is a function of a vector of m quantitative characteristics $X = \{X_1, X_2, \ldots, X_T\}$, whose parameters have been estimated by a training sample, previously knowing the number and type of existing classes. (b) Unsupervised or cluster classification techniques, which allow grouping individuals (different areas of seabed) in a variable number of classes or groups according to different similarity criteria applied to the vectors of features X that define each individual. Cluster techniques can be applied without previously knowing the existing number of classes or their number. Therefore, in this work we propose unsupervised classification tool application, taking into account that there is usually no previous information about what types of seabed exist in a specific region.

The application of cluster classification methodologies of time series is proposed, from the fact that the feature vector X is composed of T observations of sound intensity as a function of time. It is important to note that, in this context, the differences between individuals (described by time series) may depend both on the proximity of each of the observations in a grid of times and on the dynamic nature of the series (dependency structure, correlations, etc.) [14]. Therefore the first step of the procedure is the definition of an appropriate measure of dissimilarity that provides a definition of the distance between two curves or time series. Once chosen, it is applied to each pair of time series obtaining as a result the nxT dimensional dissimilarity matrix, where n is the number of observations of the seabed that are studied, and T the number of different times in which it is evaluated (points where each time series is constituted). Finally, a cluster classification method [14], preferably hierarchical [13], is applied to the dissimilarity matrix, obtaining the assignment of a group to each of the seabed observations.

In this work, the performance of six different dissimilarities belonging to the "model free" typology [14] has been studied, given its simplicity and fast runtime. To illustrate them, $\boldsymbol{X}_T = (X_1, X_2, \ldots, X_T)$ and $\boldsymbol{Y}_T = (Y_1, Y_2, \ldots, Y_T)$ have been defined as the particular realizations of the real value processes $X = \{X_t, t \in \mathbb{Z}\}$ and $Y = \{Y_t, t \in \mathbb{Z}\}$, respectively, following the indications of Montero and Vilar [14]. The simplest is the Euclidean dissimilarity, d_{L_2}, defined by

$$d_{L_2}(\boldsymbol{X}_T, \boldsymbol{Y}_T) = \left(\sum_{t=1}^{T} (X_t - Y_t)^2 \right)^{\frac{1}{2}} \tag{1}$$

that only takes into account the point-to-point distances between the time series. An alternative is the dynamic time warping distance d_{DTW}, which takes into account both the point-to-point distance and the shape differences of each pair of series. In fact, one of its main advantages is that it is capable of recognizing similar forms in curves even after they have been transformed (displaced and scaled) [14], which is the case of the present database. For computing the d_{DTW} it is necessary to define $r = ((X_{a_1}, Y_{b_1}), \ldots, (X_{a_m}, Y_{b_m}))$, where $ai, bj \in \{1, \ldots, T\}$ so that $a_1 = b_1 = 1, a_m = b_m = T$, and $a_{i+1} = a_i$ or $a_i + 1$ and $b_{i+1} = b_i$ or $b_i + 1$, with $i \in \{1, \ldots, m-1\}$, and m all possible pairs of observations.

$$d_{\mathrm{DTW}}(\boldsymbol{X}_T, \boldsymbol{Y}_T) = \min_{r} \left(\sum_{i=1,2,..,m} |X_{a_i} - Y_{b_i}| \right) \tag{2}$$

These two distances d_{DWT} and d_{L_2} do not take into account the dependency structure within each time series. They assume that each observation within the series is independent of the others. One way to measure this dependence is by calculating distances based on the autocorrelation functions (ACF) [14],

$$d_{\mathrm{ACF}}(\boldsymbol{X}_T, \boldsymbol{Y}_T) = \sqrt{(\hat{\rho}_{X_T} - \hat{\rho}_{Y_T})^T \Omega (\hat{\rho}_{X_T} - \hat{\rho}_{Y_T})} \tag{3}$$

And, on the other hand, on the partial autocorrelation functions (PACF) d_{PACF}, whose distance is defined in a similar way, where $\Omega = I$ is the weight matrix and $\hat{\rho}_{X_T} = (\hat{\rho}_{1,X_T}, \ldots, \hat{\rho}_{L,X_T})$ and $\hat{\rho}_{Y_T} = (\hat{\rho}_{1,Y_T}, \ldots, \hat{\rho}_{L,Y_T})$ the autocorrelation coefficients of X_T and Y_T, with a delay L for which we obtain $\hat{\rho}_{j,X_T} = 0$ or $\hat{\rho}_{j,Y_T} = 0$ with $j > L$ [14].

Another free-model alternative is provided by the use of dissimilarity based on the difference between the periodograms (estimates of the spectral density of the time series) of each pair of time series [14]. Defining the periodograms of $\boldsymbol{X}_T$ and $\boldsymbol{Y}_T$ as $I_{X_T}(\lambda_k) = T^{-1}|X_t e^{-i\lambda_k t}|^2$ and $I_{Y_T}(\lambda_k) = T^{-1}|Y_t e^{-i\lambda_k t}|^2$, respectively, with

$$\lambda_k = \frac{2\pi k}{T}, \quad k = 1, \ldots, \frac{T-1}{2},$$

the Euclidean distance between two periodograms can be defined as

$$d_{\mathrm{P}}(\boldsymbol{X}_T, \boldsymbol{Y}_T) = \frac{1}{\frac{T-1}{2}} \sqrt{\sum_{k=1}^{\frac{T-1}{2}} (I_{X_T}(\lambda_k) - I_{Y_T}(\lambda_k))^2} \tag{4}$$

whereas its normalized version with logarithmic transformation has the expression

$$d_{\mathrm{LNP}}(\boldsymbol{X}_T, \boldsymbol{Y}_T) = \frac{1}{\frac{T-1}{2}} \sqrt{\sum_{k=1}^{\frac{T-1}{2}} (\log NI_{X_T}(\lambda_k) - \log NI_{Y_T}(\lambda_k))^2} \tag{5}$$

where

$$NI_{X_T}(\lambda_k) = \frac{I_{X_T}(\lambda_k)}{\hat{\gamma}_{0,X_T}},$$

and $\hat{\gamma}_{0,X_T}$ is the sample variance of $\boldsymbol{X}_T$ (analogous expressions for $\boldsymbol{Y}_T$). The distance $d_{\mathrm{LNP}}(\boldsymbol{X}_T, \boldsymbol{Y}_T)$ is useful when only interested in measuring differences in the correlation structure, not in the scale.

Once the dissimilarity matrices have been calculated, it is proposed to apply the hierarchical cluster classification algorithm of maximum distance [13–15].

4 Results and Discussion

In the present section, in order to identify the type of seabed according to its substrate, the hierarchical classification algorithm of maximum distance is applied to the dissimilarity matrices obtained from the calculation of the distances described in Sect. 3. To evaluate the performance of the classification method according to the dissimilarity matrix used, we propose the calculation of three indices commonly used in this field: correct classification ratio [16], cluster evaluation index (CEI) [14], and kappa index [17]. One of the main contributions of this study is the identification of the time intervals corresponding to the acoustic data that best define each bottom class. Namely, those time intervals in which the difference between curves of the same class is minimal and the difference between acoustic curves of different seabeds is maximum. Therefore, the proposed classification methodology is evaluated in curve segments comprised in time intervals of four different lengths (3.2, 4.8, 6.4, and 8.0 ms) with moving initial time (from 0 to 40 ms, with a step of 1 ms).

Figure 1 shows the correct classification proportion obtained (using the distance d_{DTW}) for each interval of the acoustic curves as a function of the interval start time. The best classifications are obtained from the intervals located in the tail of the first echo, particularly those of shorter length (3.2 or 4.8 ms). In Fig. 2 we observe the time interval at which the highest proportion of correct classification is obtained, the most informative one, represented on the real acoustic intensity curves corresponding to the three classes of seabed. The differences in the first echo are related to the seabed roughness.

Table 1 shows almost perfect classifications applying the hierarchical cluster algorithm of maximum distance to dissimilarity matrices constructed from DTW

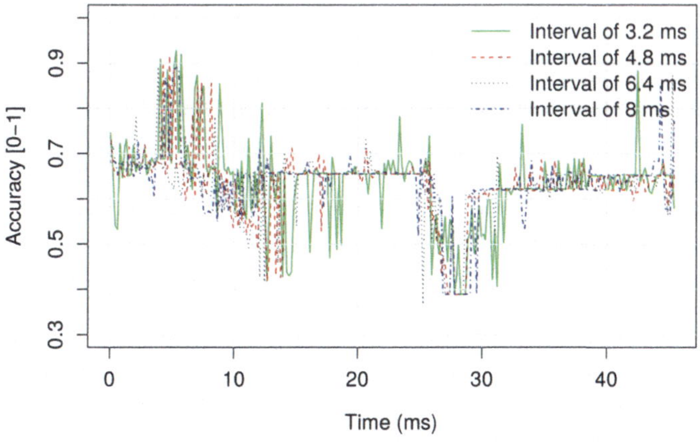

Fig. 1 Proportions of correct classification for each studied time interval, depending on its starting time. It is the result of applying the hierarchical cluster algorithm to the DTW dissimilarity matrix corresponding to four different interval lengths

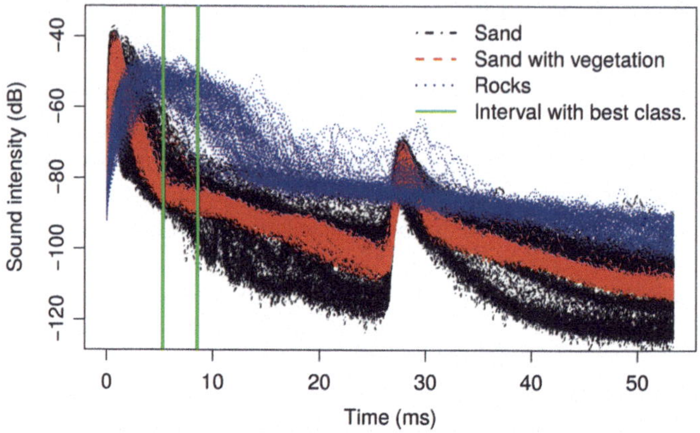

Fig. 2 Sound intensity curves as a function of time (first and second echo) corresponding to each of the three bottom classes. It highlights the time interval for which a better identification is obtained by applying the hierarchical algorithm to the d_{DTW} dissimilarity matrix

Table 1 Indices that measure the performance of the different cluster classification alternatives: proportion of correct classification, CEI, and confidence interval (95%) for kappa index

	Correct classification proportion	CEI	Confidence interval for kappa (95%)	Time interval (ms)
L_2	0.92	0.92	(0.86, 0.92)	(4.99, 8.19)
DTW	0.92	0.92	(0.85, 0.92)	(5.76, 8.96)
ACF	0.65	0.66	(0.42, 0.54)	(0.06, 3.26)
PACF	0.85	0.84	(0.72, 0.81)	(0.256, 8.26)
P	0.77	0.72	(0.59, 0.69)	(25.22, 33.22)
LNP	0.81	0.81	(0.65, 0.75)	(4.67, 12.67)

and Euclidean distances. In fact, by using the d_{DTW} distance, a correct classification ratio of 92% is obtained. This result is consistent with the 95% confidence interval for the kappa index of (0.85, 0.92), within the limits of the almost perfect classification that is commonly defined for values between 0.81 and 1. The fact that these dissimilarities are precisely the ones that provide the best classification result is due to the main differences between the acoustic curves are point-to-point and shape differences. The results show that the corrections applied to echoes have been efficient, because the effect of depth displaces and dilates the echo within the curve. On the other hand, if the d_{LNP} (81% of correct classification) and d_{PAFC} (84%) results are observed, it is verified that the seabed can also be differentiated by the autocorrelation structure of their corresponding acoustic time series. Taking these results into account, the use of other metrics that combine point-to-point distance and dependence structure, such as the d_{COR} and d_{CORT} dissimilarities, is justified [13].

Table 2 Confusion matrices resulting from applying hierarchical cluster methods to each dissimilarity matrix (in the intervals shown in Table 1)

Real	SP1	SP2	SP3	SP1	SP2	SP3
	d_{L_2}			d_{DTW}		
	Estimated			Estimated		
SP1	0.84	0.14	0.02	0.85	0.14	0.01
SP2	0.04	0.96	0.00	0.06	0.94	0.00
SP3	0.00	0.00	1.00	0.00	0.00	1.00
	d_{ACF}			d_{PACF}		
	Estimated			Estimated		
SP1	0.61	0.39	0.00	0.79	0.18	0.03
SP2	0.11	0.89	0.00	0.24	0.76	0.00
SP3	0.44	0.05	0.51	0.01	0.00	0.99
	d_{P}			d_{LNP}		
	Estimated			Estimated		
SP1	0.91	0.03	0.06	0.89	0.10	0.01
SP2	0.68	0.30	0.02	0.16	0.84	0.00
SP3	0.01	0.00	0.99	0.30	0.03	0.67

Proportions are shown as relative frequencies

Finally, Table 2 shows the confusion matrices corresponding to the use of each type of dissimilarity, in the interval for which the highest proportion of correct classification is obtained. In the diagonal of each confusion matrix, the proportion of curves to which the correct class has been assigned is shown, whereas the misclassification proportions are located outside the diagonal. All the types of sea bottom are correctly classified (values of the diagonal next to 1) by using the d_{DTW} dissimilarity or the Euclidean distance, even those very similar as the labeled SP1 and SP2. On the other hand, if the dissimilarity PAFC is applied, there is a small confusion between these classes, and the use of the d_{LNP} dissimilarity leads to confusing SP3 with SP1.

5 Conclusions

The use of time series cluster techniques has been proposed to identify the seabed class from acoustic data provided by echo sounders. The proposed methodology provides an alternative for the automatic identification of seabed types, with the advantage of not requiring expert personnel or an in-depth knowledge of the context of the problem.

Based on various indices that evaluate the performance of the classification methods used (correct classification ratio, CEI, and kappa), almost perfect classifications have been obtained by applying the hierarchical cluster algorithm of maximum distance to the matrices of dissimilarities constructed from the Euclidean or the d_{DTW} distances. In fact, by using the DTW or d_{L_2} distance, a correct classification proportion of 92% and an interval for the kappa of (0.86, 0.92), in the case of Euclidean distance, is obtained; that is, an almost perfect classification is obtained. These results indicate that the differences between the time series corresponding to the different seabed classes are mainly due to the point-to-point distance and the shape they present. Furthermore, it should also be noted that the dependency structure of each series also allows distinguishing the different types of seabed (to a lesser extent). The latter is supported by the results obtained from the dissimilarities based on the partial autocorrelation function (85% of correct classification) and on the normalized periodogram (81%).

Acknowledgements This work has been subsidized by the Xunta de Galicia through the project "Numerical simulation of high frequency hydroacoustic problems in marine environments—SIMNUMAR (EM2013/052)". In addition, the research of Javier Tarrío Saavedra has been supported by MINECO grants MTM2014-52876-R and MTM2017-82724-R, and by the Xunta de Galicia (Grupos de Referencia Competitiva ED431C-2016-015 and Centro Singular de Investigación de Galicia ED431G/01), all of them through the ERDF.

References

1. Kenny, A., Cato, I., Desprez, M., Fader, G., Schüttenhelm, R., Side, J.: An overview of seabed-mapping technologies in the context of marine habitat classification. ICES J. Mar. Sci. J. du Conseil **60**, 411–418 (2003)
2. Lurton, X.: An Introduction to Underwater Acoustics: Principles and Applications. 2nd edn. Geophysical Sciences Subseries, XXXVI, Springer Praxis Books (2010) (724 p.)
3. Sánchez-Carnero, N., Aceña, S., Rodríguez-Pérez, D., Couñago, E., Fraile, P., Freire, J.: Fast and low-cost method for VBES bathymetry generation in coastal areas. Estuar. Coast. Shelf Sci. **114**, 175–182 (2012)
4. Snellen, M., Siemes, K., Simons, D.G.: Model-based sediment classification using single beam echosounder signals. J. Acoust. Soc. Am. **129**, 2878–2888 (2011)
5. Sánchez-Carnero, N., Rodríguez-Pérez, D., Couñago, E., Aceña, S., Freire, J.: Using vertical sidescan sonar as a tool for seagrass cartography. Estuar. Coast. Shelf Sci. **115**, 334–344 (2012)
6. Rovere, A., Vacchi, M., Firpo, M., Carobene, L.: Underwater geomorphology of the rocky coastal tracts between Finale Ligure and Vado Ligure (western Liguria, NW Mediterranean Sea). Quatern. Int. **232**(1), 187–200 (2011)
7. Rodríguez-Pérez, D., Sánchez-Carnero, N., Freire, J.: A pulse-length correction to improve energy-based seabed classification in coastal areas. Cont. Shelf Res. **77**, 1–13 (2014)
8. Orlowski, A.: Application of multiple echoes energy measurements for evaluation of seabottom type. Oceanologia **19**, 61–78 (1984)
9. Voulgaris, G., Collins, M.B.: USP RoxAnn Ground Discrimination System: A Preliminary Evaluation. Technical Report, University of Southampton (1999)
10. Henriques, V., Guerra, M.T., Mendes, B., Gaudêncio, M.J., Fonseca, P.: Benthic habitat mapping in a Portuguese marine protected area using EUNIS: an integrated approach. J. Sea Res. **100**, 77–90 (2015)
11. Somerton, D.A., McConnaughey, R.A., Intelmann, S.S.: Evaluating the use of acoustic bottom typing to inform models of bottom trawl sampling efficiency. Fish. Res. **185**, 14–16 (2017)
12. Rodríguez-Pérez, D., Sánchez-Carnero, N., Freire, J.: ECOSONS software. Available at http://www.kartenn.es (2014). Accessed 27 July 2017

13. Tarrío-Saavedra, J., Sánchez-Carnero, N., Prieto, A.: Supervised and Unsupervised Seabed Classification Using Functional Data Analysis and Time Series clustering from Acoustic Curves. Submitted
14. Montero, P., Vilar, J.A.: Tsclust: an R package for time series clustering. J. Stat. Softw. **62**, 1–43 (2014)
15. Everitt, B. S., Landau, S., Leese, M., Stahl, D.: Hierarchical clustering. In: Cluster Analysis, pp. 71–110, 5th edn. Wiley (2011)
16. Francisco-Fernández, M., Tarrío-Saavedra, J., Naya, S., López-Beceiro, J., Artiaga, R.: Statistical classification of early and late wood through the growth rings using thermogravimetric analysis. J. Therm. Anal. Calorim. **127**, 499–506 (2017)
17. Cook, R.J.: Kappa. In: Encyclopedia of Biostatistics, vol. 4. Wiley (2005)

Sustainable Development of Estuary Ports—Study Case: Bahía Blanca Estuary, Argentina

Daniela Escudero, Olga Cifuentes and Silvina Medus

Abstract Estuaries are coastal wetlands. Their surfaces are permanently or temporarily flooded by the effect of the tides. They serve as habitat for numerous species and the refuge of migratory fauna. At the same time, they provide a natural shelter against currents and waves, being therefore sites especially suitable for the installation of ports. The need to expand these facilities to meet the growing maritime transport demands requires a careful evaluation of the works to ensure environmental sustainability, in order to avoid irreparable consequences if the design and location are not adequate. The objective is to identify which of the areas of an estuary is most suitable for new sustainable port developments. Analyzing the case of Bahía Blanca, the deepwater port complex (Argentina), and considering the current location of the facilities, the impacts that would derive from the construction and port operation are identified. They are evaluated from their possible location in the internal and external areas of the estuary. SWOT matrices are developed and compared to identify the most convenient location. As a result, the external zone of the estuary would be environmentally more favorable for new port developments. However, it is necessary to make a resilience map that avoids conflicts when decisions are made. In that moment, some of the social actors can exercise power relations on technical justifications. Particular interests would prevail that it would not always be consistent with sustainable development.

Keywords Port developments · Estuary · Environmental sustainability

D. Escudero (✉) · O. Cifuentes · S. Medus
Facultad Regional Bahía Blanca, Universidad Tecnológica Nacional,
Bahía Blanca, República Argentina
e-mail: daniela_escudero@yahoo.es

A. Vega Sáenz et al. (eds.), *Proceedings of the 25th Pan-American Conference of Naval Engineering—COPINAVAL*, https://doi.org/10.1007/978-3-319-89812-4_28

1 Introduction

Estuaries are coastal, partially closed, bodies of water that arise from the mixture of fresh water from rivers or streams with seawater. They constitute transition zones between the land area and the sea. Although they are influenced by the tides, these areas are protected from waves and winds by reefs, barrier islands, or peninsulas [1]. They are especially suitable for the installation of ports because they provide natural shelter against storms. Their wetlands and marshes are unique environments that provide a variety of resources, benefits, and services, being one of the most productive areas of the planet. Their occupation for port uses leads to a modification of the natural equilibrium conditions that consequently modify the remaining channels, wetlands, and marshes to find a new equilibrium position (Medina Santamaría 2002 in [3]).

This is the case of the Bahía Blanca port complex (Argentina), located in the interior of the estuary of the same name. Its geographical conditions, its depth (which allows the navigation of vessels up to 45 ft draft), and its strategic location make it a natural outlet for regional production towards international markets [2–4]. This estuary has wetlands and marshes that are habitat for numerous species and a source of natural and economic resources (Melo et al. 2002; Petracci 2011 in [3]). In recent years, the pressures to increase the port spaces in response to growing production and energy demands, as well as the needs of urban settlers in their surrounding area make it necessary to plan these expansions taking into account environmental sustainability [3]. For this reason, this work aims to identify which is the most convenient estuary area for new sustainable port developments, in order to avoid irreparable consequences if the design and location are not adequate.

2 Methodology

The research begins with surveys and bibliographic searches that allow the socioecological characterization of the Bahía Blanca estuary, port facilities, and main urban centers that surround them. For the present work, only the current situation is synthesized, but the results of the planning processes for the sector of the last 20 years for the evaluation process are included.

Based on the information obtained, surveys were prepared for interviews with specialists in environmental, port, transport, and urban planning issues to complement data and deepen concepts. The potential impacts of the port developments were identified, both in the construction and operational phases. The most relevant aspects (land use, location, natural resources, and urban centers, existing infrastructure, biological, and activities) were evaluated comparing them through a SWOT matrix (strengths, opportunities, weaknesses, threats), in order to suggest the more favorable location (internal and/or external) for the expansion of the port complex in this estuary.

The final considerations on the sustainability of the socioecological system were elaborated (Calvente 2007 in [3]).

3 Area of Study

The Bahía Blanca estuary is located in the southwestern area of the province of Buenos Aires, Argentina, between 38° 42′S and 39° 15′S, and between 61° 45′ and 62° 29′W. It comprises an area of approximately 3000 km^2. Their extensive wetlands and tidal plains are connected by a series of side channels that flow to the main channel that runs on to the NW–SE direction. Argentina's largest deepwater port system sits on its northern coast (Melo 2007 in [3]). For this study, the estuary was divided into two zones: an internal one, from the headwaters to the mouth of the Napostá stream and an external one, from the aforementioned mouth to its outer limit. Only the north coast is considered for new port developments (see Fig. 1).

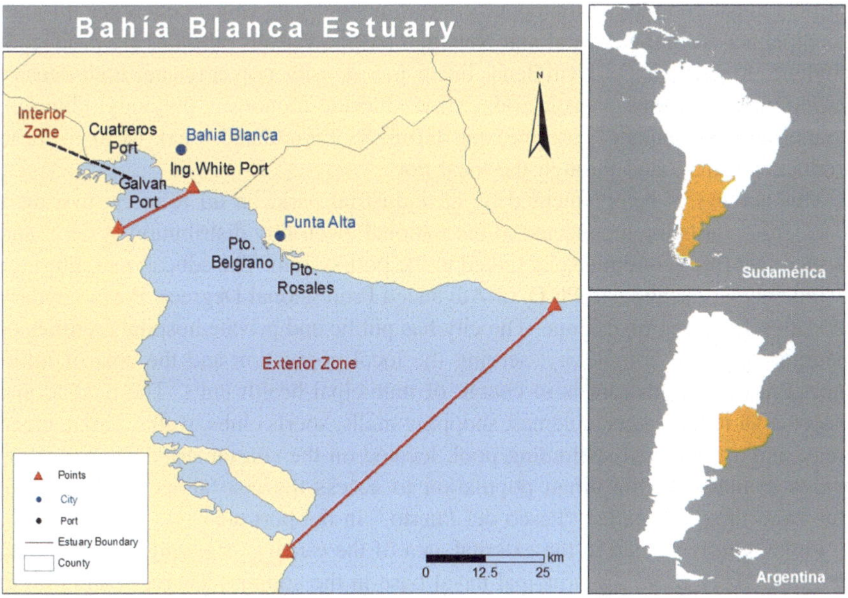

Fig. 1 Bahía Blanca estuary (Argentina), ports, and main urban centers

3.1 Urban Centers in the Estuary

At the head of the estuary, on the north inland coast are General Cerri, Ingeniero White, and Bahía Blanca. In the most external zone, the Puerto Belgrano Naval Base and Punta Alta city are located.

The General Cerri town is located at the head of the estuary, about 15 km from Bahía Blanca. It has 8715 inhabitants [5]. In its surroundings, the only fruit and vegetable sector of the Bahía Blanca district developed. Its economic activity is related to the food industry (an oil mill, a flour mill, a dairy factory, and a refrigeration plant). A gas treatment and transportation plant, and another one for the production of inert gases are installed in this area. The ethane produced in the mentioned plant is raw material for the Bahía Blanca Petrochemical Hub and the inert gases are used in medicine and the production of soft drinks.

Bahía Blanca is one of the most important maritime port cities of Argentina. Although its urban center is almost 10 km from the coast, it is integrated with Ingeniero White, a town bordering the port of the same name. Its population is about 301,500 inhabitants [5]. It constitutes an outstanding transport and communications node. It links the center and south of the Pampean region with the Patagonian north through road and rail connections. It has the most important petrochemical complex in the country. From natural gas, which converges through gas pipelines from the Austral and Neuquén basins, various products are produced: ethylene, low-density polyethylene, linear low-density polyethylene, high-density polyethylene, chlorine, caustic soda, vinyl chloride monomer, polyvinyl chloride, ethane, propane, butane, gasoline, and fertilizers (granulated urea). Most of these products are marketed through the local port.

This activity is complemented by an industrial park, an oil refinery, two thermoelectric plants interconnected to the national electricity distribution system, and an incipient free economic zone linked to the port system. The educational offerings include nursery schools to Ph.D. or Advanced Professional Degrees. Postsecondary education has a regional scope. The city has public and private hospital facilities of medium and high complexity, serving the local population and the area of influence. Primary health care is in charge of municipal health units. The recreational aspect is varied: theaters, cinemas, shopping malls, sports clubs, walks, parks, green areas, and a municipal swimming pool, located on the coastal strip. There are few spaces available to the urban population to access the coastal sector: two yacht clubs and the place called "Paseo del Puerto," in the port area.

Punta Alta is located in the external area of the estuary, near and around Puerto Belgrano Navy Base, the principal naval base in the country. There, in addition to tasks related to the strictly military, drydocking facilities for commercial ships are offered. The city commercial sector is oriented to the satisfaction of local demand for goods and services. There are two flour mills and a dairy industry. It has a medium complexity hospital, supplemented with eight nursing rooms for primary care. The naval base has a hospital too, and specialized naval workshops, public schools, a printing press, a banking center, civil registry, post office, and a hotel, among other facilities. The population has access to the coastal sector for recreational purposes in Villa del Mar, where the so-called Paseo del Humedal has been developed, in the Arroyo Pareja beach. In the summer season, the navy base sector called Punta Ancla, is also habilitated for civilian uses.

3.2 Port Developments in the Estuary

Describing its location from the headland towards the external zone of the estuary, on the north coast and along 60 km, are the ports Cuatreros, Galván, Cangrejales, and Ingeniero White (with its industrial area) and in the most external zone, the military installations of the Puerto Belgrano Navy Base and Rosales Port. This complex constitutes the deepsea port of Argentina [3, 4].

Because of its draft, Cuatreros Port today is not used for commercial activities, and its facilities are used by a fishing club [3].

The facilities known as Bahía Blanca Port are formed by Galván, the Cangrejales area, and Ingeniero White. All of them constitute a set of facilities scattered along 25 km on the north coast of the estuary. The Management Consortium of the Port of Bahía Blanca (its acronym in Spanish, CGPBB) is the nonstate public entity that is in charge of its administration and exploitation, as of its creation, after the sanction of the National Ports Law, in 1992. The maritime jurisdiction of this Consortium includes the access route to the entire port complex. It is constituted by a channel, recently deepened, 190 m wide and 97 km long, which allows the navigation of vessels with a maximum draft of 45 ft. It has a modern beaconing system and traffic monitoring, which gives good safety conditions for day and night navigation.

Puerto Galván is made up of two wharves destined for grains, by-products, and general cargo; to which are added the Flammable Berths 1 and 2 intended for chemical products, and liquid and gaseous combustibles. In 2016, a third Post was added to supply fuel oil to the new thermoelectric plant, which also allows the decongestion of the current sites for flammables. In the area known as Cangrejales, there is an export terminal for grain and terminal products associated with a hydrocarbon fractionation plant and an industry that manufactures fertilizers and ammonia. This sector was won to the sea with the filling of the first deepening dredging of the access channel. Currently, a regasification vessel is moored at the hydrocarbons wharf, through which imported gas enters through methane tankers to compensate part of the country's energy deficit.

There are three terminals specialized in the loading of grains and by-products located in Ingeniero White Port. Those are the main export items of the port. A multipurpose dock, for container movement, and a general loading dock are also situated there. In addition, it has sites assigned to service vessels: coastguard, mooring, pilots, dredging, and tugboats, as well as the dock for coastal fishing vessels [3, 4]. The growth of port facilities has been significant in the last 25 years. Seven of the terminals described have been built in this period. Therefore, the availability of land for port use in the sector is limited. The land accesses have not had the same evolution. This generates congestion and annoyance in the urban sector that surrounds the port, mainly in peak periods of merchandise inflow. To this is added the lack of accessibility of the population to the coastal sector, which also generates discontent. The CGPBB has tried to correct this last conflict by materializing two projects: "Paseo del Puerto" and "Balcón al Mar."

In the external zone, there is Puerto Belgrano Navy Base. Its port facilities make up a shipside of 243,000 m^2, surrounded by several berths that together total 2472 m. The dry docks N° 1 and N° 2, of 215 and 205 m long, respectively, stand out. In these, large-scale naval repairs can be carried out, assisted by their specialized workshops [4].

Rosales port is towards the estuary mouth. It is managed by the Puerto Rosales Delegation, which depends on the government of the province of Buenos Aires. It has a continuous wharf of 300 m in length and 30 ft depth. It is mainly used by pilots, small craft, and fishing vessels. In the maritime sector, in depths that reach 60 ft, are installed two buoys SPM (Punta Ancla and Punta Cigüeña) for the operation of oil tankers, linked to land by a submarine pipeline of 2000 m in length that connects them with a pumping system, tank park, and an oil pipeline that takes the product to various refineries [3, 4]. The road connectivity to Puerto Rosales is confused. It should be circulated through urban sectors of the Punta Alta locality to join the port sector with the national route. The railway connection is nonexistent. Formerly it was about 2700 m from the wharf, but now it is unusable [3].

3.3 Estuary Ecological Value

Over the tidal flats of the Bahía Blanca estuary, wetlands and marshes develop. They constitute ecosystems of significant ecological value for their primary production, food source, provision of habitats, stabilization of sediments, and biofiltration (Parodi 2007 in [3]), [6]. The intertidal areas are the habitat of diverse communities of invertebrates, resting and wintering areas of migratory birds, and a nesting and feeding site important for various species of waterfowl [6]. The internal sector is the only flowering place for winter phytoplankton, which starts the trophic chain by providing food for many commercial species (fish, shrimp, and prawns), which are the basis of local artisanal fishing. There are also several species of sharks, sea lions, and Franciscan dolphins, in danger of extinction. Whales and killer whales are common at the estuary (Fidalgo 2007 in [3]).

The wildlife and the different vegetal communities of the estuary constitute an interesting ecotourism resource, as well as for sport fishing and wildlife observation (Nebbia and Zalba 2007; Parodi 2004; Massola 2005; Rosake 2005 in [6]).

The relevance of the estuary is reflected in the creation of the Natural Reserve of Multiple Uses "Bahía Blanca, Bahía Falsa and Bahía Verde" by the Province of Buenos Aires in 1991, as well as other reserves of municipal scope [3].

4 Potential Environmental Impacts Due to Port Developments

In this section, selected aspects are evaluated according to the problems identified in the local estuary and the needs that a port development requires, both in its construction (see Table 1) and operational phases (see Table 2). The information included in them is a synthesis of the complete evaluation, which can be consulted in [3].

5 SWOT Matrix Elaborated from the Location of Port Developments in Internal and External Zones

In the last decade, several factors were combined to be instilled in the local society: the importance of planning port developments, linked to urban planning and environmental sustainability. Programs, documents, and reports were prepared with the participation of social actors specialized in the aforementioned topics. Among

Table 1 Potential impacts in the construction phase of a port industrial site

Action: opening of work sites (land area)		
Milieu	Interior zone impacts	Exterior zone impacts
Atmosphere	Dust, noise, vibrations, pollution (use of equipment, vehicles)	Dust, noise, vibrations, pollution (use of equipment, vehicles)
Flora and fauna	Destruction of vegetal cover	Destruction of terrestrial habitats
Soil	Low availability of land for port uses The landscape is already impacted Surface consolidation reclaimed from the sea by contributions from quarries	Change in land use (rural to industrial port) Difficult integration with the landscape
Waterbodies	Similar to existing ones	Possibility of saline intrusion by uptake of groundwater
Socioeconomic	New jobs (temporary) and growth of service areas	Jobs (temporary) and growth of service areas
Action: opening dredging (extraction, sand filling)		
Milieu	Interior zone impacts	Exterior zone impacts
Atmosphere	Possible bad odors circumscribed to the work area	
Flora and fauna	Modification and/or destruction of coastal and marine habitats	
Soil (seabed and coasts)	Modification of current speeds. Modification of sedimentological processes. Incorporation of land reclaimed from the sea	
Water	Seawater pollution by resuspension of different parameters. Turbidity increasing	
Socioeconomic	Dredging companies only demand goods and/or services	

Table 2 Potential impacts in the operational phase of the port industrial site

Milieu	Interior zone impacts	Exterior zone impacts
Atmosphere	Similar to those already existing	Generation of emissions and increase of particulate material
Flora and fauna	Decrease in the internal productivity of the estuary	Affecting of habitats and species by increasing maritime traffic
Freshwater	Impacts similar to existing ones	Pollution from effluent spills, spills
Seawater	Contamination by resuspension of bed solids (maintenance dredging), spillage of effluents or accidental spills. Increased turbidity and other parameters in suspension by ship propellers	
Soil (seabed and coasts)	Changes in the hydrodynamics of currents and speeds that would modify erosion and sedimentological patterns	
Socioeconomic	Loss of access to the coast and recreational coastal spaces	

them stand out: the URB-AL Program (2006); the Franco-Argentine Cooperation Program ADEFRANCE (2008–2010); the Development Plan of Bahía Blanca (2010); and the ENARSA-YPF Project (2011). The latter was developed for the installation of a regasification plant in the inner area of the estuary. It triggered social mobilization, requesting reports by the Justice to the local universities and scientific research institutions related to the subject [4]. Based on these documents and complementing the information with interviews with specialists in environmental, port, transport, and urban issues, the SWOT matrix was built. The most relevant aspects of the location of new port developments in the internal versus external zone of the estuary are compared (see Tables 3, 4, 5 and 6).

Table 3 Strength factors of port developments according to their location

Interior zone	Exterior zone
Current facilities allow operational logistics planning for an expansion in its vicinity	Location of urban centers allows the planning of access works in order to minimize conflicts with urban traffic
Operational rail network	Less ship navigation time
Existence of environmental control and monitoring programs	Amplitude ship maneuvering areas that minimize nautical accidents (greater amplitude of the estuary)
The hydrodynamics of the estuary facilitate the maintenance of the depths achieved with dredging, in both areas	
In both locations there are qualified human resources for the port operation and associated industrial developments. The distance between both zones allows the affecting of the human potential for any of them	

Table 4 Weakness factors of port developments according to their location

Interior zone	Exterior zone
Increase in the transit of trucks and railroads in densely populated areas	Lack of terrestrial access to the port that implies its complete design
Internal zone depth to imply important volumes of dredging to bulk carrier navigation	The characteristics of the waves in the sector imply the need for breakwaters
Protected species (e.g., Olrogs gull) would be affected by the proximity of new harvests to their breeding grounds	Populations of migratory birds rest and feed in the lagoon of Arroyo Pareja, near Puerto Rosales
The recovery of benthic populations would be affected by dredging	
Lack of strategic national, provincial, or local transportation plans that involve the port of Bahía Blanca and its development (investments linked to power relations)	
Overlapping jurisdictions (national, provincial, and municipal) complicate development planning and environmental management (dilution of responsibilities)	

Table 5 Opportunity factors of port developments according to their location

Interior zone	Exterior zone
Port industrial development would improve the economy of the General Cerri town	Port industrial development would improve the economy of Punta Alta
Land reclaimed from the sea by filling with sediments from dredging would allow the installation of port terminals	
It is possible to develop an adequate and efficient interurban communications system	
The revival of railway lines would relieve road traffic	

Table 6 Threat factors of port developments according to their location

Interior zone	Exterior zone
Dredging and ship traffic would affect phytoplankton bloom in the sector, with consequences throughout the food chain	The location of the developments could close Arroyo Pareja and/or Punta Ancla beach
The dredging of the navigation channel would affect the area of protected crabs	
Increase in the probability of nautical incidents due to increased vessel traffic and the natural narrowing of the estuary in the sector	
Increased probability of introduction of exotic species (increase in maritime traffic)	
The increase in maritime traffic would affect the distribution of fish and bird populations	

6 Final Remarks

From the ecological point of view, the estuary is one of the most productive environments on the planet, a key to the survival of many species, but vulnerable to environmental changes caused by the construction and operation of ports.

From the socioeconomic aspect, the ports, industrial activities, and transit of merchandise by land and sea, would originate a positive impact on the regional and national economy, resulting in a direct benefit for a sector of society and indirect for the whole of it. However, they would also modify the current functioning of the estuary, losing places of recreation, which could lead to loss of wealth in biodiversity and natural resources such as fishing. For this reason, the sustainability of the project will be given by the resilience of the site which is chosen.

From the evaluation of the information in the Bahía Blanca estuary, it appears that the most suitable sector for the location of new port developments is the outer zone. Nevertheless, the interaction of the different social actors with their interests and strategies over others (fishermen: by their source of work, population: by recreational places, industries: by production, port terminals: by coastal areas), the overlapping of jurisdictions that it dilutes responsibilities, and the lack of state policies on transport with a strategic vision of regional integration, sometimes does not lead to sustainable development of the estuary. Making a resilience map in advance of these developments would avoid conflicts that occur when decisions are taken at turbulent times and when relations of strength and power are imposed on technical justifications. If these conditions do not materialize, the ports and the logistical chains that integrate would grow anarchically. Particular interests, that would not always be consistent with the common good, would prevail, motorizing a discontent that generally leads to social conflicts.

Acknowledgements Thanks to the Management Consortium of the Port of Bahía Blanca, for their collaboration.

References

1. USEPA—United State Environmental Protection Agency—Homepage: Basic Information About Estuaries. National Estuary Program (NEP). https://www.epa.gov/nep/basic-information about-estuaries#whatis. Accessed 27 June 2017
2. Programa UR-BAL: Proyecto común R7 P1 02, Impacto del Crecimiento del Sector Industrial Químico y Petroquímico en la Gestión Urbana de Ciudades Portuarias. Bahía Blanca, República Argentina (2006)
3. Mujica, G., Escudero, D., Cifuentes, O.: Desarrollos portuarios ambientalmente sostenibles en el estuario de Bahía Blanca, Environmental Engineering Master's Thesis, Universidad Tecnológica Nacional Facultad Regional Bahía Blanca, 338 p, (2015). Repositorio Institucional Abierto (RIA), http://ria.utn.edu.ar/handle/123456789/996?show=full. Accessed 10 July 2017
4. CGPBB—Consorcio Gestión Puerto de Bahía Blanca—Homepage: Institucional/Servicios. http://puertobahiablanca.com/descripcion-general.html. Accessed 10 July 2017

5. INDEC—Instituto Nacional de Estadísticas y Censos—Homepage. Población. http://www.indec.gob.ar/. Accessed 14 Mar 2015
6. Universidad Nacional del Sur, Departamento Biología, Bioquímica y Farmacia: Resolución unánime del Consejo Departamental registrado bajo N° CDBBYF 611/1, Revista de la Asociación Argentina de Sedimentología. Vol. 10 N°1 ISSN 1853-6360. La Plata, Argentina (2011)

Considerations Regarding Ship Recycling

Natalia de Souza Ribeiro, Euler Sanchez Ocampo and Newton Narciso Pereira

Abstract Ship dismantling is a common practice at the end of the ship's life for the recovery of materials used in its structure. However, the main countries involved in this activity are located in underdeveloped countries that use the worst techniques, presenting hazards to workers and the environment. This work presents the opportunities that this activity can offer to the shipyards and the steel mills in the country. It is intended to address information on the practice of ship recycling, as the main countries involved, the main differences between practices in underdeveloped countries, and in developed countries. Information about the legislation created in recent years is presented in order to make this activity more sustainable, as well as the opportunities that this activity offers. In this context are shown the various materials present on the ship as well as their possible applications. It also addresses the current situation of Brazilian shipyards, which have experienced a severe economic crisis, which has led to the closure of yards and the dismissal of several workers. As such, the recycling of Brazilian vessels presents a potential market for shipyards, as well as offering a secondary steel market to the steel industries present in the country. The research work was done through a bibliographical study, surveys, mapping, and data collection, which allowed us to show the possibility of carrying out this activity in Brazil.

Keywords Recycling · Dismantling of vessels · Shipyards

N. de Souza Ribeiro · E. S. Ocampo · N. N. Pereira (✉)
Department of Production Engineering, School of Industrial and Metallurgical Engineering at Volta Redonda - EEIMVR, Federal Fluminense University, Volta Redonda, Rio de Janeiro, Brazil
e-mail: newtonpereira@id.uff.br

N. de Souza Ribeiro · E. S. Ocampo · N. N. Pereira
Center for Sustainable Systems Studies - CSSE(Eng.) - CESS(Port), Av. dos Trabalhadores 420 - Sala C77, Vila Sta. Cecília, Volta Redonda, RJ 27255-125, Brazil

A. Vega Sáenz et al. (eds.), *Proceedings of the 25th Pan-American Conference of Naval Engineering—COPINAVAL*, https://doi.org/10.1007/978-3-319-89812-4_29

1 Introduction

Currently shipping has become the main means of transport between countries and accounts for more than 90% of international trade in goods. According to data from [7] in 2016, 862 oceanic vessels were dismantled and 668 of these vessels were shipped to countries in South Asia.

The dismantling of ships around the world has generated concern due to its high environmental and social impact. Most of the ships arriving at the end of their useful life are sent to underdeveloped countries. The major problem is that environmental and worker protection laws are practically nonexistent or ignored in these places, thus dismantling ships is a dangerous practice [6].

Shipbreaking or breaking ships is an activity that occurs after the life of the ship in order to recover materials used in its construction, especially steel. This process is difficult and dangerous, so it is necessary to be carried out correctly, because the ships have in their structure various materials and wastes that are dangerous and that if not handled correctly present risks to the environment and to man.

The ship recycling industry is a potential source of steel, in addition to a multitude of items that can be recovered from the ship, such as machinery, equipment, and household accessories. The ship recycling industry allows almost all products to be reused, recycled, and resold in addition to providing employment opportunities directly and indirectly in the countries involved [4].

Despite the opportunities, the problems associated with this practice have led to the creation of regulations aimed at curbing risky practices. Among these regulations are the Hong Kong Convention (HKC), established by the International Maritime Organization (IMO), and the Regulation of the European Union, which follows the same standards as the HKC; however, this is aimed at vessels with a European flag.

This research presents a panorama of the recycling of ships around the world, the main countries involved in this practice both Asian and European, the social and environmental impacts of this activity, as well as the opportunities that this market can bring to Brazilian shipyards.

2 Methodology

To reach the research objectives, an exploratory research was carried out. The exploration research was done through a bibliographical survey of different international articles that deal with ship dismantling practices in the world. The impacts from the economic, social, and environmental points of view related to this global operation were identified, as well as the by-products and processes that involve the recycling of ships.

In addition, research was carried out on the current situation of Brazilian shipyards considering the resources available for the vessel recycling operation.

The shipyards were mapped based on their location using Google Earth. Likewise, the main steel mills existing in the country were mapped in order to verify the distance between the two facilities, which allowed us to verify the radius of influence of the shipyard with the steel mills.

To estimate the potential of this market in Brazil, a survey was carried out on the website of the National Agency for Waterway Transportation (ANTAQ), on the vessels registered in the country, and based on these data, a projection was made on the year in which vessels can be recycled. Firstly, the vessels were separated by type, and after this division the boats that had the same characteristics were grouped in the same category. After that, we added 25 years to the year in which each vessel was registered and with these data the vessels were divided by year of recycling from the year 2017.

In addition, based on the materials present on the ships, a search was made to identify which materials can or cannot be recycled and their possible destination. Finally, research was carried out on the national steel industry.

3 Results

3.1 Top Asian Dismantlers

Throughout the investigation we find two types of actors who do the dismantling of ships in the world, the yards that do the dismantling of ships on beaches and the shipyards that have practices specialized in ship dismantling. Every year, the NGO Shipbreaking Platform provides a list containing information about vessels that were scrapped during the year. This list has information such as the name of the vessel, the type of vessel, the country of origin, and the country of destination for the clearing. Based on these lists a survey of the data obtained in the last six years (2011–2016) was performed (Fig. 1).

According to these data, a total of 4258 vessels were dismantled in the world in the last six years, 2013 being the year in which it had the largest number of ships

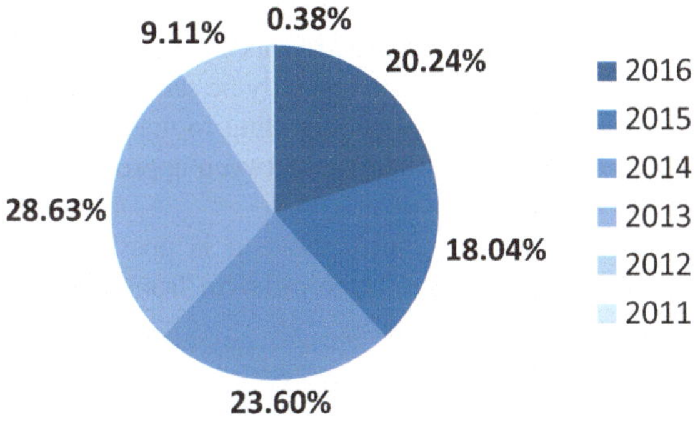

Fig. 1 Ships scrapped per year. *Source* Adapted from shipbreaking platform [5–7]

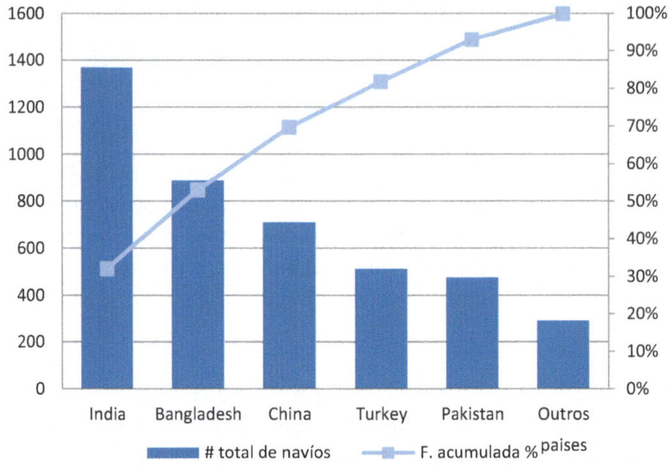

Fig. 2 Ships dismantled by destination country. *Source* Adapted from shipbreaking platform [5–7]

dismantled, representing 28.63%. It can also be observed that most of these vessels were sent to Asian countries. This information is shown in Fig. 2.

More than 80% of vessel recycling activity was concentrated in five countries: India, Bangladesh, China, Turkey, and Pakistan, India being the country that received the largest number of vessels. About 20% of this activity was carried out in other countries, with Denmark, Belgium, Lithuania, the United States, the Netherlands, and Indonesia as examples. Figure 3 shows the total dismantling in tons of light displacement (LDT). This weight is approximately equal to the weight of the ship's steel, disregarding cargo, fuel, ballast water, and others.

Alang in India was the location in Asia where most shipwrecks expressed in LDT in the last six years in Asian countries represented 59% of the total. When a more detailed analysis was carried out regarding the main break points of ships in each country, it was obtained (Fig. 4).

It has been found that the main destinations are Alang, India; Chitagong, Bangladesh; Aliaga, Turkey; and Gaddani, Pakistan. These sites represent the main areas of clearing in these countries. Figure 5 details these locations.

On the other hand, by the European Union Regulation for the Recycling of Ships the flagships of this union can only be sent for dismantling in shipyards certified by the European community, according to the 18 yards approved in December 2016. Figure 6 shows the difference between approved shipyards and Asians dismantling on the beaches.

In the figures obtained above it is possible to observe the organization and structure of these sites, quite different from what can be observed in the clearing sites on beaches in Asia.

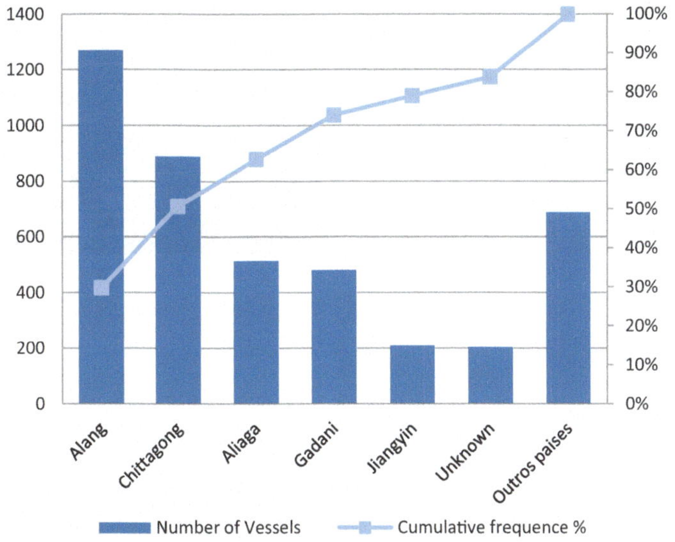

Fig. 3 Total in LDT dismantled in main shipyards. *Source* Adapted from shipbreaking platform [5–7]

Fig. 4 Dismantled ships by shipyard destination. *Source* Adapted from shipbreaking platform [5–7]

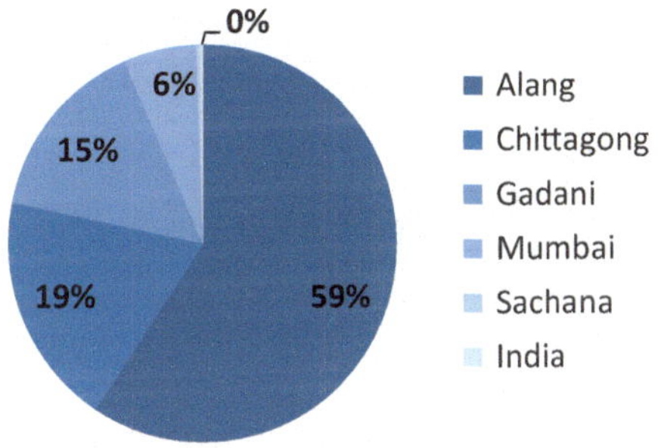

3.2 *Dismantling Materials*

A survey was carried out in order to know what materials may be present on the ship that may or may not be recycled and their possible reuse. This information is cited in Table 1.

Based on the information presented in Table 1, it can be inferred that most of the waste from ships can be recycled later. A small share of this waste does not have reuse capacity. Materials such as asbestos and PCBs are extremely hazardous and are not recyclable. The Brazilian shipyards need knowledge for correct handling, as well as the practices used by countries specialized in ship dismantling for the disposal of these materials.

Alang - India

Aliaga, Turkey

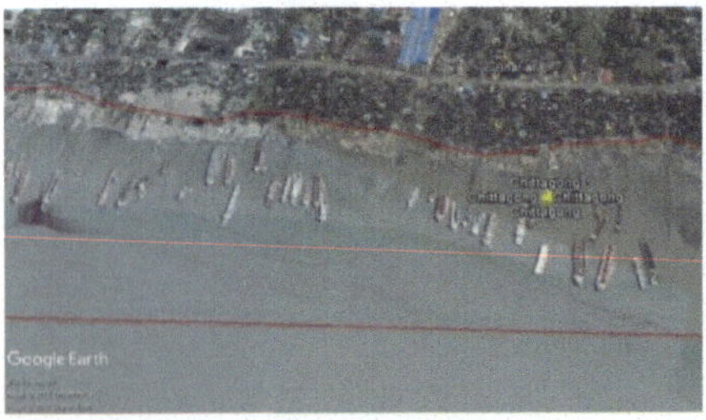

Chittagong, Bangladesh

Gaddani, Pakistan

Fig. 5 Main points of ship dismantling in Asia at beaches. Adapted from Google Earth®

Able UK Limited Teesside Environmental Reclamation and Recycling Centre

Fornaes APS, Denmark

NV Galloo Recycling Ghent, Belgium

Keppel – Verolme Prof, Holland

Fig. 6 Example of some yards approved by the European Union in December 2016. Adapted from Google Earth

Table 1 Ship dismantling materials

Type of material	Name of material	Recyclable?	Use
	Antimony (Sb) Berílium (Be)		1. Steelmakers 2. Electronics industry 3. Industry: household utensils, tools, auto parts, building structures, food and beverage cans; food cans, auto parts, steel for construction 4. Industry of: car batteries, seals; roofs and batteries
Metals and Alloys	Cadmium (Cd) Lead Mercury Tellurium (Te) Arsenic Hexavalent chromium Zinc Mixture of batteries and accumulators Electrical and electronic components	Yes	
Inorganic components	Glass, cathode ray tube Asbestos (powder and fibers)	Yes No	Identified by the acronym CRT (cathode ray tubes), they are recyclable materials. There are different ways to recycle this material, also, applications such as bricks, tiles, ceramic manufacturing, glass foam for insulation of new CRT and other. Of difficult reutilization besides being highly contaminating and dangerous for human health
Organic components	Mineral oils Nonhalogenated solvents Halogenated organic solvents Components with PCB	Yes No No Yes	Generation of electricity by burning Distillation and refinement of oil in diesel fuel or marine fuel Manufacture of lubricant N/A N/A 1. Electronics 2. Printed circuit boards
Substances with other organic and inorganic compounds	1. Oily waters and emulsions 2. Wastes of an explosive nature 3. Waste packaging 4. Chemical waste from ship operation; organic waste from ship operation	Yes Yes Yes Yes	Reverse logistics: internal and external markets, covering other market segments
Others	1. CFC 2. Halons 3. Radioactive material 4. Microorganisms/ sediments	No No Yes No Yes	They must be destined for incineration They must be destined for incineration

(continued)

Table 1 (continued)

Type of material	Name of material	Recyclable?	Use
	5. Liquid and gaseous fuels		Reuse of uranium, knowing that radioactive uranium contains 97% of reusable radioactive material Note: difficult to reuse, with 96% of uranium and 1% of plutonium N/A Reverse logistics: Generating electricity through burning

The table was prepared on the basis of [2, 3, 8, 9]

Likewise, it is fundamental knowledge of the regulations in force in the country on how to carry out the disposal of materials considered dangerous and the bodies responsible for such activity. In addition, a more detailed study of the possible destination for the residues from the dismantling, such as a possible market in the country that could absorb these materials.

3.3 Opportunities for the Brazilian Naval Sector

The following information was obtained through the database of the National Agency of Waterway Transportation (ANTAQ) [1]. The site offers information on the more than 2000 different vessels registered in Brazil. Among these vessels we can see oil tankers, bulk carriers, tugboats, and marine platform suppliers, among others (Fig. 7).

To obtain this graph, a survey was made of the year in which the vessel was registered and to this year was added 25 years, which is an estimated useful life of a vessel. From this projection was created the graph showing the quantity of vessels that have the potential to be recycled in the coming years starting in 2017. A large number of vessels may be recycled between the years 2032 and 2040. Ships smaller than 500 DWT were disregarded in the analysis due to the low capacity of total LDT input resulting from the structure of the ship.

The data presented in Fig. 7 show a potential market for different vessels. These vessels will reach the end of their useful life and will have to be dismantled, thus presenting an opportunity to the national shipyards to do the dismantling of Brazilian flag vessels. In this analysis all other registered vessels were discarded, whose date for recycling is prior to 2017, as it was not possible to identify whether these vessels are still in operation.

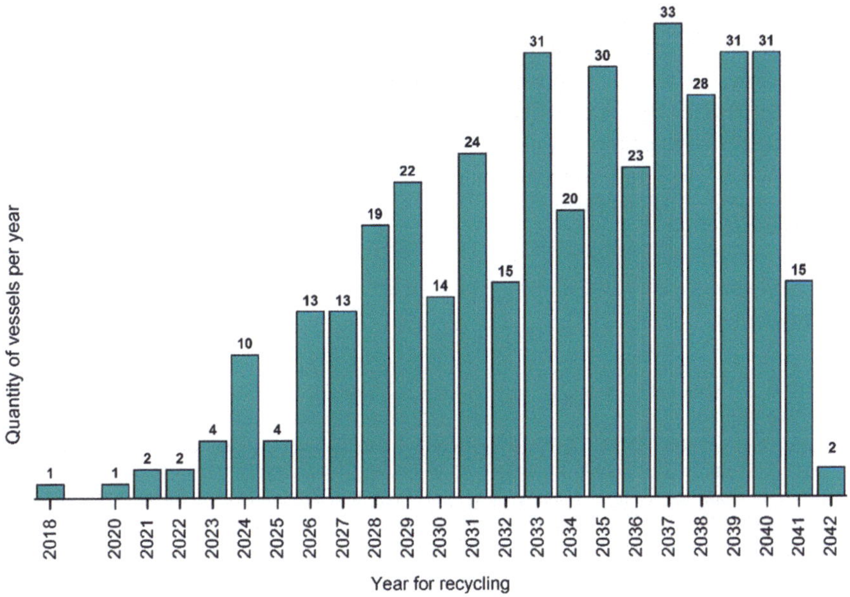

Fig. 7 Number of ships with potential to be recycled per year. *Source* Adapted from [1]

4 Final Thoughts

The recycling of ships is of extreme importance for the disposal and reuse of materials present in the vessels. Despite the advantages, this activity is still practiced in a dangerous way in Asian countries, involving several environmental and social problems, which generate concern in the international scenario. This activity allows tons of steel and other materials to be recycled or reused, thus offering an economic opportunity to these sites.

As seen, there are a considerable number of vessels registered in Brazil that could be recycled in the national shipyards. The recycling of ships can offer opportunities to Brazilian shipyards, as well as the national steel sector that has the capacity to obtain the main component of this activity, which is steel.

However, it is necessary for national shipyards to be prepared to handle hazardous materials from the process as well as to dispose of them correctly. Likewise, a better knowledge of current national legislation is needed on the correct destination of hazardous materials and the environmental aspects related to recycling. It is also necessary to carry out further research for the knowledge of markets that can obtain the other materials from the dismantling, such as the reuse of materials and what to do with the recycling materials.

In addition, a study on the operational costs of ship recycling at shipyards, such as labor, logistics, and materials costs to assess the feasibility of this activity, is indispensable. Similarly, it is important to know the technical demand required to carry out ship dismantling and how the shipyards should prepare to maintain the ship dismantling and construction activities.

Therefore, the recycling of national vessels may be the beginning of a promising market for Brazil. New legislation in force such as the Hong Kong Convention and the European Union Ship Recycling Regulation may allow for future international recognition of national shipyards and the possibility of recycling vessels from different parts of the world.

References

1. ABENAV. Associados. Available: http://abenav.org.br/web/dados-do-setor/associados/. Accessed 28 Apr 2017
2. Basel Convention, IMO, ILO: Hong Kong International Convention for the Safe and Environmentally Sustainable Recycling of Ships, pp. 1–41 (2009)
3. Eropean Maritime Safety Agency: EMSA's Best Practice Guidance on the Inventory of Hazardous Materials. European Maritime for Safety Agency, pp. 1–39 (2016)
4. Hiremath, A.M., Pandey, S.K., Asolekar, S.R.: Development of ship-specific recycling plan to improve health safety and environment in ship recycling yards. J. Clean. Prod. **116**, 279–298 (2015). https://doi.org/10.1016/j.jclepro.2016.01.006
5. NGO Shipbreaking Platform. List of ship dismantled worldwide in 2012. Available: http://www.shipbreakingplatform.org/annual-lists-of-scrapped-ships/. Accessed 12 Oct 2016
6. NGO Shipbreaking Platform. List of ship dismantled worldwide in 2015. Available: http://www.shipbreakingplatform.org/annual-lists-of-scrapped-ships/. Accessed 12 Oct 2016
7. NGO Shipbreaking Platform. List of ship dismantled worldwide in 2016 Available: http://www.shipbreakingplatform.org/annual-lists-of-scrapped-ships/. Accessed 12 Mar 2017
8. Secretariat of the Basel Convention: Basel Convention on the Control of Transboundary Movements of Hazardous Wastes and Their Disposal Adopted by the Conference of the Plenipotentiaries on 22 March 1989. Basel (1989)
9. UNEP: Technical Guidelines for the Environmentally Sound Management of the Full and Partial Dismantling of Ships. Sweden (2003)

Review of Renewable Energies for Naval Propulsion and Its Application in the National Fleet

P. La Paz and J. Freiria

Abstract For naval propulsion, as with other power generation plants, the use of renewable energies is a subject that demands more and more attention as fossil fuel reserve is in the process of being depleted in the medium or long term. On the other hand, no less important, the generation of energy from these fuels is one of the main sources of emissions of greenhouse and other toxic gases, such as sulfur oxide (the main cause of acid rain). It is necessary to reduce its use and increase the number of sources of green energy, minimizing the damage to the environment. Special attention is given to the application of wind and solar energy as renewable energy sources. The problem is to manage them efficiently and economically profitwise, allowing the cost to be below the cost of production with fossil fuels, taking into account the initial investment, maintenance, and needs of the boat.

Keywords Propulsion · Renewable energies · Solar energy · Wind energy

1 Introduction

Currently, 90% of world trade transportation is done by vessels. Although compared with other means of transport, the maritime one is among the most environmentally friendly, the magnitude of operations makes this industry an important emitter of greenhouse gases. It is estimated that in 2007, international maritime transport emitted 870 million tons, that is, around 2.7% of global emissions.

One way to reduce these emissions is to use renewable energy. This type of energy has not yet been generalized, given that it is in a period of improvements in technology to make them more attractive in relation to efficiency costs than the technologies currently used. At this moment, the use of renewable energies is being

P. La Paz (✉) · J. Freiria
Facultad de Ingeniería, Universidad de la República, Montevideo, Uruguay
e-mail: plapaz@fing.edu.uy

J. Freiria
e-mail: jfreiria@fing.edu.uy

A. Vega Sáenz et al. (eds.), *Proceedings of the 25th Pan-American Conference of Naval Engineering—COPINAVAL*, https://doi.org/10.1007/978-3-319-89812-4_30

promoted as an improvement for environmental conditions and not as an economic improvement, which entails the process [1].

As the production technology is being improved the negative aspects that led in the past to change wind energy by the use of steam from the combustion of coal first, and then fossil fuels, can be remedied.

Current technologies contain programs that can determine optimal routes depending on weather conditions, reorient the ship if deemed necessary, and calculate approximate arrival times. At the same time the mechanization and automation included allows keeping limited crews.

2 Naval Propulsion Through Wind Energy

Propulsion assisted by the wind is one of the oldest forms of ship propulsion. For centuries, maritime propulsion was carried out using wind power, with low load volumes and high labor contents. Then most of the ships began to use fossil fuel energy; coal steam turbines began to be used in 1800. Towards the end of the twentieth century interest in this energy returned due to the oil crisis. At present, the interest in renewable energies goes beyond an economic solution, with the aim of reducing emissions. Several countries and zones of influence have incorporated regulations on permitted emissions, a situation that requires accelerating improvements at the risk of being limited and losing competitiveness.

Propulsion assisted by the wind is mainly used as auxiliary energy using different systems: rigid sails, rotors, comets, and wind turbines among others. This type of energy allows the reduction of the amount of emissions of combustion gases because they maintain the same speed of the ship decreasing the necessary power of the engine or because, while maintaining the power of the engine, it increases the speed, which delays the journey less [1].

2.1 Rigid Sails

2.1.1 Generalities

Rigid sails are aerodynamic profiles built with different materials, such as metal plates or composite materials.

The amount of rigid sails to be placed on the deck depends on the type of vessel, its dimensions, and cargo stowage among others. The sails can be integral or have a vertical telescopic system that can be lowered in adverse weather conditions or in restricted operating conditions. However, this solution adds considerable mechanical complexity and the possible occupation of spaces that could affect the profitability of the vessel [1, 2].

2.1.2 Basis

The wing profile of the sails grants the fundamental characteristics of the aerodynamic performance, being used in a generalized NACA profiles way.

The main forces developed in a wing profile when a fluid with a certain angle of attack falls on it are: the lift force (drag) and the drag force; the total aerodynamic force is the vector sum of both forces, as shown in Fig. 1.

The lift force is produced by pressure differences on both sides of the profile and is perpendicular to the apparent wind speed direction. The equation that describes this force is:

$$L = \frac{1}{2} \cdot \rho \cdot C_{\mathrm{L}} \cdot A \cdot V_{\mathrm{A}}^{2} \tag{1}$$

where ρ is the density of the air, A is the area, V_{A} is the apparent wind speed, and C_{L} is the lift coefficient.

The drag force is parallel to that of the apparent wind flow opposing the advance of the profile. The equation that describes this force is:

$$D = \frac{1}{2} \cdot \rho \cdot C_{\mathrm{D}} \cdot A \cdot V_{\mathrm{A}}^{2} \tag{2}$$

where C_{D} is the drag coefficient; it depends on both the geometry of the profile and the angle of attack.

Aerodynamic efficiency is defined as the ratio between the lift coefficient and the drag coefficient. It is a useful measure when quantifying the behavior of the various systems at certain angles of attack and apparent wind speed.

The sails are symmetrical because they have to be oriented in both bands. They are accompanied by a system of ailerons that seek optimal configuration through the optimization of the angle of attack to increase performance. The adjustment mechanism must be automated, keeping control from the bridge in order to make decisions manually.

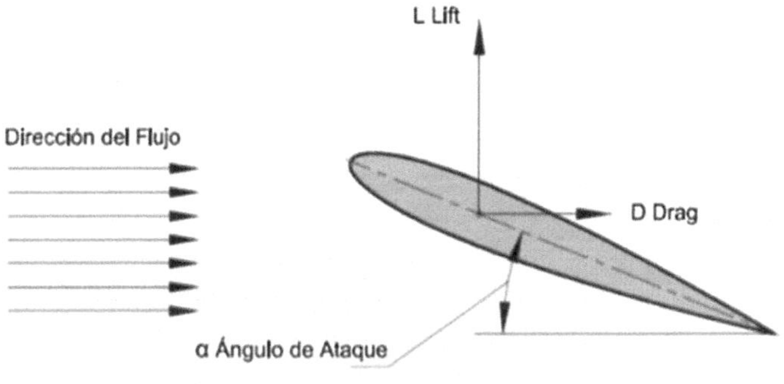

Fig. 1 Forces that act on the profile when influencing a fluid on it [12]

The number of candles to be installed must be studied through CFD programs in order to verify the interference between them that could result in lesser use [1, 7].

2.2 Flettner Rotors

2.2.1 History

In the early 1920s, the German engineer Anton Flettner invented the rotor that bears his name, vertical cylinders whose operation is based on the Magnus effect, which is based on a cylinder in rotation, where the air current through generates a force that translates into a lifting force and a drag force.

In 1924, under the supervision of Flettner, the Buckau ship (Fig. 2) in Germany was adapted with this technology, being the first to use it, making its first trip in 1925 by the North Sea. It consisted of two rotors of 3 m diameter, 15 m high driven with an electrical system of 37 kW, rotating the rotors at 120 rpm. According to studies carried out, the power provided by the rotors was approximately 1000 hp, achieving a speed of 8 knots. At that time it was much more efficient than the sails, because the weight of the sails and the rigging were greater than the weight of the rotors and at the same time these generated more power than the sails with the same exposed area [3, 13].

The second vessel to adopt this technology was in the United States and the third in Germany; in 1926 the German navy ordered the construction of the ship "Bárbara" of 3000 tons with three rotors, sailing between 1926–1929 as a freighter in the Mediterranean Sea [3, 13].

Currently, a German company dedicated to the manufacture of wind turbines called Enercon built the ship E-Ship 1, which made its first trip in the year 2010. This vessel uses a hybrid diesel–electric propulsion system and contains four Flettner rotors with automatic control systems; the rotors have a height of 27 m and

Fig. 2 The ship Buckau, the first using Flettner rotors [3]

a diameter of 4 m. The gases produced by the combustion of diesel engines are used to power a turbine that generates the electricity used to rotate the four rotors. The savings are approximately 25–30% of fuel, of which 15% is directly attributable to the use of wind energy [6].

At present, international attention is focused on large-scale maritime transport for the use of this technology, but it is also viable in medium and small-sized vessels, provided they have the room available on the deck to place the rotors [9].

2.2.2 Magnus Effect

To understand the operation of the rotors it is necessary to explain the "Magnus effect". It is given by the rotation of an object in a medium where there is a fluid in motion. By turning the cylinder immersed in a stream of viscous fluid, differential accelerations are generated in opposite sides (Bernoulli), which gives rise to a lift and drag force.

2.2.3 Variables

For the efficient use of the rotor (Fig. 3) there are some parameters that can be controlled to vary the relationship between the lift coefficient and the drag coefficient, some of design and other operatives; they are: spin ratio, aspect ratio, and Thom's disc [4].

The spin ratio (SR) is the ratio between the speed of rotation (Ω), the diameter of the rotor (d), and the apparent wind speed (V_A) expressed as

$$\mathrm{SR} = \frac{\Omega d}{2V_A} \tag{3}$$

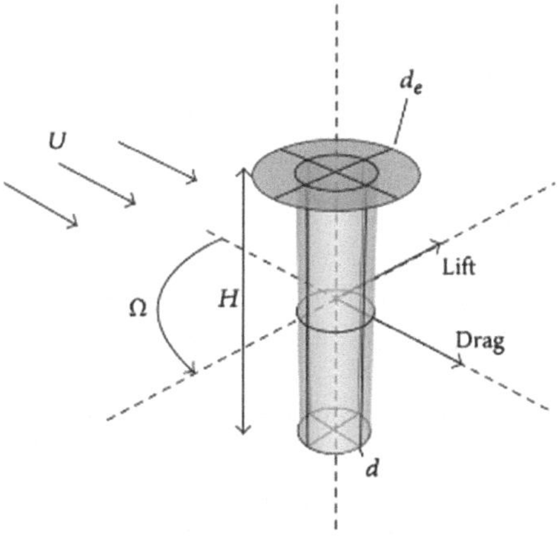

Fig. 3 Principal parameters for Flettner rotors [4]

Generally, the values of interest for the maritime application are in the range of $1 \leq \text{SR} \leq 3$.

The aspect ratio (AR) is the ratio between the rotor length and rotor diameter. For high relationships, the rotors behave like wings, generating vortices that help lift. The lower the values of AR, the lower the lift and the lower the speed ratio reached [4].

Thom's disc is used to optimize aerodynamic efficiency. The presence of this plate modifies the flow in the upper part of the rotor, increasing its effective aspect ratio. For high speeds you can increase the lift to nearly double. When increasing (d_e/d) the rotor generates a higher lift and this value also occurs in higher SRs [4].

It was presented in a Seifert study how the diameter of the Thom disc is related to SR to obtain optimal rotor performance. He observed that at low SR the smaller plates generate less drag; for moderate SRs larger disks are preferred and for smaller SR ratios smaller plates are preferred [4].

In the graphs shown in Figs. 4, 5 and 6, the behavior of lift coefficients, drag coefficients, and the relation between them with respect to the variations of the aforementioned parameters are studied [4].

By doing an analysis, it can be noted that the point of greatest efficiency is defined by SR = 2, AR = 8, and $d_e/d = 3$.

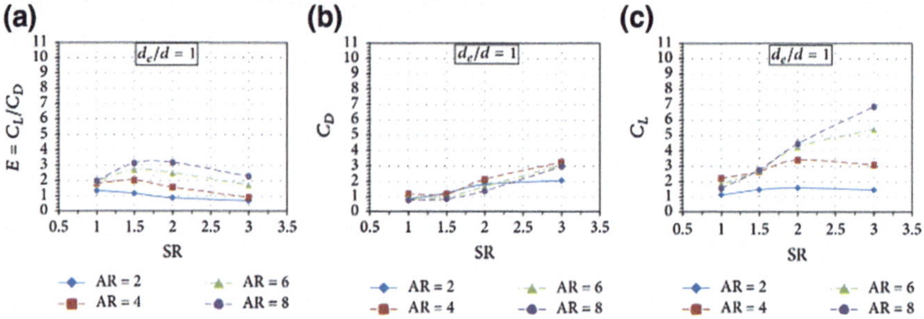

Fig. 4 Behavior of Flettner rotor without Thom disc [4]

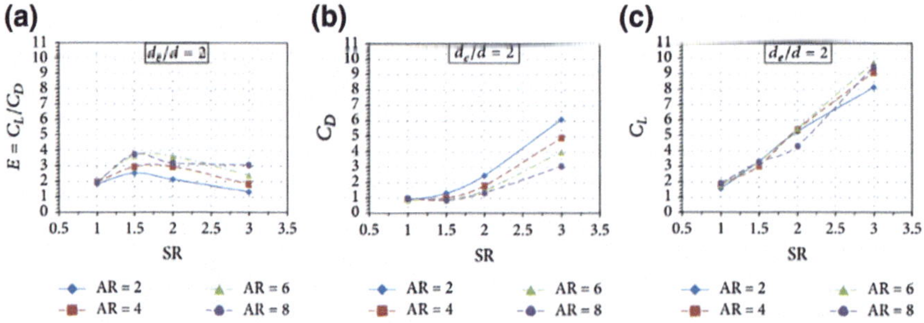

Fig. 5 Behavior of Flettner rotor with Thom disc [4], $d_e/d = 2$ [4]

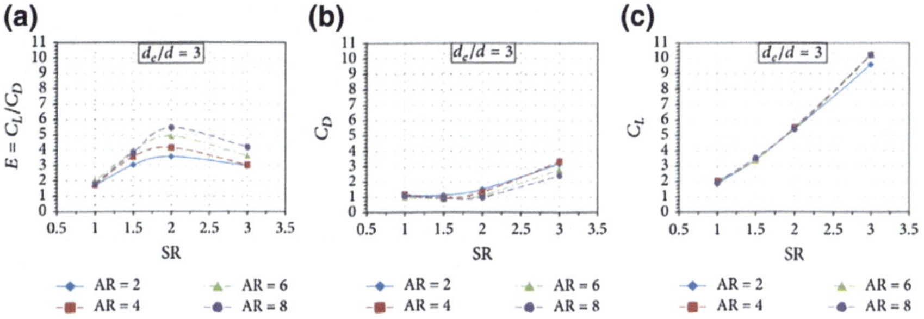

Fig. 6 Behavior of Flettner rotor with Thom disc [4], $d_e/d = 3$ [4]

Other conclusions that can be obtained are that an increase of AR translates into an increase in efficiency, and as the ratio between the diameter of the Thom disc increases and the diameter of the cylinder, when the lift coefficient varies according to SR, the curves for the different ARs become similar.

2.3 *Kites*

The tow kites can be treated as a surface of a wing, so you can use the aerodynamic concepts seen above.

The main idea is the use of high winds, which are higher than on the surface, consequently with greater energy input.

The most outstanding advantages of this system are indicated below [2]:

- Stowage facility with minimum occupation of spaces.
- The height of flight allows generating up to 25 times the energy provided by conventional sails, achieving savings quantified between 10 and 35%.
- They have relatively low investment costs.
- They have automatic control, are easy to handle, and have high reliability.
- They generate minor heeling moments over conventional candles.

As disadvantages, it can be highlighted that:

- They cannot be used with winds whose directions do not generate lift.
- They cannot be used with low-speed winds.
- They cannot be used in dense maritime traffic areas for reasons of safety.

The Skysails system represented in Fig. 7 consists of three parts: a tow kite with its rigging, a launch and recovery system, and a control system for automated operation. The equipment is placed in the forecastle. The kite is made of highly resistant textile fibers. For launch, the kite is lifted by the telescopic mast; the winch releases the tow rope until the kite has reached its operating altitude [11].

Fig. 7 Ship kite

The control console directs the device in order to perform different regular dynamic flight maneuvers in the air to generate propulsion. This force, as well as the cables of the control system, is transmitted to the ship by means of a high-resistance shaft [11].

2.3.1 Flight Envelope

The flight envelope is defined as the locus where the kite can fly; it is schematized in general with a quarter of a sphere (Fig. 8).

A kite can be arranged to perform a static or dynamic flight. In the dynamic flight, the kite is in continuous movement describing a trajectory in the form of eight inside the flight envelope, in order to increase the tension in the rope and therefore the power delivered to the ship [8].

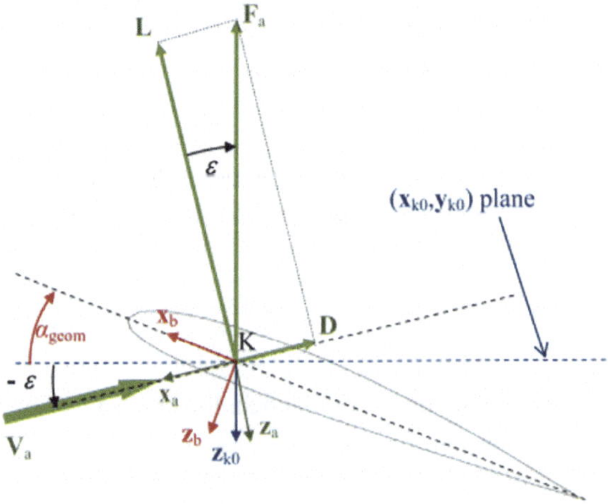

Fig. 8 Reference systems for kite forces [8]

The reference systems in which it will be working and the speed relationships are described below.

1. Kite reference system (R_{KO}): Originates from the center of comet K; the unit vector X_{KO} corresponds to the direction of the comet's speed [8].
2. The aerodynamic reference system (R_{A}): Originates from the center of the kite, is defined in relation to the apparent wind speed and aerodynamic forces, the X_{A} is parallel and opposite to the vector of apparent wind speed and force of drag, and the Z_{A} is parallel and opposite to the lift force. It is assumed that the drag and lift forces are in the plane of symmetry of the kite [8].

To calculate the tension in the ropes, the method of zero mass is used, which implies that the weight of the kite is disregarded against the other forces (tension, drag, and lift); this is a valid hypothesis for this type of job:

$$\vec{T} + \overrightarrow{F_{\mathrm{A}}} = 0 \tag{4}$$

$\vec{T}$ is the tension in the strings and $\overrightarrow{F_{\mathrm{A}}}$ the aerodynamic force.

Both forces are aligned on the same axis that goes from the point of grip of the ropes to a point belonging to the kite.

Another equation that governs the movement of the kite is:

$$\overrightarrow{V_{\mathrm{A}}} = \overrightarrow{V_{\mathrm{WR}}} - \overrightarrow{V_{\mathrm{K}}} \tag{5}$$

where $\overrightarrow{V_{\mathrm{A}}}$ is the apparent velocity, $\overrightarrow{V_{\mathrm{WR}}}$ is the relative speed of the wind, and $\overrightarrow{V_{\mathrm{K}}}$ is the speed of the kite. The relative wind speed can also be written as $\overrightarrow{V_{\mathrm{WR}}} = \overrightarrow{V_{\mathrm{WT}}} - \overrightarrow{V_{\mathrm{S}}}$, where $\overrightarrow{V_{\mathrm{WT}}}$ is the actual wind speed and $\overrightarrow{V_{\mathrm{S}}}$ the speed at which the ship is sailing.The aerodynamic force can be defined as the sum of the lift force and drag force; using Eqs. (1), (2), and (4) you get:

$$\vec{T} = \frac{1}{2} \cdot \rho \cdot C_{\mathrm{L}} \cdot A_{\mathrm{k}} \cdot V_{\mathrm{A}}^{2} \widehat{z_{\mathrm{A}}} + \frac{1}{2} \cdot \rho \cdot C_{\mathrm{D}} \cdot A_{\mathrm{k}} \cdot V_{\mathrm{A}}^{2} \widehat{x_{\mathrm{A}}}$$

Because ε is the angle between the aerodynamic force and the lift force, the drag force can be defined as $D = L \bullet \tan \varepsilon$; therefore the tension would be:

$$\vec{T} = \frac{\frac{1}{2} \cdot \rho \cdot C_{\mathrm{L}} \cdot A_{\mathrm{k}} \cdot V_{\mathrm{A}}^{2}}{\cos \varepsilon} \widehat{z_{\mathrm{ko}}} \tag{6}$$

In this way the developed tension is defined; the component that interests us is the collinear to the direction of the ship; the force projected on the other axis would be a drift force [8].

Equation (6) is determined when the apparent speed is constant throughout the flight, which is more approximate in the static flight kites, but in the case where the kite is out of dynamic flight these forces are integrated with respect to the time along the flight path to obtain average values in a given trajectory. This allows comparing the efficiency of dynamic and static flight with the average propulsion force [8].

3 Naval Propulsion Through Solar Energy

Photovoltaic solar energy is one that is obtained from the transformation of solar energy into electrical energy. The photovoltaic solar panels are constituted by photovoltaic cells that consist of a semiconductor material, generally silicon, with impurities in their faces that provoke opposite charges in these. When solar irradiation hits the panel, the electrons are forced to move from one face to the other, causing a direct current and a potential difference between both faces of the panel to be generated.

Photovoltaic solar energy can be used in at least two different ways.

3.1 Direct Energy Transformation

The panels distributed on the deck and other parts of the ship are used for charging batteries and their use for powering the electric propulsion engines.

The panels can be placed on the roof or in a flat sky exposed to solar radiation; the system usually includes a DC to AC converter and a frequency converter that allows them to vary the speed of rotation. They can have a battery charger to charge them in case of being in port, according to needs and availability of the energy [10].

3.2 Indirect Application: By Obtaining Hydrogen Gas

Water is divided into its chemical components, generating gases with high combustion power: hydrogen and oxygen.

This system can either be placed on board the ship or on land, and the hydrogen obtained is either stored directly on the ship or transported to the ship to be used as fuel [5].

Another alternative to be implemented is the use of solar energy in conjunction with wind energy, from the placement of solar panels on rigid sails. The sails can rotate in two axes: one of them allows the elevation (the angle 0° corresponds to the panel in an axis parallel to the water, and 90° in a plane perpendicular to it) and the other one allows it to change the angle of attack depending on the apparent wind direction. This allows them to function as a sail, a solar panel, or both at the same time.

As efficiency of the solar panels is less than 25%, a lot of space is needed to obtain the required power for electrical motors that propel the ship; they are used for auxiliary services only both during navigation or port applications [1].

4 Application in National Fleet

The Uruguayan fleet is of medium and low size; it would be composed mainly of traffic vessels, port tugboats, and fishing vessels.

The technologies that would be most suitable to install in national vessels would be the rigid sails combined or not with solar energy, Flettner rotors, and solar energy as the objective of producing hydrogen through hydrolysis.

The first two options require changing the layout of the ship, because it is necessary to place them on deck, thus it is a solution for new construction.

The other option would be viable in the case of being able to transport and carry out the safe combustion of hydrogen. It is necessary to make changes in the internal combustion engine and in how the fuel is stored, but the changes would not affect the structure of the ship.

Kites are not applicable in the Uruguayan fleet because these vessels need great maneuverability and are used for long-distance trips; they generally sail in areas of great maritime traffic, therefore it is not justified to deploy thc system.

References

1. Cachaza, L., Pérez López, G., Villa Caro, R., Fraguela, J.: Reducción del Consumo de Combustible Mediante Velas Rígidas y Paneles Solares. In: Conference: XXIV Congreso Panamericano de Ingeniería Naval, Transporte Marítimo e Ingeniería Portuaria, At Montevideo-Uruguay, Volume: COPINAVAL (2015)
2. Colomar Lix, G.: Medios de Propulsión Eólica Alternativos a la Vela Rígida. Tesis de Grado, Facultad Náutica de Barcelona (2012)
3. Craft, T.J., Iacovides, H., Launder, B.E.: Dynamic Performance of Flettner Rotors with and without Thom Disc. University of Manchester, Manchester, UK (2011)
4. De Marco, A., Mancini, S., Pensa, C., Calise, G., De Luca, F.: Flettner Rotor Concept for Marine Applications: A Systematic Study. Department of Industrial Engineering, University of Naples Federico II, Via Claudio 21, 80125 Naples, Italy. Received 19 January 2016; Revised 31 May 2016; Accepted 15 June 2016. Academic Editor: Ryoichi Samuel Amano (2016)
5. Jiménez Ariza, J.M.: Modelado de un Sistema Híbrido con Energías Renovables para Propulsión Eléctrica en Embarcaciones. Universidad Politécnica de Cartagena. Aplicación a una Plataforma Flotante Robotizada, Tesis de Grado (2016)
6. National Port Administration (ANP) Homepage. http://www.anp.com.uy/inicio/novedades/noticias/e_ship_uno. Accessed 19 Jan 2018
7. Ouchi, K., Uzawa, K., Kanai, A.: Huge Hard Wing Sails for the Propulsor of Next Generation Sailing Vessel. Graduate School of Engineering, The University of Tokyo, Japan. ACT Corporation, Japan (2011)

8. Leloup, R., Roncin, K., Behrel, M., Bles, G., Leroux, J. B., Jochum, C., Parlier, Y.: A continuous and analytical modeling for kites as auxiliary propulsion devoted to merchant ships, including fuel saving estimation. France (2015)
9. Nuttall, P., Kaitu, J.: The Magnus Effect and the Flettner Rotor: Potential Application for Future Oceanic Shipping (2016)
10. Salas, M., Luco, R., Ekdahl, H.: Energía Solar para Propulsión de Embarcaciones Fluviales. Universidad Austral de Chile, Valdivia, Chile (2013)
11. SkySails Homepage. http://www.skysails.info. Accessed 19 Jan 2018
12. Torres Saez, D.: Diseño y Análisis Aerodinámico de Sistemas de Propulsión Mediante Vela Rígida. Tesis de Grado. Universidad Austral de Chile, Facultad de Ciencias de la Ingeniería. Escuela de Ingeniería Naval (2013)
13. Traut, M., Gilbert, P., Walsh, C., Bows, A., Filippone, A., Stansby, P., Wood, R.: Propulsive Power Contribution of a Kite and a Flettner Rotor on Selected Shipping Routes. University of Manchester, UK (2013)

The Abandonment of Ships: Consequences for the Crew and for the Ship

Asunción López Arranz, Raul Villa Caro, José Angel Fraguela Formoso and José de Troya Calatayud

Abstract The purpose of this treatise is to analyze what happens with those ships that remain for more than six months berthed, moored, or anchored in the same place without noticeable activity and how they are regulated by national and international laws. All this has important consequences for both the crews and the ship. Undoubtedly the abandonment puts the crew in a difficult situation, cutting off their income and increasing their expenses. On the other hand, with regard to the consequences for the ship, once the abandonment procedure has been resolved, it will be sold in public auction and the proceeds of the sale will be deposited in the public treasury. In short, the consequences of abandonment can be personal and material not counting situations of an environmental nature. The methodology used in the analysis involves the use of scientific, legal, and social methods.

Keywords Abandonment · Ship · Crew

1 Introduction

In Spain the abandonment of ships is a widely known issue both for its problems and the number of ships abandoned in their ports. As an example, currently in Spanish ports there are 11 ships that are abandoned as a result of failure to pay the crew or other types of administrative problem or other difficulties, as shown by the International Transport Workers Federation (ITF). Among the ports are Algeciras with four abandoned ships, Ceuta with two and Tenerife, Tarragona, Valencia, Las Palmas de Gran Canarias, and Almeria, all of them with one each. At the same time and in the matter of recreational-type boats in Spain there are more than 10,000 abandoned boats. All this generates an important problem that must be resolved within different economic, personal, security, and environmental areas. The problem has been accentuated notably by the economic crisis that we have been suf-

A. L. Arranz (✉) · R. V. Caro · J. A. Fraguela Formoso · J. de Troya Calatayud
Universidade da Coruña, Campus de Esteiro, 15401 Ferrol, A Coruña, Spain
e-mail: a.larranz@udc.es

A. Vega Sáenz et al. (eds.), *Proceedings of the 25th Pan-American Conference of Naval Engineering—COPINAVAL*, https://doi.org/10.1007/978-3-319-89812-4_31

fering since 2008, which has affected all countries given its globalized aspect. It can be said that the economic crisis has caused a clear phenomenon of reduced flow of containerized cargo and what flow remains is distributed among fewer and fewer ports. In many cases of conventional general merchandise, shipping companies have suffered large economic losses that result in ships being abandoned [1].

In this discussion only one kind of abandonment is being analyzed, that which is produced by the serious breach of the obligations assumed by the shipowner and proprietor. Others exist such as the case that occurs as a result of a serious accident [2]. Our study is carried out on ship abandonments produced in most cases by guilty or negligent causes of the shipowners and owners, usually as a result of an economic crisis.

The approach to this problem must be multidisciplinary, given that the abandonment of a ship has administrative, commercial, labor, technical, and environmental consequences. In addition there is a variety of parties involved, such as state administrators, workers (i.e., the crew), boat owners, technicians, and so on.

It is evident that the most serious repercussions of a ship abandonment are for the crew given that this is the most vulnerable group and therefore the one that can have greater difficulties to solve as a result of the abandonment, even to the point of endangering their subsistence and that of their families [3]. The abandonment of a ship is very similar to the abandonment of a work center by a businessman. The big difference here is that there are national and international laws that may come into play as well as the loopholes provided by flags of convenience [4]. Thus the consequences for this group of workers are those from nonpayment of wages and also, in most cases, lack of food or fuel for the ship and the inability to be repatriated, all of which can be considered as violations of human rights [5].

The objective of this work is to analyze the abandonment of a ship and its consequences for all parties under a legislative and doctrinal analysis.

2 Abandonment Concept and Procedure

The Royal Legislative Decree 2/2011 of September 5, which approves the Revised Text of the Law of State Ports and the Merchant Navy, defines ***abandoned*** in Article 302.2 as "those ships that remain for more than three months berthed, moored or anchored in the same place inside a port without externally appreciable activity, and without having paid the corresponding fees or tariffs, and is so declared by the Board of Directors of the Port Authority." But the concept of abandonment in a legal sense will be that situation of serious noncompliance by one of the parties, specifically the proprietor and shipowner, of the obligations agreed upon with its workers and in general with all the commitments made. Generally this noncompliance translates into nonpayment of wages, not repatriating the crew, and not covering the basic needs of the crew, such as food, ship supplies, and the like [6].

The declaration of abandonment will require the processing of the corresponding procedure. This declaration will establish the circumstances of the assumed abandonment and the owner, the shipowner, the captain of the vessel, or, where appropriate, the consignee, will be called to a hearing as provided for in Law 39/2015, of October 1, of the Common Administrative Procedure of the public administrations. This is usually a difficult procedure, because the Port Authority must collect the corresponding contact information from the flag countries in order to make the notifications, and in most cases their whereabouts cannot be determined [7]. As the Law indicates in articles 40–45, the procedure begins once the abandonment is accredited, calling the owner to a hearing. If the notification attempt is unsuccessful, whether it be because the address is unknown or if the notification is ignored or if the process of localization is unable to be carried out, the notification will be made through an announcement published in the "Official State Bulletin." Thus, previously and with an optional nature, the Administrations may publish an announcement in the Official Autonomous Community or the Province Bulletin, on the bulletin board of the town hall of the last known address of the interested party, or in the consulate or consular section of the corresponding embassy. The Public Administrations may establish other forms of complementary notification through the other means of dissemination, which shall not exclude the obligation to publish the corresponding announcement in the "Official State Bulletin." The notification will open a period of 15 working days in which the notified parties can appear before the corresponding port authority, make claims corresponding to their rights, and present the required documentation.

Once the abandonment of the ship is declared by the Board of Directors of the Port Authority, the process will continue. Either the ship will be sold in public auction or sunk. In the case of the public auction, the proceeds will be transferred to the Public Treasury after deducting the credits accrued in its favor for the corresponding charges and port fees [8], as well as the costs of the procedure. But if the ship is in such bad condition that it endangers maritime safety, it will be sunk. This situation and the need to be able to carry out auctions to pay off expenses accrued in accordance with the Geneva Convention of May 6, 1993 [9], led almost all the Port Authorities in the clauses of the auction documents to determine the criteria for distribution of the Agreement as well as the notifications of the auction. This procedure has been established in Law 14/2014, of July 24, on Maritime Navigation [10], which in its Article 480 states that.

[T]he forced sale of the ship will comply with the provisions of the Law of Civil Procedure or in the administrative regulations that result from the application for the auction of movable assets subject to registration publicity in everything not covered by the International Convention on Maritime Privileges and the Naval Mortgage, made in Geneva on May 6, 1993, and in this law [11].

And it establishes the requirements to be met in Article 481, so that before proceeding to the compulsory sale of the vessel, the competent judicial or administrative authority will provide notifications of said sale to the following parties.

(a) to the registered owner of the Registry of Movable Property, as well as to the competent authority responsible for registering the ship in the State that would have authorized it to fly its flag temporarily, if applicable; (b) to the person who has registered the ownership of the vessel in his favour; (c) to all the holders of registered mortgages or liens that have not been constituted to the bearer; (d) to all the holders of registered mortgages or liens constituted to the bearer and of the maritime privileges enumerated in Article 4 of the International Convention on Maritime Privileges and the Naval Mortgage, provided that the judge or competent administrative body had received notification of their respective credits.

It can happen that, if there is no exhaustive compliance with the rule regarding the distribution in accordance with the provisions of the Geneva Convention (only the entry into the Public Treasury), the representatives of owners, shipping companies, or owners' banks of real rights will be allowed to make the impugnation of the same, given that if the requirements of distribution of the Agreement were not fulfilled, this Agreement could not be also invoked for the extinction of the pre-existing charges. All these extremes have already been overcome both through administrative practice and through some judicial resolutions, such as the order of the Provincial Court of Barcelona 226/2008, of June 26 (AC 2008/1711), in whose foundations it is pointed out that " …the objection made at the time by COMA Y RIBAS, SL, that these requirements had not been met because the deadline for prior notification of 30 days before the forced sale, provided for in Article 11.2 of the Convention (RCL 2004, 1066) had not been respected." To frame this objection, we must start from what is regulated in Article 12.1 of the Convention, where it indicates that the forced sale of the vessel causes the lifting of the liens held on the vessel so that "the sale has been made in accordance with the legislation of that State and with the provisions of the article 11 and in the present article." Article 11 of the Convention, in its first paragraph, provides that.

Before the forced sale of a ship by a State Party, the competent authority of that State Party shall ensure that notifications are made in accordance with the provisions of this article: (a) to the authority responsible for registration in the State of registry; (b) to all the beneficiaries of obligations, mortgages or registered liens that have not been constituted to the bearer; (c) to all the beneficiaries of the registered mortgages, obligations or encumbrances constituted to the bearer and to all the holders of the maritime privileges listed in Article 4, provided that the competent authority in charge of proceeding with the forced sale receives notification of their respective credits; and (d) to the person who has the ownership of the vessel registered in his favour.

It is not being denied that this notification was made or that it contained the information required by the second section of this same Article 11. The objection is that it has not been done with the minimum notice provided in said second section, "at least thirty days before the forced sale …." This notification, according to Sect. 3 of Article 11, must be in writing, "either by registered mail, or by any means of electronic communication or other suitable means that provides a receipt acknowledgment, to the interested persons indicated in paragraph 1, if known"; and in any case, "by edicts published in the newspapers of the State in which the forced sale is made and, if the authority that proceeds to the forced sale deems it convenient, in other publications." It is argued that the publication in the BOE, on July 5,

2007, is not earlier than 30 days before the date of the auction, on July 30, 2007, which presupposes that the publication as imposed by the Convention (RCL 2004, 1066) must be by official newspapers. In fact it is not like that because Article 11.3 refers to newspapers of the state in which the sale is made, and that purpose can be served by publications that provide general widespread information and by those specialized in maritime information in order to ensure a broad knowledge of the announced sale. The Port Authority has accredited documentary evidence that the edict announcing the sale of the ship and its conditions were published in the following newspapers: in *El País*, on June 29, 2007; in *Diario del Puerto*, on June 29, 2007; and in *Maritimes* on June 29, 2007 (pages 127 to 129). With this, the Port Authority complied with the advertising requirements imposed by Article 11 of the Agreement, so that the publication in the BOE, of the announcement of the auction on July 5, 2007, does not vitiate, for the purposes of Article 12 of the Convention, the compulsory sale nor does it prevent deploying its effects on the levies that weigh on the ship. It is not up to us now to determine the order in which the credits of the creditors concurrent with the proceeds of the sale of the vessel must be satisfied, because no ruling was requested from the judge in that sense, nor, in principle, was the same in that proceeding. The only thing that is clarified in this case is the origin or not of lifting the preventive arrest agreed by the Commercial Court, based on the invocation of a maritime credit, which may also be maritime privilege. And the lifting proceeds as a result of the forced sale of the vessel, in application of Art.12 of the 1993 Geneva Convention (RCL 2004, 1066). The forced sale not only causes the embargo, as encumbrance of the vessel, to be without effect (Article 12.1 of the Agreement) but also that does not allow for the right of reipersecutoriety (Latin: ius persequendi) that the Article 8 recognizes as a maritime privilege. Therefore, regardless of whether the maritime credit invoked to agree to the seizure were a maritime privilege, the verification of the compulsory sale of the vessel must result in the lifting of said encumbrance and that the maritime privilege is not required against the beneficiary.

3 Consequences for the Crew

The abandonment of a ship in the vast majority of cases is caused by the economic problems of the proprietor and shipowner, which leaves both the ship and its crew in a totally vulnerable situation. In order to understand the fate that these workers tend to have, it is necessary to start from the analysis of this peculiar activity, which is work at sea and their working conditions, to then assess the serious damage they suffer when they are left to their own devices in any port. In many cases, we would be talking about the violation of human rights [12].

The crew of a ship, excluding warships, is composed of workers in the service of a shipowner generally linked by an employment contract that meets all the characteristics of the contract, as established in Article 1 of the Royal Legislative Decree 2/2015, of October 23, which approves the revised text of the Workers' Statute Act, which states that "it will apply to workers who voluntarily provide paid services for third parties and within the scope of organization and direction of another person, physical or legal, called employer or businessman" [13]. If there is no doubt about the nature of the contract and the type of work of the crew, the third fundamental element is the place where the service is performed. The principal characteristic of this element is its mobility, because its location varies, and that makes it a special work center.

Another peculiarity of this activity is that the provision of services is carried out under different jurisdictions, which produces situations of normative conflict and lack of protection for these workers. In addition to having to carry out a job in conditions of isolation, hardship, and danger, they are poorly covered because of the conflict of the norms of applications of the states. This is especially problematic when navigating with flags of convenience, which determine the application of the so-called "ship's flag law," built on the fiction that the ship is an integral part of the state. That leads, in principle, to the application of labor and social security legislation of the nationality of the vessel to the seamen aboard [14]. The harsh working conditions of seafarers are exacerbated by the difficulties in applying national, international, or community standards. This is why it is important to apply international standards that oblige shipowners to maintain decent working conditions and protect the health and well-being of sea workers [15]. Currently it is estimated that 83% of seafarers from all over the world come from developing countries that allows us to know that their working conditions and salaries are not adequate. There is also an important difference between officers and subordinates. The latter represent approximately 67% of the total of the seafarers who are miserably paid in most cases (Table 1).

It is estimated that around 60% of the international fleet, including all types of vessels, fly flags of convenience and 40% fly a national flag. The ITF calculates that close to 30% of the BDC ships have subscribed to one of their collective agreements, which protects the workers.

The legal relationship between shipping businessmen and the employees on board is formalized through the boarding contract, which defines rights and duties between both parties. This particular contract has its own characteristics that, together with that of compensation, make up a series of rights such as the expenses incurred to the crew by reason of boarding, navigation, and disembarkation, or more specifically, accommodation, meals, and repatriation [16]. This obligation is recognized in Spain in the collective agreements of the sector and through the ILO Conventions, specifically in No. 22 on the contract for the enlistment of seafarers, of 1926, No. 23 of 1926, on the repatriation of seafarers, and No. 166, on repatriation of seafarers revised in 1987, and finally the Maritime Labour Convention of 2006.

The abandonment of a crew causes a situation of need, both for lack of payment of wages and for an increase in expenses, so that specifically Spanish legislation has established assistance services abroad for the support and repatriation of employees on board and Spanish or foreigners who need it [17]. Currently, Law 47/2015, of October 21, regulating the social protection of workers in the maritime fishing

Table 1 Development of salaries in 2015, in dollars per month of ITF data service

	Captain		Sailor	
	Bulk-carriers	Tankers	Bulk-carriers	Tankers
United Kingdom	9300–11,000	11,000–15,800	3500–4200	4200–4900
Denmark	8760–10,820	9790–11,850	3810–4120	4330–4640
Spain	7300–7500	8300–10,000	2100–2300	2200–2400
Pakistan	4800–5200	5300–5600	1300–1400	1300–1400
India	4700–4900	5500–5900	1160–1400	1215–1400
Croatia	5000	6000–8000	1300–1400	1300–1400
South Korea	4500–5000	5500–6500	1400–1700	1700–2300
Latonia	4500–5000	5500–5800	1300–1400	1300–1400
Russia	4500–4800	5000–5500	1300–1400	1350–1450
Poland	4500–4700	5900–6300	1200–1400	1350–1550
Mexico	4330–4530	4750–4950	1150–1400	1200–1450
Montenegro	4100–4300	6150–6350	1300–1400	1390–1490
Egypt	3880–4080	4490–4690	1400–1450	1400–1450
Rumania	3800–4000	5050–5250	1150–1400	1350–1550
Philipines	3600–4000	3800–4300	1050–1400	1350–1550
Ukraine	3400–3660	3980–4190	870–1000	970–1200
China	2700–3300	3550–4150	820–1000	970–1200
Burma	2150–2350	2550–2750	370–420	450–500

Source Own elaboration

sector, recognizes that abandonment is a professional risk in the work of the salaried employees. This is because these workers have a significant exposure to the dangers and difficulties both at sea and in port, especially in foreign ports (Article 40).

The provision established in Spanish legislation, in cases of abandonment, will depend on whether it occurs in a Spanish or foreign port. In the case of needing assistance abroad, because of the abandonment of the ship by insolvent companies, and so on, the maintenance and restitution to their place of residence will be carried out without precluding the responsibility that falls to the shipping company, the shipowner, or the legal representative. In the second case, it will be the assistance to the sea workers, who for whatever reason are in national territory and may require urgent attention, until the competent authorities of the foreign country can take care of them.

In short, the purpose of the provision of services abroad is for the Naval Social Institute to advance the economic amount of the expenses that are appropriate for the maintenance and repatriation to the place of residence of those employees who prove the abandonment situation in a foreign port. Naturally they must be Spaniards and natives of the nations that make up the European Union who have a Spanish residence, or Spaniards even though they are not included in the field of application of the social security dedicated to the special regime of sea workers. An administrative procedure with the formalization of the corresponding request to the Naval Social Institute will need to have been initiated.

An important advance in the coverage of these risks for the crews has been the entry into force on January 18, 2017, of the rules of the Maritime Labour Convention (MLC 2006) [18], with regard to the abandonment of crews In Port. These rules, now in force, establish the obligation of every state signatory of the agreement to carry on board the necessary official document that certifies having insurance covering crew abandonment. This certificate must be in English and in an accessible location aboard the vessel. Two years after being approved by an overwhelming majority at the 103rd Session of the International Labour Conference (ILC), it has been confirmed that the amendments to the Code of the Maritime Labour Convention (MLC, 2006), adopted in 2014, came into force on January 18, 2017. The ratifying members had until July 18, 2016 formally to express their disagreement with the 2014 amendments. The new provisions were widely supported, and only two governments declared that they would not be bound by the amendments, unless they notify that they accept them at a later stage. The 2014 amendments establish new binding international legislation to protect seafarers better from abandonment and provide for compensation in case of death or prolonged disability [19].

The most important novelty is the protection of the crews. They can now go and request directly from the corresponding insurance company the expenses of repatriation, as well as the salaries owed to them.

The current financial guarantee rules establish the necessary requirements to guarantee the establishment of a fast and effective financial guarantee system to assist seafarers in case of abandonment.

Standard A2.5.2, in Sect. 2, states that

> [F]or the purposes of this rule, a seafarer must be considered to have been abandoned when, in violation of the requirements of this Convention or the conditions of the employment agreement of the seafarers, the ship-owner: (a) do[es] not bear the cost of the repatriation of seafarers; (b) has left seafarers without the necessary support and maintenance, or (c) has unilaterally broken ties with seafarers and has not paid contractual wages at least for a period of two months. This obligation is for all Member States which will be obliged to ensure that there is, for ships flying their flag, a financial guarantee system that meets the requirements stipulated in this standard.

The financial guarantee system may consist of a social security scheme, insurance, or a national fund or other similar system. The member will determine the modality of the system, after consultation with the organizations of shipowners and seafarers concerned [20]. In addition, the financial guarantee system shall provide direct access, sufficient coverage, and prompt financial assistance, in accordance with this standard, to all abandoned seafarers on board a vessel flying the member's flag.

It clearly and conclusively shows that for the purposes of paragraph 2 (b) of this standard, the necessary maintenance and support for seafarers must include: adequate food, shelter, potable water supply, fuel essential for survival on board the ship, and necessary medical attention.

In addition, each member state should require that ships flying its flag, and to which paragraphs 1 or 2 of regulation 5.1.3 apply, carry on board a certificate or other documentary evidence of the financial guarantee issued by the supplier of the same. A copy of these documents must be displayed in a conspicuous place and accessible to seafarers. When there is more than one financial guarantee provider that provides coverage, the document issued by each provider must be carried on board. The certificate or other documentary evidence of the financial guarantee shall contain the information required in Annex A2-I. They must be written in English or accompanied by an English translation.

The assistance provided by the financial guarantee system must be provided without delay at the request of the seafarer or his designated representative, and must be accompanied by the necessary documentation that justifies the right to the benefit. Taking into account rules 2.2 and 2.5, the assistance provided by the financial guarantee system should be sufficient to cover the following: (a) the salaries and other outstanding benefits that the shipowner has to pay to seafarers under the agreement of employment, the relevant collective agreement or the national legislation of the flag state; the amount owed shall not exceed four months of outstanding wages and four months of other outstanding benefits; (b) all expenses that seafarers have reasonably incurred, including the cost of repatriation mentioned in paragraph 10; (c) the essential needs of seafarers include: adequate food, clothing, if necessary, accommodation, drinking water supply, essential fuel for survival on board the vessel, necessary medical attention, and any other cost or reasonable expense that derives from the act or omission constituting the abandonment until the arrival of the seafarers to their home.

The cost of repatriation will include travel made by appropriate and expeditious means, normally by air transport, and include the provision of food and shelter to seafarers from the moment they leave the ship until their arrival at home, the necessary medical, travel, and transportation of personal effects, as well as any other cost or reasonable charge arising from the abandonment.

The financial guarantee shall not end unless the provider has notified it 30 days in advance to the competent authority of the flag state.

If the insurance provider or other financial guarantee mechanism has made any payment to a seafarer in accordance with this standard, said provider, in accordance with the applicable legislation, will acquire by subrogation, assignment, or by other means, for an equivalent amount as maximum to the sum paid, the rights that the seafarers would have enjoyed.

No provision of this standard shall impair the right of appeal of the insurer or the provider of the financial guarantee against third parties.

The provisions of this regulation are not intended to be exclusive nor to undermine any other right, claim, or measure of reparation to commemorate seafarers who have been abandoned. National legislation may provide that any amount payable under this rule is deducted from amounts received from other sources and derived from any right, claim, or measure of reparation that may give rise to compensation under this rule.

If more time is required to verify the validity of certain aspects of the application submitted by the seafarer or his designated representative, this should not prevent seafarers from receiving immediately the part of assistance requested which has been recognized as being justified. The certificate or other documentary evidence mentioned in paragraph 7 of Standard A2.5.2 must include the following information:

(a) the name of the vessel; (b) the port of registration of the vessel; (c) the ship's call sign; (d) the IMO number of the vessel; (e) the name and address of the supplier or the providers of the financial guarantee; (f) contact details of the persons or entity responsible for processing seafarers' assistance requests; (g) the name of the owner; (h) the period of validity of the financial guarantee; and (i) an attestation from the financial guarantee provider, indicating that this guarantee meets the requirements of standard A2.5.2.

The national legislation should provide that the financial guarantee system intended to ensure that the payment of compensation provided in paragraph 1, (b), of this rule resulting from contractual claims, defined in standard A4.2.2, complies with the following minimum requirements: (a) the contractual indemnity, when provided for in the employment agreement of seafarers and without prejudice to the provisions of paragraph (c) of this paragraph, shall be paid in full and without delay; (b) no pressure should be exerted for the acceptance of a payment below the contractual sum; (c) when the characteristics of a sailor's prolonged disability make it difficult to evaluate the total compensation to which he may be entitled, a payment or several provisional payments must be made to avoid being in a situation of undue precariousness; (d) in accordance with regulation 4.2, paragraph 2, seafarers must receive payment without prejudice to other legal rights, but the shipowner may deduct such payment from any other resulting compensation, or from any other claim made by the people of the sea against the ship-owner and related to the same incident; and (e) any contractual claim for compensation may be presented directly by the seafarers concerned, or their closest relative, a seafarer representative, or a designated beneficiary.

National legislation should ensure that seafarers receive prior notification that the shipowner's financial guarantee will be cancelled or rescinded. National legislation should ensure that the provider of the financial guarantee notifies the competent authority of the flag state that the shipowner's financial guarantee has been cancelled or rescinded.

4 Consequences for the Vessel

When a ship is abandoned, in addition to the known problems that affect the crew, the port suffers a setback, because it is given a ship that has hardly any value. In addition, if the ship is exempt from maintenance during the time it remains moored to the dock, and additionally, without crew, its low value will still be reduced even further after the occurrence of adverse phenomena such as, for example, marine corrosion. Given this scenario it is difficult to believe that the ship can be repaired before a hypothetical judicial auction of its sale, if it occurs. Scrapping will begin to become the only possible destination for these vessels, but even so this option will also be complicated to carry out, because the hypothetical tow necessary to move the ship to other countries could exceed even the cost of the ship itself [22].

To the hypothetical devaluation of the ship due to corrosion, it should be added that it will be difficult for the port where the ship is abandoned to pay for the pending berth, inasmuch as when shipowners leave the ship, they do so because they are generally in an economically critical situation. In addition, this situation of abandonment of the ship could lead to a generation of possible risk of sinking or fire, for which the port authorities should try to ensure that these ships do not prolong their time of stay in port more than what can be considered a prudent interval, such as one year. Therefore, and to avoid this situation of unwanted abandonment of the ship, after five or six months of abandonment, the Port Authorities should start taking action. And this type of problem is even more acute in small ports, where there are a limited number of berths, in which the abandonment of the ship can prevent other hypothetical ships that arrive at the port from docking due to lack of space.

In summary, the negative environmental effects produced by the abandonment of a ship in a port could be classified into two types, those related to ports, and those related to the ships themselves as detailed in the following lines [20].

- Relative to the Ship
  - Loss of the value of the vessel due to aging and deterioration thereof
  - Lack of maintenance of the propulsion equipment of the ship, and the structure in general
  - Absence of renewal of certificates, insurance, and classification of the vessel
  - Visible emergence of dirt and corrosion
- Relative to the Port
  - Lack of payment of tariffs and port taxes
  - Occupation of space
  - Problems of atmospheric pollution due to discharges
  - Difficulties to identify and locate the shipowner
  - Nonpayment to local suppliers

To deal with these problems that are generated in the port and on the ship, the following considerations should be taken into account.

- Study of waste treatment and ship dismantling procedures
- Regulation of the procedures by which a ship whose owner is missing can be scrapped or moved

The scrapping procedures that are created must guarantee the lowest possible environmental impact from the point of view of transport, such as energy consumption, noise generation, and emissions, as well as the logistics of the transport of the boats to the dismantling centers. There is another important issue to be defined, which is the reference to the regulation and capacity of a port to move vessels with unaccounted-for owner or owners. It should be noted that a recently abandoned boat is not a boat in poor condition, but will soon be if the relevant actions are not taken. Many times, when a ship is finally declared abandoned, a scenario in which the port could auction it, it turns out that after months of lack of maintenance, the poor condition of the ship detracts from potential buyers, who sometimes desert the sale process. Should this occur, neither the creditors nor the Administration will manage to recover their debts [23].

The imminent appearance of dirt and corrosion already commented on will add to the bad image that these abandoned ships give to the ports, not to mention that they will become a danger to the ships that dock next to them. There are cases in which ships coming from operations for drug trafficking, moor at a dock, and spend years in that position, waiting for the trial, and deteriorating, while losing some of their elements (fenders, etc.).

And who should be blamed for the existence of derelict ships? It should be assumed that the Port Authorities (AAPP) are the first ones interested in removing these ships as quickly as possible. The legal process to be able to get rid of a boat in a state of neglect is, for the moment, and as Javier Adame, technical advisor of the FEADPT, says, is so long and complex that the permission to act arrives when the ship hardly has any value [21].

5 Conclusions

First. The abandonment of a ship has consequences in many areas: legal (such as administrative, commercial, and labor), economic, technical, and environmental.

Second. The weakest and most vulnerable part when abandonment occurs is, and has always been, the crew. Given that, as has been seen, the work is in some cases precarious, due to its low salaries, when a ship is abandoned, this situation is markedly accentuated.

Third. Probably there are no ports that have not suffered, or are suffering, the problems resulting from the abandonment of ships. The structural problem itself will be compounded by possible breach of contractual obligations, without forgetting the appalling effect abandonment produces and the danger that it represents for nearby vessels.

Fourth. The lack of maintenance of the boats will lead to their deterioration, which will affect their seaworthiness and stability. Such conditions could lead to the ship sinking, burning, and so on. It is a matter of great importance.

Fifth. The legal avenues to move a boat under these conditions are well defined, but they imply generally long execution times. It could be the case that when the time finally arrives in which the embargo of a ship could be executed, it has already sunk.

References

1. Pais Montes, C.: La nueva geografía económica del transporte marítimo, In: (Tesis) Repertorio de la Universidad de, Directores Freire Seoane, pp. 17–745. Mª Jesús y González Laxe, A Coruña. (A Coruña), Fernando (2014)
2. Richards, P., Banigan, J.J.: How to abandon ship, p. 19. Cornell Maritime Press, Maryland (1988)
3. Carril Vázquez, X.M.: Una cuestión de Derechos Humanos: la Protección de Seguridad Social de nuestros trabajadores del mar emigrantes en Noruega, Editorial Aranzadi, p. 31–51. Pamplona (2015)
4. López Arranz, A.: Conflictos na aplicación da lexislación das condicións laborais e de seguridade social aos traballadores de plataformas marítimas, vol. 6, pp. 136–145. Anuario da Facultade de Ciencias do traballo (2015)
5. Basurco, OF.: El convenio refundido sobre el trabajo marítimo de 2006 ¿introduce satisfactoriamente las directrices de la OMI obre el abandono de marinos?, en Olga Fotinopoulou Basurco (Coordinadora) Derechos del hombre y trabajo marítimo: los marinos abandonados, el bienestar y la repatriación de los trabajadores del mar, EuskoJaurlaritza, (Vitoria), pp. 116–118 (2009)
6. Joyanes, G.: Domingo, Abandono de buques y tripulación, pp. 21–22. Marge Books, Valencia (2009)
7. Ribes, D.: Sebastian, Los buques abandonados en la Ley de navegación marítima, p. 470. Dykinson, Madrid (2015)
8. Palomar, A.: La ordenación administrativa de la navegación marítima, en Enrique San Juan y Belen Campuzano (Coordinadores) Comentarios a la Ley de navegación marítima, pp. 106–110. Tirant lo Blach, Valencia (2016)
9. García, G., Luis, J.: El Convenio internacional sobre privilegios marítimos e hipoteca naval, Ginebra, 6 de mayo de 1993, en Civitas, Iglesias Prado Juan Luis, (coord.), pp. 3728–3765. Estudios jurídicos en homenaje al profesor Aurelio Menéndez Iglesias Prada, Juan (1996)
10. Gabaldón García, J.L y Ruíz Soroa, J.M, Manual de Derecho de la Navegación Marítima, Marcial Pons. Ediciones Jurídicas SA, Madrid, pag. 240–249, (2006)
11. Domínguez Cabrera, M.P.: El buque como objeto de Registro de bienes Muebles, Revista de Derecho Mercantil, febrero/2004, pp. 7–10. (2004)
12. Carril Vázquez, X.M.: Una cuestión de Derechos Humanos: la Protección de Seguridad Social de nuestros trabajadores del mar emigrantes en Noruega, pp. 31–51. Editorial Aranzadi, Pamplona (2015)
13. Martínez Girón, J. Arufe Varela, A.: Derecho Crítico del Trabajo, pp. 51–57, netbiblos, A Coruña. (2012)
14. Carril Vázquez, X. M.: Aspectos laborales y de seguridad social de los pabellones de conveniencia. In: Revista Española de derecho del Trabajo, número, vol. 108, p. 2. (2001)
15. Carril Vázquez, X. M.: La seguridad social de los trabajadores del mar, pp. 197. Civitas, Madrid (1999)
16. Carril V, Xosé M.: El régimen jurídico español de los trabajadores del mar en casos de abandono, en Olga Fotinopoulou Basurco (Coordinadora)Derechos del hombre y trabajo marítimo: los marinos abandonados, el bienestar y la repatriación de los trabajadores del mar, pp. 221. EuskoJaurlaritza, Vitoria (2009)
17. Martínez Girón, A.: Arufe Varela y Xosé Manuel Carril Vázquez, Derecho de la Seguridad Social, 2ª edición, pp. 3–4. Netbiblo, ACoruña (2008)
18. De La Villa de la Senrra, P.: Convenio de la Organización Internacional de trabajo sobre el Trabajo Marítimo. In: Ratificación y vigencia. Texto normativo, Revista del Ministerio de Trabajo e Inmigración, vol. 82, pp. 401 (2006)

19. Oficina Internacional de Trabajo, UE, "Compilación de instrumentos sobre trabajo marítimo", pp. 127–133, (2015)
20. González, Joyanes D.: Abandono de buques y tripulaciones. Margebooks, Valencia (2009)
21. Sanlorenzo C., Pérez G.: Barcos abandonados, pp. 1–6. Un problema para todos. Skipper (2010)
22. Salinas, M, El mundo del Marine Surveyor. Obtenido de EL ABANDONO DE BUQUES EN PUERTO: http://captmsalinas.blogspot.com.es/2012/09/el-abandono-de-buques.html. (2013)
23. Fayos, J. L.: TransEurope Marinas. Obtenido de Embarcaciones abandonadas en los puertos: http://www.transeuropemarinas.com/. (2017)

Green Artificial Reef PROARR: Repopulation of Coastal Ecosystems and Waste Recycler of the Maritime Industries

Luis Manuel Carral Couce, María Jesús Rodríguez Guerreiro, José Angel Fraguela Formoso, Jose Carlos Alvarez-Feal and Almudena Filgueira Vizoso

Abstract The extractive work of shellfish harvesting and fishing in the autonomous community of Galicia is a strategic economic activity. In this sense, the action of artificial reefs is of great interest as they act as instruments of arrangement and protection of the marine resources and ecosystems. The PROARR research project has led to the design of independent modules of production and protection reefs, aimed at the construction of polygons of artificial reefs. One of the aims of the research work has looked for the reutilization of inert waste materials from industrial processes linked to the maritime area: naval construction, fishing, and shellfish harvesting, incorporating them into the production process of the reef. In this way, existence of dumps is reduced and the value chain of those processes is increased. The actual sustainability within the activity of maritime industries must promote the absorption of this inert waste by the sea, thus closing the activity cycle. In the case of the utilization of shells in the manufacture of artificial reefs (as an additive to the mortar used in its construction) it represents the double advantage of the elimination of a residue, with the subsequent increase in the value chain of these aquaculture processes and, on the other hand, returning to its origin a product which is part of a structure that can achieve the development of resources and favour the conservation of the ecosystems.

Keywords Artificial reef · Ecosystems regeneration · Waste recycling

L. M. Carral Couce (✉) · M. J. R. Guerreiro · J. A. Fraguela Formoso
J. C. Alvarez-Feal · A. F. Vizoso
Universidade da Coruña, A Coruña, Spain
e-mail: lcarral@udc.es

A. Vega Sáenz et al. (eds.), *Proceedings of the 25th Pan-American Conference of Naval Engineering—COPINAVAL*, https://doi.org/10.1007/978-3-319-89812-4_32

1 Introduction

Artificial reefs (AA) are defined as "submerged structures placed deliberately on the marine bottom to imitate some of the characteristics of a natural reef" [1]. With this we aim to reach several varied objectives such as: improvement in the production of fishing, conservation of resources in diminishing the pressure on the existent reefs, restoration of the marine bottom because of the settlement in them of communities of invertebrates and seaweeds, and even new applications that include tourism and recreational use of the sea [2–4].

The idea of submerging structures to "improve" the marine resources arises from the casual discovery that fishing in the vicinity of shipwrecks increased in relation to nearby zones [5]. From this knowledge, a wide variety of objects/substrates has been used as AA. Concrete structures made for the purpose as well as "opportunity materials" such as logs, tires, subproducts from quarries, cars, and even offshore structures have been utilized [6, 7].

AAs present marked differences depending on the geographical area of application and goal to be achieved [5]. Japan was a pioneer in the construction of artificial reefs, with the active leadership of the government in the promotion and financing of projects as a way to improve commercial fishing [6, 8]. In North America this instrument has been used for various purposes, but notably emphasizing the enhancement of its recreational use in sport fishing and diving [7]. Hong Kong has recently promoted an innovative program of artificial reef deployment linked to the creation of protected marine areas, thereby seeking to reverse the effects of overfishing [9, 10]. In the Mediterranean, AAs are commonly used in support of actions to avoid trawling, seeking the protection of marine habitats of Posidonia meadows [11]. In other parts of Europe, artificial reefs are considered a management tool to sustain artisanal fishing and compensate for the effects of stock depletion, which is all the more important when this type of traditional fishing is an active means of sustaining local communities [1, 5].

The concept of sustainable development includes the controlled use of natural resources as well as the reduction in the consumption of energy with emission of carbon dioxide and other greenhouse gases (GHG). Regarding the construction of AAs, it can be estimated that the most commonly used material in the construction of AAs is concrete [12]. From the production of their components [13], it is indicated that Portland cement production releases large amounts of carbon dioxide and other greenhouse gases into the atmosphere. With regard to natural aggregates (sand, gravel), which account for 60–80% of the volume of concrete, their high consumption generates enormous pressure on the surrounding ecosystems [14].

The result is that the environmental impact the extraction of additions of the concrete produces is particularly severe in terms of aggregates, as there are problems associated with the different extraction or production technologies.

2 Objectives of the PROARR Project

The PROARR Research Project is aimed at the design of production and protection modules for the construction of artificial reef polygons in the autonomous community (AC) of Galicia. The Project, whose results have been the subject of a patent (Fig. 1), ended in the year 2012 having reached the objectives determined in its memorandum:

- Design of a reef production module of the alveolar type adapted to the geomorphology of the Galician sea bottom having a high ratio of apparent volume compared to its weight, which easily allows reaching the aim of the increase of the biomass of the fishery resources based on the increase in their survival, their growth, and their reproduction.
- Design of a reef protection module aimed at the protection of marine ecosystems against illegal bottom trawling and other physical aggressions. The stability of these modules was calculated adapted to the power and typologies of the Galician bottom trawlers that work in our fishing grounds.
- Determination of an optimized manufacturing, transport, and installation process, linked to the geographic and logistical reality of Galicia, so that the industrial activity of production and transport presents favorable economic costs.
- The recycling of inert materials from other industrial processes linked to maritime, naval, and extractive (fishing and shellfish) areas developed in our community, incorporating them into the reef production process. In this way, the existence of landfills is reduced and the value chain of these processes is increased.

This project establishes the objective of determining the waste that can be recycled as a consequence of its integration as an aggregate of the concrete in the production of AAs.

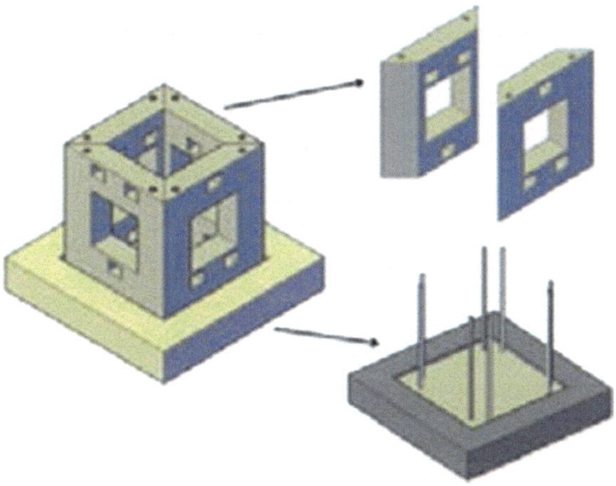

Fig. 1 PROARR artificial reef module. Own source

3 Socioeconomic Effects of the Installation of the AA

The assessment of the impact that the AA produces on a community must be carried out in social and economic terms, more than based only on criteria of general or particular economic profit [5].

In any case, the valuation is relatively simple when the effects can be quantified (a clear example would be to evaluate the income based on considering the catch rate and its market value; Fig. 2). However, it will be more difficult in those cases in which the benefits are not commercialized (e.g., the enjoyment of sport fishing, the increase in biodiversity), or when the induced changes are complex or difficult to perceive (e.g., the links between habitat and fishing). Likewise, it will be difficult to assess the effects on the community by the achievement of the sustainability of the fishing activity, on which a significant number of direct jobs depend. All these results that are difficult to evaluate form, largely, the social benefits of AA.

Other difficulties that should also be evaluated derive from the need to consider negative externalities, such as when a reef attracts fish out of their current habitat and reduces the catch rate of vessels operating in those areas [12].

The result of the analysis conducted to consider the socioeconomic benefits derived from the implementation of the green type AA [15] establishes a basic distinction between the relationship of its implementation and the type of benefit derived from it. In this analysis, the benefits obtained are presented directly, indirectly, and passively.

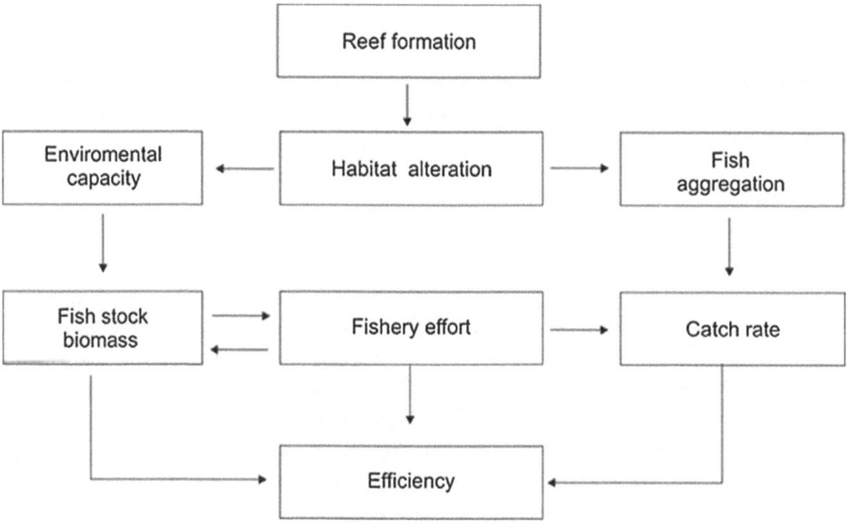

Fig. 2 Bioeconomic implications of AA's establishment. *Source* Own source based on [5]

In this way [5] it determines as direct benefits those derived from "extractive" uses such as fishing and aquaculture, as well as those derived from "recreational" activities.

We must add the employment of inert materials derived from the recycling of waste from the maritime industries in the AA [16].

Indirect-type benefits cover a wider range, because they now include increased production and protection of young fish, deviation of effort from fisheries, protection of coastal areas, and improvement in water quality as the AA acts as a substrate for the elimination of nutrients (biofiltration). As an indirect effect derived from the consideration of the green type AA, it would be necessary to take into account the reduction that occurs in the environmental impact of related industries in the maritime field. Finally, the benefits of the "passive" type as a result of its implementation would be those in which the community obtains the benefit derived from "knowledge" of the fact itself of its implementation. As possible cases, we would have those that occur after the success of the AA by increasing the diversity of species or by conserving a unique habitat intact for future generations. In the case of the consideration of the AA under the green alternative, the passive effect will derive from the achieved sustainability of the maritime industries in the realization of their productive activities. In any case, the relative importance of these three types of benefits varies from one project to another according to the different aims assigned to the AAs [5]. However, the bioeconomic implications contained in Fig. 2 should be expanded under consideration of the socioeconomic effects of recycling, the reduction in waste, and sustainability in the performance of industrial activities in the maritime industries.

4 Waste Recycling of Maritime Industries

With an estimated total annual production of 10 billion cubic meters [17] in the general field of construction and in particular in the production of AAs, concrete is the most widely used material in the world. The consumption of natural aggregates for their production—sand and gravel—is of great importance, because they correspond to a greater proportion of the final volume.

The use of these two aggregates of concrete corresponds to a severe environmental impact derived from the problems associated with the different technologies of extraction or production. The extraction of sand in coastal areas [18–21] affects local ecosystems [22–24]. Finally, the production of fine aggregates from the mechanical crushing of coarse aggregates presents a high energy cost and poses difficulties with respect to their incorporation into concrete, as stated by Elavenil and Vijaya [24], Sreebha and Padmalal [25]. In any case, the use of aggregates in concrete, as it is a nonrenewable resource, can be reduced by replacing it with waste from the maritime industries.

Associated with any production process, and in particular in maritime activities, the generation of waste constitutes an inefficiency of the processes because it means a loss of materials and additional energy consumption. At the same time, it leads to an additional activity of collection, treatment, and elimination with generation of new economic and environmental costs for society. The adequate management of the productive process in order to make it environmentally clean requires reducing the generation of waste and the reuse of materials and products, or otherwise recycling or enabling other types of recovery [16]. The goal is to manage waste in order to reduce the amount of waste which is to be disposed of.

In our analysis of the waste coming from the industrial processes of the maritime sector, only nonhazardous ones can be taken into consideration, because the dangerous ones cannot be recycled without going through an authorized manager. In this consideration, we must also consider biological waste: mollusc shell waste such as mussel or oyster shells. It is important to point out that these shells must be burnt (to eliminate all remaining organic matter) and crushed, obtaining a grain size suitable for their use [16].

An analysis of the waste generated by a shipyard of new construction and repairs was carried out in order to determine what could be recycled for the manufacture of concrete in the construction of AAs. In addition to the quantities generated, it is necessary to take into account the possible treatment that must be carried out according to the applicable environmental legislation. After the evaluation of the waste studied, grit, rock wool, and fiberglass wool were selected as suitable materials to be included in the preparation of the specimens of the concrete object of this study. At the same time [26] the use of oyster shells and mussels in the manufacture of concrete has been studied and validated. These shells constitute important residues of the activity of the canning industry [12].

5 Specimen Samples and Tests

Once the materials have been selected, the methodology followed to analyze the incorporation of each of them as an aggregate of the concrete has consisted of determining an optimum dosage for its elaboration. Looking for the validation of the results, a set of blank specimens was prepared, without adding the recycled materials, in order to put them to tests that would provide the necessary information to characterize the concrete (Table 1). The specimens offering the best results were the ones that determined which dosage to use for the rest of the tests with the recycled material.

Then, starting from the optimal dosage, new dosages were calculated by substituting different quantities of the fine aggregate for each of the recycled materials. The code [27] recommends replacing quantities not higher than 20% in weight of the coarse aggregate. In this case, due to the granulomere of the recycled material, the fine aggregate was replaced in different proportions depending on the characteristics of the recycled material.

The specimens obtained were subjected to the compression test to determine the characteristic resistance of the concrete for 7 and 28 days of curing period and also

Table 1 Results test to compression blank test specimens. Own source

	Mass (kg)	Load (kN)	Time (s)	Strength (MPa)
Sample	Values in 7-day period			
I	12.4	413.2	51	23.38
III	12.4	456.3	61	25.82
VI	12.3	595.9	72	33.72
VIII	12.3	595.9	72	33.72
Sample	Values in 28-day period			
II	12	464.1	57	26.26
V	12	520.6	53	29.46
VII	11.9	652	78	36.89
IX	11.9	716.1	85	40.53

to the water absorption test. In this way, all the results obtained could be compared and conclusions and recommendations were drawn for each material.

5.1 Concrete Processing

The suitable dosage to obtain concrete with the desired characteristics, both for the blank samples and for the samples with recycled materials, was calculated starting from a theoretical characteristic resistance of 30 MPa and plastic soft theoretical consistency, satisfying the recommendations for concrete intended for the construction of artificial reefs.

With each of the calculated dosages, a mixture of 20 l was prepared, from which two cylindrical test tubes with a diameter of 15 cm and a height of 30 cm and a cube measuring 15 cm on each side were extracted in all cases. All the molds were previously impregnated with a release agent to avoid the adherence of the concrete to them and facilitate their demolding after 24 h. The product used was Sika® release agent–LN [28].

Approximately one minute after adding the additive, the mixture was homogeneous and the consistency test known as Abrams Cone was carried out, whose method is described in the standard [28]. Knowing the consistency of the concrete, the specimens were elaborated according to those specified in the standard [29], vibrating with a needle vibrator, and they were demoded 24 h later. Then they were introduced into a water bath at 20 °C where they were cured for 7 or 28 days, according to the tests to which they were subsequently subjected.

5.2 Tests Carried Out

Compression test. Simple compression resistance is the most important mechanical characteristic of concrete, and its determination is carried out by testing specimens according to standardized operating methods. The compression test must be carried out in accordance with the standard [29] and consists of compressing the specimens of hardened concrete until they break in a standardized compression machine.

Absorption test by the Fagerlund method. This method [29] is used to determine the coefficient of water absorption by capillarity of the hardened concrete.

Given the purpose of the concrete to be processed, it is useful to know or estimate the total water absorption of this concrete, which will always be completely submerged, as this influences the durability of the concrete. The absorption of water depends on the interconnection of the existing pore network inside the concrete.

5.3 Test Results

As a summary and in order to observe the results obtained in the compression test more clearly, two bar graphs are presented in Fig. 3a, b in which the resistance of the test pieces of each mixture is compared, both in blank specimens and with the recycled materials, in 7 days and to 28 days of the curing period. Under no circumstances, neither in 7 or 28 days of the curing period, is the resistance of the blank specimens reached, obtaining the best results for the concrete with oyster shell, mussel, and grit, offering higher resistance the higher the amount of material replaced and repeatedly exceeding the starting strength of 30 MPa.

The concrete with glass fiber and rock wool has very low resistance values, the more the greater the amount of material included. Therefore, taking into account this and the disadvantages when preparing the material to make the mixture, it is

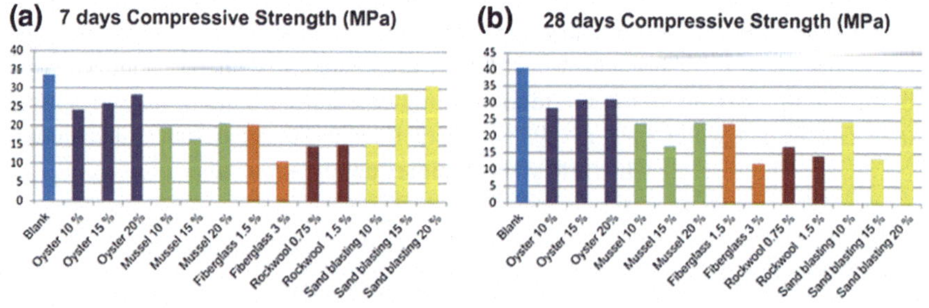

Fig. 3 Comparison of the strength of the concrete specimens obtained after 7 and 28 days of curing period. Own source

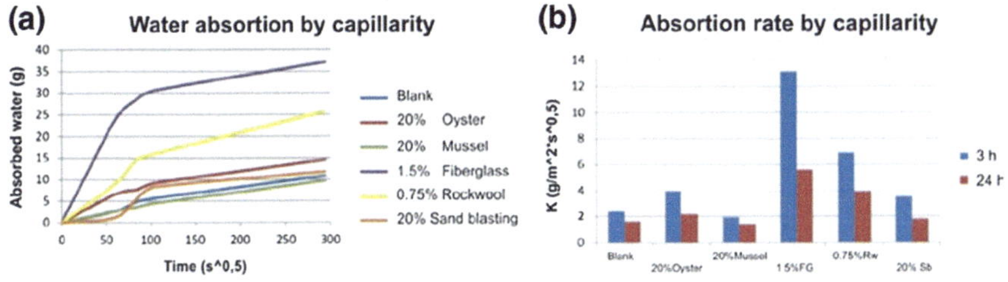

Fig. 4 Comparison of the values obtained by water capillarity absorption. Own source

concluded that both fiberglass and rock wool are not suitable materials to be included in the concrete.

In order to compare the results more clearly and draw conclusions, a graph comparing the best results obtained for the absorption of each concrete is presented (Fig. 4a, b).

As can be seen in both figures, the concrete with greater absorption is that which contains glass fiber followed by that containing rock wool. On the other hand, the best values are offered by concrete with mussel shell, and even lower than those obtained for concrete in blank specimens. The absorption of concrete with oyster shell and grit can also be considered acceptable.

6 Conclusions

The substitution of any amount of fine aggregate by any of these materials supposes a decrease in the characteristic resistance. However, with the oyster shell and the grit it was possible to reach and even exceed the theoretical target strength of 30 MPa.

Fiberglass and rock wool are not suitable materials to be included in the concrete: first, due to the inconvenience derived from the preparation of the material, and second, due to the poor results obtained in the tests of compressive strength and water absorption by capillarity. These results are a consequence of the low workability involved in the incorporation of these materials in the concrete, because it is practically impossible to obtain a homogeneous and correctly vibrated mass, even with small quantities of those materials and maximum water/cement ratio and fluidizing additive.

The capillary absorption test was carried out according to the Fagerlund method, by means of which information is obtained about the absorption capacity of the concrete over time and not only at the initial moment. Overall, water capillary absorption increases when substituting fine aggregate for other materials, with the exception of the mussel shell. However, water absorption is lower the greater the percentage of substitution, without taking into account the fiberglass and the rock wool.

Acknowledgements The work carried out is part of the research project "Design and production of an artificial reef module for the protection and rehabilitation of coastal ecosystems in the autonomous community of Galicia." CN-10MMA003CT; Programme INCITE Programas Sectoriais PEME I + D and R & D SUMA Xunta de Galicia.

References

1. OSPAR Convention.: OSPAR Commission. https://www.ospar.org/convention/agreements?q=reef (2015). Accessed 25 Apr 2015
2. Pears, R.J., Williams, D.M.: Potential effects of artificial reefs on the Great Barrier Reef: Background paper **60**, 1–33 (2005)
3. Seaman, W.: Unifying trends and opportunities in global artificial reef research, including evaluation. ICES J. Mar. Sci. **59**, S14–S16 (2002)
4. Seaman, W.: Artificial habitats and the restoration of degraded marine ecosystems and fisheries. Hydrobiologia **580**(1), 143–155 (2007). https://doi.org/10.1007/s10750-006-0457-9
5. Whitmarsh, D., Santos, M.N., Ramos, J., Monteiro, C.C.: Marine habitat modification through artificial reefs off the Algarve (southern Portugal). Ocean Coast Manage. **51**(6), 463–468 (2008)
6. Baine, M.: Artificial reefs: a review of their design, application, management and performance. Ocean Coast Manage. **44**, 241–259 (2001)
7. Seaman, W., Sprague, L.M.: Artificial habitat practices in aquatic systems. In: Seaman, W., Sprague, L.M. (eds.) Artificial habitats for marine and freshwater fisheries, pp. 1–29. Academic Press Inc, San Diego (1991)
8. Stone, R.B., McGurrin, J.M., Sprague, L.M., Seaman, W.: Artificial habitats of the world: synopsis and major trends. In: Seaman, W., Sprague, L.M. (eds.) Artificial habitats for marine and freshwater fisheries, pp. 31–60. Academic Press Inc, San Diego (1991)
9. Pitcher, T.J., Buchary, E.A., Hutton, T.: Forecasting the benefits of no take human made reefs using spatial ecosystem simulation. ICES J. Mar. Sci. **59**, S17–S26 (2002)
10. Wilson, KDP., Leung, AWY., Kennish, R.: Restoration of Hong Kong fisheries through deployment of artificial reefs in marine protected areas. ICES J. Mar. Sci. **59**:S157–63. Bruce Anchor Group. (2002)
11. Jensen, A.C.: Artificial reefs of Europe: perspectives and future. ICES J Mar. (2002)
12. Carral, L., Alvarez-Feal, J.C., Tarrio-Saavedra, J., Rodriguez-Guerreiro, M., Fraguela Formoso, J.: Social interest in developing a Green modular artificial reef structure in concrete for the ecosystems of the Galician rías. J. Clean. Prod. **172**, 1881–1898 (2018)
13. Malhotra, V.M., Bilodeau, A.: High-volume fly ash system: the concrete solution for sustainable development. In: Mehta, P.K. (ed.) Proceedings of the International Symposium on "Concrete Technology for Sustainable Development in the Twenty-First Century", pp. 43–64. Radha Press, New Delhi, India (1999)
14. De Brito, J., Saikia, N.: Recycled aggregate in concrete: green energy and technology. Springer-Verlag, London (2013)
15. Fundación EOI.: Sectores de la nueva economía 20–20– Economía verde. Available in: https://books.google.es/books?isbn=8415061048. Accessed May 2017. (2010)
16. Ferreño, S., Carral, L., Guerreiro, M., Fraguela, J., Villa, R.: Estudio, análisis y valorización de residuos industriales procedentes de la industria de la construcción naval, para su reciclaje y uso en otras aplicaciones marinas. VII Simposio Marítimo Symtechnaval—La Habana—Cuba (2012)
17. Meyer, C.: The greening of the concrete industry. Cem. Concr. Compos. **31**(8), 601–605 (2009)

18. Boyd, S., Limpenny, D., Rees, H., Cooper, K.: The effects of marine sand and gravel extraction on the macrobenthos at a commercial dredging site (results 6 years postdredging). ICES J. Mar. Sci. **62**, 145–162 (2005)
19. Desprez, M.: Physical and biological impact of marine aggregate extraction along the French coast of the Eastern English Channel: short and long-term postdredging. (2000)
20. Phua, C., van den Akker, S., Baretta, M., van Dalfsen, J.: Ecological effects of sand extraction in the North Sea. Stichting De Noordzee, Utrecht (2002)
21. Van Lancker, V.R.M., Bonne, W., Velegrakis, A.F., Collins, M.B.: Aggregate extraction from tidal sandbanks: is dredging with nature an option? Introduction J. Coast. Res. **2010**(51), 53–62 (2010)
22. Kori, E., Mathada, H.: An assessment of environmental impacts of sand and gravel mining in Nzhelele Valley, Limpopo Province, South Africa. 3rd International conference on biology, environment and chemistry, IPCBEE, Bangkok. Thail. **2012**, 137–141 (2012)
23. Padmalal, D., Maya, K., Sreebha, S., Sreeja, R.: Environmental effects of river sand mining: a case from the river catchments of Vembanad lake Southwest coast of India. Environ. Geol. **2007**(54), 879–889 (2007)
24. Elavenil, S., Vijaya, B.: Manufactured sand, a solution and an alternative to river sand and in concrete manufacturing. J. Eng. Comput. Appl. Sci. **2**, 20–24 (2013)
25. Sreebha, S., Padmalal, D.: Environmental impact assessment of sand mining from the small catchment rivers in the southwestern coast of India: a case study. Environ. Manage. **47**, 130–140 (2011)
26. Martinez-Garcia, C.: Estudio del comportamiento de la concha de mejillón como árido para la fabricación de hormigón en masa. Trabajo Fin de Grado. Repositorio Universidade da Coruña. http://hdl.handle.net/2183/17489. (2016)
27. Code on Structural Concrete EHE-08.: Ministerio de Fomento. Spain. (2010)
28. EN 12350-2.: Testing fresh concrete. (2009)
29. EN 12390-3.: Testing hardened concrete. (2009)

Factors Affecting Microbial Enhanced Oil Recovery (MEOR)

Miguel A. Hernández Rivera, Jose Marcio Vasconcellos and Marcia E. Ojeda Morales

Abstract Oil and natural gas are the primary sources of energy used by humanity, as renewable energy is not yet available to the entire population. After an oil field is abandoned, or its production declines significantly, an estimated two-thirds of the oil originally onsite remains in the subsoil, which is difficult and expensive to extract by conventional methods. Enhanced oil recovery techniques are used to increase the recovery of oil from these sites, but they are expensive and often harmful to the environment. A technically and economically viable alternative is the use of microorganisms, which are generally obtained from the oil reservoirs themselves. These microbes are resistant to oil reservoir conditions, including high pressures, temperatures above 85 °C, high salt concentrations, and extreme pH values. During microbial growth, when supplied with appropriate nutrients, these microbes generate biosurfactants, solvents, gases, and organic acids, and in some cases, biopolymers. In addition, such microbes are capable of breaking up long chains of paraffins and asphaltenes, modifying the wettability of the storage rock, decreasing the viscosity and density of the oil, and increasing the pressure within the pore network of the rock. The successful use of microbial enhanced oil recovery has been reported in which the growth of native bacteria from a reservoir was stimulated by supplying nutrients, with most treated reservoirs having shown positive results.

Keywords MEOR · Biosurfactants · Wettability

M. A. H. Rivera (✉) · M. E. O. Morales
Biotechnology Laboratory, Academic Division of Engineering and Architecture, Juarez Autonomous University of Tabasco, Km 1 Carretera Cunduacán-Jalpa, Col. La Esmeralda, Cunduacán 86690, Tabasco, Mexico
e-mail: miguel.hernandezr@ujat.mx

J. M. Vasconcellos
Naval Architecture and Ocean Engineering Department, COPPE, Federal University of Rio de Janeiro, Av. Pedro Calmon, 550-CEP 21941-901, Rio de Janeiro, Brazil

A. Vega Sáenz et al. (eds.), *Proceedings of the 25th Pan-American Conference of Naval Engineering—COPINAVAL*, https://doi.org/10.1007/978-3-319-89812-4_33

1 Introduction

Oil is one of the most important sources of energy and raw materials for the global chemical industry. However, available oil reserves have decreased considerably in recent years, partly due to the population growth of the planet but also due to the difficulty of finding new oil deposits. Although a great deal of oil remains in numerous mature and abandoned fields around the world, the extraction of such oil by conventional methods is not economically attractive.

Approximately 1 trillion barrels of oil are estimated to have been extracted globally, leaving an untapped surplus calculated to be between 2 and 4 trillion barrels of hydrocarbons [1], representing approximately 67% of the total known oil reserves in the world [2].

In general, between 5 and 10% of the oil originally onsite is recovered during the first stage of oil extraction [3]. It is possible to increase the efficiency of this stage by decreasing the hydrostatic load, which is accomplished by injecting a gas into the production column (gas lift). When the natural energy of the reservoir has decreased, the reservoir is stimulated by injecting water or gas under pressure to sweep the oil to the producing well or to maintain the pressure.

Although this second stage increases hydrocarbon recovery to between 10 and 40% [3], an estimated 50–67% of the oil originally onsite remains in the reservoir [4, 5]. This oil is trapped in the pores of the reservoir structure by viscous and capillary forces, highlighting the relative inefficiency of the primary and secondary stages [6].

After the previously described extraction stages, the residual oil is composed of heavier oil fractions and is therefore of high viscosity. In addition, this oil exhibits a high water–oil formation interfacial tension, resulting in large capillary forces that retain the oil in the pores of the formation, inhibiting their mobility. To increase recovery, enhanced oil recovery (EOR) methods have been developed, consisting of a set of extraction techniques that use external sources of energy. These processes are more complicated and expensive than those of the previous stages [7], limiting their application [8]. Therefore, the oil industry requires alternative methods that are more economically favorable and are environmentally friendly [9].

2 Microbial Enhanced Oil Recovery (MEOR)

Microbial enhanced oil recovery (MEOR) is a technology that can be implemented with considerably low operating costs [10]. MEOR has several advantages over other conventional EOR processes, as it does not consume the large amounts of energy that thermal methods require and does not depend on oil prices, unlike chemical methods [11].

The MEOR process is an EOR method that uses microorganisms to improve the recovery of oil. Microbes influence the process in several ways (Table 1).

Table 1 Several microorganisms, their metabolic products, and their applications in the MEOR process [3]

Microorganisms	Microbial product	Application in MEOR
Bacillus, *Leuconostoc*, and *Xanthomonas*	Biomass	Selective packing and alteration of wettability
Acinetobacter, *Arthrobacter*, *Bacillus*, and *Pseudomonas*	Surfactants	Reduction of surface and interfacial tension
Bacillus, *Brevibacterium*, *Leuconostoc*, and *Xanthomonas*	Polymers	Selective packing, modification of viscosity and flow patterns
Clostridium, *Zymomonas*, and *Klebsiella*	Solvents	Increase in mobility and permeability, reduction in viscosity, and density
Clostridium, *Enterobacter*, and *Acidogenic mixed consortium*	Acids	Increase permeability and cause emulsification
Clostridium, *Enterobacter*, and *Methanobacterium*	Gases	Increase pressure, interfacial tension, and viscosity reduction

Bacteria are the only microorganisms used in MEOR due to their small size and ability to generate metabolic products, such as gases, solvents, organic acids, biosurfactants, and biopolymers as well as the biomass itself, increasing the recovery of oil from low-producing, nonflowing, or abandoned wells to extend the productive life of an oil field.

Biosurfactants lower the interfacial tension of water–oil systems, reduce the capillary forces that keep oil trapped in the pores of the formation and alter the wettability of the storage rock [3, 12]. In addition, biosurfactants decrease the density and viscosity of the hydrocarbon medium, increasing its mobility [13].

Bacteria have the ability to survive in the hostile environments of oil reservoirs with respect to temperature, pressure, pH, and salinity. This ability makes their use especially attractive for EOR processes.

The average size of bacterial cells is between 0.50 and 5.0 μ, facilitating their entry into the pores of the formation. Many oil deposits are located in basins of sedimentary rocks, in which the lithography is composed primarily of sandstones or carbonates that have pore sizes greater than 30 μ, with the narrowest sections not less than 10 μ wide [11]. The microbes injected into the formation should be small and spherical, with sizes less than 20% of the pore opening [4].

In 1926, Beckman proposed the possibility of using bacteria to improve oil production for the first time [4], observing that certain microorganisms or their metabolic products could be used to increase the amount of oil recovered. However, it was not until 1946 that Zobell et al. performed the first experimental work to confirm Beckman's theory [4].

The MEOR process is considered cost effective and can be applied with minimal modifications to existing facilities in the field. However, despite the positive experimental results and field tests, the MEOR process has not been widely accepted in the oil industry. This is primarily due to a negative perception about the use and handling of bacteria in the MEOR process regarding a potential risk to people, animals, plants, or the environment. There is inconsistent information available on the technical performance of the MEOR process, as there is a lack of standardization of field tests, treatment of results, and posttreatment follow-up. However, much progress has been made in the microbiological aspect of the process to select bacteria that can be used, including their survival and effectiveness inside the reservoir as well as possible competition or synergistic effects with native bacteria.

3 Microorganisms and Their Effect on Oil Recovery

3.1 Production and Effect of Biosurfactants

Biosurfactants offer several advantages over surfactants of chemical origin. For example, most chemical surfactants are deactivated in the presence of 2–3% NaCl (w/v) and are toxic to the environment, whereas biosurfactants retain their properties in environments with moderate to high salinity, are biodegradable and have low toxicity [14, 15]. Biosurfactants can also be obtained from low-cost and renewable carbon sources. They can increase the recovered oil fraction by reducing the water–oil interfacial tension and by altering the wettability of the rock, which is considered one of the most important mechanisms of the MEOR process [11, 12, 16, 17]. Figure 1 shows that the biosurfactant produced by an *Azospirillum brasilense* strain is capable of reducing the surface tension of water from 72 to 38 mN/m, a 53% reduction.

These results are consistent with other studies that showed similar decreases in water surface tension [18, 19].

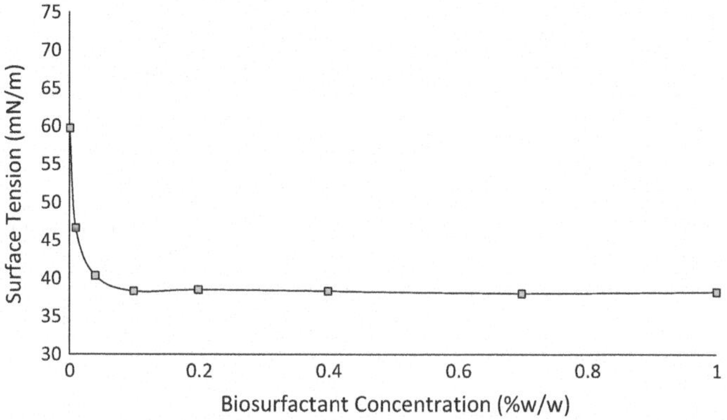

Fig. 1 Water surface tension reduction by biosurfactant produced by *Azospirillum brasilense*

3.2 Production of Solvents and Gases

The MEOR process uses microorganisms that metabolize low-cost substrates from renewable sources to provide products that are similar to those used in chemical methods [8, 17, 20]. This process has the added benefit of being environmentally friendly because the generated products are easily biodegradable and have lower toxicity than the alternative chemical surfactants used in waterflooding [15, 21]. During stationary growth, the bacterial genus *Clostridia* and a number of other anaerobic bacterial strains produce solvents, such as acetone, butanol, ethanol, and isopropanol as well as others at lower concentrations [22].

When these liquid and gaseous substances are produced in the reservoir, they mix with the crude oil and decrease the viscosity and the capillary forces that oppose its movement. An additional and important effect of the gases produced inside the pores is that they increase the pressure in the reservoir, while solvents and acids dissolve the carbonate rock, increasing its porosity and permeability.

3.3 Selective Packing

The stimulation of native bacterial growth with the appropriate nutrients or the injection of exogenous bacteria causes an increase of cellular biomass that helps to change flow patterns. Nutrients and injected bacteria preferentially flow through the areas of highest permeability, and as a result of cell growth, the generated biomass creates plugs in these highly permeable areas, forcing the fluid to flow through less permeable areas with greater oil saturation [23]. This same effect is caused by bacteria that produce water-insoluble biopolymers, polysaccharides secreted by some bacteria as a means of protection and to facilitate their adhesion to a surface [24].

3.4 Degradation of Hydrocarbons

One of the most important functions of the bacteria used in MEOR is their ability to break down heavier hydrocarbon chains, generating lighter products with lower viscosity. It has been reported that some bacterial genera preferentially degrade long chains of alkanes with more than 20 carbon atoms [25, 26].

3.5 Production of Biosurfactants Ex Situ

When biosurfactants are produced ex situ, the bacteria are isolated from oil obtained from the well to be stimulated or from water from the formation [25, 27].

In other cases, they are obtained from a site contaminated by oil spills [8, 15, 17, 28]. Next, in appropriate facilities outside the well to be intervened, the bacteria are provided the nutrients and appropriate conditions for their growth. Once the maximum population growth stage is reached, the culture broth containing the biosurfactants, with or without the cells, is injected into the reservoir through the injector wells [7, 13, 21].

Microbial metabolites remove hydrocarbons that adhere to the rock pore walls as well as paraffin deposits. This process is generally recommended for reservoirs rich in paraffins [13]. The application can also be performed directly in the producer well for cleaning purposes. Most of the experimental data available on the supply of biosurfactants have been obtained in the laboratory, simulating well conditions using sand-packed columns or Berean stone plugs and oil-impregnated sandstone or carbonates [12, 13, 17, 26, 28, 29].

3.6 Production of Biosurfactants for the MEOR Process in Situ

The in situ production of biosurfactants involves supplying specific nutrients to stimulate the growth of native bacteria in the reservoir capable of synthesizing biosurfactants and degrading hydrocarbons [30]. Native oleophilic bacteria grown in the laboratory can also be introduced. The introduction of nutrients is performed through the injection wells.

The in situ production of biosurfactants by stimulating native bacteria or by the introduction of exogenous bacteria has several advantages over ex situ production. The bacteria produce biosurfactants, organic acids, solvents, and gases (primarily CO_2 and CH_4) that increase the pressure in the reservoir, favoring the displacement of oil. Additionally, these oleophilic bacteria are able to break down the heavier hydrocarbon chains, generating lighter products with lower viscosity [25, 26].

However, the supply of nutrients or the inoculation of exogenous bacteria has some drawbacks. For example, the temperature of the reservoir is a critical factor for the selection of microorganisms. The bacteria that have been most studied for MEOR in situ processes are thermophilic bacteria of the genus *Bacillus*, which are generally isolated from the formation water. The nutrients can cause the multiplication of sulphate reducing bacteria that produce H_2S. The selection and adaptation of the microorganisms that will be inoculated in the reservoir is crucial, and the resistance of the inoculated microorganisms to temperature, pressure, high salinity, pH, low oxygen content, and the presence of heavy metals must be analyzed. Additionally, cheap and easily accessible carbon sources are required [7, 14].

4 Application of MEOR: Field Results

Many field applications have reported on the use of MEOR, most of which have been successful, substantially increasing oil production and decreasing the water cut. Gao and Zekri [10] summarized the application of MEOR in 11 oil fields in four countries (Malaysia, the United States, Argentina, and China) between 1999 and 2008. In total, 209 wells were involved (185 producers and 24 injectors). In some cases, bacteria and nutrients were injected, whereas in others, only nutrients were injected. The production of 144 wells increased (78% of the total wells studied). Approximately 70% of the treated reservoirs had temperatures below 55 °C, more than 60% of the wells that showed successful results had temperatures below 55 °C, and almost all successful cases had temperatures below 85 °C. Many microorganisms do not survive at high temperatures, thus it seems reasonable to conclude that the MEOR methods are more effective at temperatures below 85 °C. The permeability of the reservoirs fluctuated between 0.70 and 4000 mD, suggesting that the success of the treatments does not depend on permeability. MEOR has been applied successfully in reservoirs with salinities of up to 140,000 ppm and densities between 41° and 16° API [31].

Currently, the MEOR process is no longer a promising technology and has become a commercial technology. This technology has been commercially applied in 48 commercial oil fields, the results of which are summarized in Table 2 [32].

According to published information [30–35], the appropriate native microorganisms from the reservoir were carefully selected and stimulated by providing adequate nutrients, which were fed into the reservoir through injection wells or directly into the producing wells.

Table 2 Application of the MEOR process in oil fields [32]

Type of treatment	Number of treatments	Number of wells	Number of successful treatments	Success percentage (%)	Increase in recovery (%)
In situ testing in producing wells	49	47	36	73	140
Producer wells	19	18	17	89	133
Injector wells	238	81	234	98	54
Total	306	146	287	94	92

5 Conclusions

(a) MEOR is an environmentally friendly and safe technology.
(b) The field results show a significant increase in the recovery of oil in wells with a high water cut.
(c) There does not seem to be any significant relationship between the permeability of the reservoirs and the success of MEOR.
(d) MEOR provides the best results in reservoirs with temperatures below 85 °C, although it has been used in reservoirs with higher temperatures [13].
(e) MEOR is applicable in reservoirs with salinities of up to 140,000 ppm.
(f) It is not essential to introduce bacteria into the reservoir for MEOR, as a number of native species can be stimulated using the appropriate nutrients.
(g) MEOR is applicable over a wide range of API densities.
(h) MEOR utilizes the surface facilities from the waterflooding process.
(i) Depending on the characteristics of the reservoir, the increase in the cost per barrel may be less than USD$10 [32]. However, lower costs of USD$2.90 and $5.10 have been reported in the Papagayos and Piedras Coloradas fields in Argentina, respectively [10].

References

1. Hall, C., Tharakan, P., Hallock, J., Cleveland, C., Jefferson, M.: Hydrocarbons and the evolution of human cuture. Nat. **426**, 318–322 (2003)
2. Shibulal, B., Al-Bahry, S.N., Al-Wahaibi, Y.M., Elshafie, A.E., Al-Bemani, A.S., Joshi, S.J.: Microbial enhanced heavy oil recovery by the aid of inhabitant spore-forming bacteria: an insight review. Sci. World J. 1–12 (2014)
3. Sen, R.: Biotechnology in petroleum recovery: the microbial EOR. Prog. Energy Combust. Sci. **34**(6), 714–724 (2008)
4. Brown, L.R.: Microbial enhanced oil recovery (MEOR). Curr. Opin. Microbiol. **13**(3), 316–320 (2010)
5. US Deparment of Energy.: Enhanced oil recovery:CO_2 injection. Available: http://www.fe.doe.gov/programs/oilgas/eor (2012). Accessed 2016/08/05
6. Bryant, R.S.: Potential uses of microorganisms in petroleum recovery technology. Proc. Okla. Acad. Sci. **67**, 97–104 (1987)
7. Marchant, R., Banat, I.M.: Microbial biosurfactants: challenges and opportunities for future exploitation. Trends Biotechnol. **30**(11), 558–565 (2012)
8. Al-Bahry, S.N., Al-Wahaibi, Y.M., Elshafie, A.E., Al-Bemani, A.S., Joshi, S.J., Al-Makhmari, H.S., Al-Sulaimani, H.S.: Biosurfactant production by *Bacillus subtilis* B20 using date molasses and its possible application in enhanced oil recovery. Int. Biodeterior. Biodegradation **81**, 141–146 (2013)
9. Youssef, N., Simpson, D.R., McInerney, M.J., Duncan, K.E.: In-situ lipopeptide biosurfactant production by Bacillus strains correlates with improved oil recovery in two oilwells approaching their economic limit of production. Int. Biodeterior. Biodegradation **81**, 127–132 (2013)
10. Gao, C.H., Zekri, A.: Applications of microbial-enhanced oil recovery technology in the past decade. Energy Sour, Part A: Recovery, Utilization, and Environ. Eff. **33**(10), 972–989 (2011)

11. Al-Sulaimani, H., Joshi, S., Al-Wahaibi, Y., Al-Bahry, S., Elshafie, A., Al-Bemani, A.: Microbial biotechnology for enhancing oil recovery: current developments and future prospects. Biotechnol. Bioinf. Bioeng. **1**(2), 147–158 (2011)
12. Al-Sulaimani, H., Al-Wahaibi, Y., Al-Bahry, S., Elshafie, A., Al-Bemani, A., Joshi, S.: Residual-oil recovery through injection of biosurfactant, chemical surfactant, and mixtures of both under reservoir temperatures: induced-wettability and interfacial-tension effects. SPE Reservoir Eval. Eng. **15**(02), 210–217 (2012)
13. Arora, P., Ranade, D.R., Dhakephalkar, P.K.: Development of a microbial process for the recovery of petroleum oil from depleted reservoirs at 91–96 oC. Biores. Technol. **165**, 274–278 (2014)
14. Al-Sulaimani, H., Al-Wahaibi, Y., Al-Bahry, S., Elshafie, A., Al-Bemani, A., Joshi, S., Zargari, S.: Optimization and partial characterization of biosurfactants produced by Bacillus species and their potential for ex-situ enhanced oil recovery. SPE J. **16**(03), 672–682 (2011-a)
15. Ojeda-Morales, M.E., Domínguez-Domínguez, M., Hernández-Rivera, M.A., Álvarez-Ramírez, J.G.: Biosurfactant Synthesized by Azospirillum lipoferum ALM1B2: Characterization and Application for Environmental Protection. Water Air Soil Pollut. **227**(6), 1–13 (2016)
16. Kowalewski, E., Rueslåtten, I., Steen, K.H., Bødtker, G., Torsæter, O.: Microbial improved oil recovery-bacterial induced wettability and interfacial tension effects on oil production. J. Petrol. Sci. Eng. **52**(1), 275–286 (2006)
17. Al-Wahaibi, Y., Joshi, S., Al-Bahry, S., Elshafie, A., Al-Bemani, A., Shibulal, B.: Biosurfactant production by Bacillus subtilis B30 and its application in enhancing oil recovery. Colloids Surf. B **114**, 324–333 (2014)
18. Ojeda-Morales, M.E., Domínguez-Domínguez, M., Hernández-Rivera, M.A., Zavala-Cruz, J.: Biosurfactant production by strains of azospirillum isolated from petroleum-contaminated sites. Water Air Soil Pollut. **226**(12), 1–15 (2015)
19. Suthar, H., Hingurao, K., Desai, A., Nerurkar, A.: Selective plugging strategy-based microbial-enhanced oil recovery using Bacillus licheniformis TT33. J. Microbiol. Biotechnol. **19**(10), 1230–1237 (2009)
20. Al-Araji, L., Rahman, R.N.Z.R., Basri, M., Salleh, A.B.: Microbial surfactant. Asia Pacific J. Mol. Biol. Biotechnol. **15**(3), 99–105 (2007)
21. Desai, J.D., Banat, I.M.: Microbial production of surfactants and their commercial potential. Microbiol. Mol. Biol. Rev. **61**(1), 47–64 (1997)
22. Bryant, R.S., Bailey, S.A., Stepp, A.K., Evans, D.B., Parti, J.A., Kolhatkar, A.R.: Biotechnology for Heavy Oil Recovery, https://www.researchgate.net/publication/236869112 (1998)
23. Cheng, J., Li, W., Zhang, J., Wu, J., Yang, Z. and Guo, C.: Studies on de Pilot Test with Microbial Profile Modification after Polymer Flooding in Da-qing Oil Field. Paper No. IPTC 11227, International Petroleum Technology Conference, Dubai, UAE (2007)
24. Stewart, T.L., Scott, H.F.: Biomass plug development and propagation in porous media. Biotechnol. Bioeng. **72**(3), 353–363 (2001)
25. Gudiña, E.J., Pereira, J.F., Rodrigues, L.R., Coutinho, J.A., Teixeira, J.A.: Isolation and study of microorganisms from oil samples for application in microbial enhanced oil recovery. Int. Biodeterior. Biodegradation **68**, 56–64 (2012)
26. Xiaolin, W., Zhaowei, H., Xumou, D., Wei, L., Rui, W., Xiaolei, W.: The Application of Microbial Enhanced Oil Recovery in Chaoyanggou Daqing Low-Permeability Oilfield. Open Pet. Eng. J. **5**, 118–123 (2012)
27. She, Y.H., Zhang, F., Xia, J.J., Kong, S.Q., Wang, Z.L., Shu, F.C., Hu, J.M.: Investigation of biosurfactant-producing indigenous microorganisms that enhance residue oil recovery in an oil reservoir after polymer flooding. Appl. Biochem. Biotechnol. **163**(2), 223–234 (2011)
28. Joshi, S., Bharucha, C., Jha, S., Yadav, S., Nerurkar, A., Desai, A.J.: Biosurfactant production using molasses and whey under thermophilic conditions. Biores. Technol. **99**(1), 195–199 (2008)

29. Jimoh, I.A.: Microbial Enhanced Oil Recovery. Luma Print, 6700 Esbjerg. Ph.D. Thesis, Aalborg University Esbjerg, Denmark (2012)
30. Jenneman, G.E., Miffitt, P.D., Young, G. R.: Application of a microbial selective-plugging process at the north burbank unit: prepilot tests. SPE Paper 27827. 1994 SPE/DOE Symposium on Improved Oil Recovery, Tulsa, Oklahoma, USA (1994)
31. Zahner, R.L., Tapper, S.J., Marcotte, B.W.G., Govreau, B.R.: What has been learned from a hundred MEOR applications. SPE Paper 145054. SPE Enhanced Oil Recovery Conference, Kuala Lumpur, Malaysia (2011)
32. Titan Oil Recovery Inc. http://titanoilrecovery.com/, Accessed 2017/07/11
33. Town, K., Sheehy, A.J., Govreau, B.R.: MEOR success in Southern saskatchewan. SPE Paper 124319. SPE Annual Technical Conference and Exhibition, New Orlans, USA (2009)
34. Zahner, R.L., Sheehy, A., Govreau, B.R.: MEOR success in Southern California. SPE Paper 129742. 2010 SPE Improved Oil Recovery Symposium, Tulsa, Oklahoma, USA (2010)
35. Akintunji, F., Marcotte, B.W.G., Sheehy, A., Govreau, B.R.: A Texas MEOR application shows outstanding production improvement due to oil release effects on relative permeability. SPE Paper 154216. Eighteenth SPE Improved Oil Recovery Symposium, Tulsa, Oklahoma, USA (2012)

Does Ecuador Prevent Pollution Caused by Ships?

Nadia Mercedes Mendieta Villalba

Abstract Maritime transport is an important source of sea pollution, which is why active prevention policies are required to minimize or prevent its effects and consequences, specifically in two relevant aspects: operational pollution caused by the manipulation of polluted merchandise and accidental pollution whose control is effected through the international maritime conventions for both the ships and crew. In this context, this investigation presents a synthesis of the issue of the lack of pollution prevention caused by ships in Ecuador, based on a study about compliance with the conventions by the International Maritime Organization, to which Ecuador is a signatory. To this effect, maritime statistics for the world and the coast of Ecuador are presented. Additionally, the paper emphasizes the International Convention for the Prevention of Pollution by Ships (MARPOL 73/78), which regulates the construction of facilities for the reception at ports in which ships can dock and unload the pollutants produced during their operations and the designations used in maritime areas under this convention. Finally, through a poll for maritime personnel, the real importance given to international maritime conventions is shown as well as the need to overcome weaknesses and suggest new initiatives on the subject of sea environmental safety and protection in Ecuador.

Keywords International Maritime Organization · International maritime conventions · Pollution prevention · Reception facilities

N. M. Mendieta Villalba (✉)
Universidad Politécnica Salesiana, Chambers # 227 and 5 de Junio,
Guayaquil, Ecuador
e-mail: nmendieta@ups.edu.ec

A. Vega Sáenz et al. (eds.), *Proceedings of the 25th Pan-American Conference of Naval Engineering—COPINAVAL*, https://doi.org/10.1007/978-3-319-89812-4_34

1 Introduction

1.1 Ecuador and the International Maritime Organization (IMO)

The sources of sea pollutants include the discharges by oil tankers, ballast water, sewage water, and garbage, among others, which are released intentionally or by accident during collisions or groundings [1]. Other sources are the emissions of sulphur oxide (SOx) and nitrogen oxide (NOx), which are used-fuel emissions, harmful for human beings. Today, mandatory technical and operational measures have been established to reduce the emissions of greenhouse gases by ships [2].

The IMO (International Maritime Organization), a specialized agency of the United Nations, establishes regulations that the crafts that perform international trips must fulfill to guarantee maritime security, protection of human life at sea, and prevention of environmental pollution [3].

Ecuador has ratified 16 conventions, international protocols, and IMO amendments, integrating them to its legislation once this takes effect [4]. However, the existence of the IMO regulatory framework does not prevent maritime accidents from happening, because to prevent accidents (and, consequently, pollution too) it is necessary to verify compliance with IMO regulations in ships and national ports through inspections, a job that must be in turn verified by the Ecuadorian maritime authority [5].

1.2 Pollution from Ships

1.2.1 Causes and Consequences

In the world, during the period from 1970 to 2016, approximately 50% of all oil spills happened while the ships were traversing open waters. Collisions and groundings represented 59% of the causes for these spills and 19 of the biggest spills happened before the year 2000 [6].

Maritime accidents produce pollution and can be caused by human error (navigation or piloting mistakes), fatigue, age of the ship, structural failure (cracking of the hull, corrosion), fires, and explosions. Additionally, the oil slick formed has a process on the ocean that depends on sea currents, meteorology, states of the sea, and properties of the oil, such as viscosity and thickness [7].

On the other hand, when the oil slick arrives at the coast, sea turtles, dolphins, or seals are poisoned; reefs and sea creatures die by suffocation. The feathers on the birds in contact with the oil lose their insulating properties which can produce their deaths by hypothermia [8]. Other consequences are a decrease in the production of fishing communities, decrease in tourism, and serious public health issues.

1.2.2 Sea Disasters in Ecuador

The statistics for ship accidents in Ecuador for the 2005–2010 period are presented in the graphs shown in Figs. 1, 2, 3 and 4, based on the statistics of maritime accidents recorded in the archive of the General Directorate of Merchant Navy and Littoral (Digmer in Spanish).

2 Measures to Prevent Pollution Caused by Ships

2.1 Legal Instrument

The legal instrument published by the IMO to prevent pollution caused by ships is the International Convention for the Prevention of Pollution from Ships (Marpol 73/78), which stipulates in its six technical annexes the ban of the release of substances

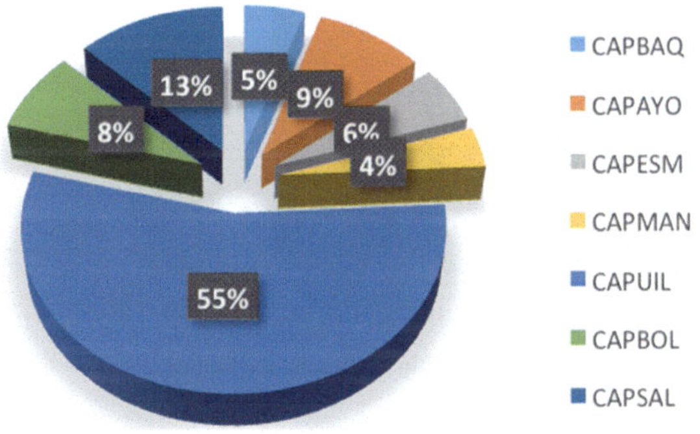

Fig. 1 Number of maritime accidents by port captaincy

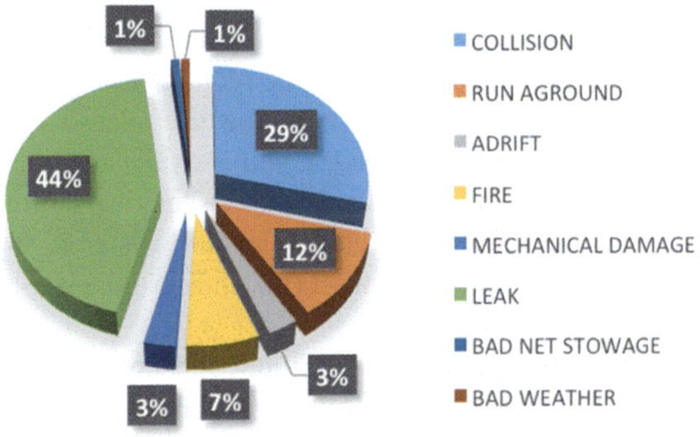

Fig. 2 Percentage of maritime disasters by cause

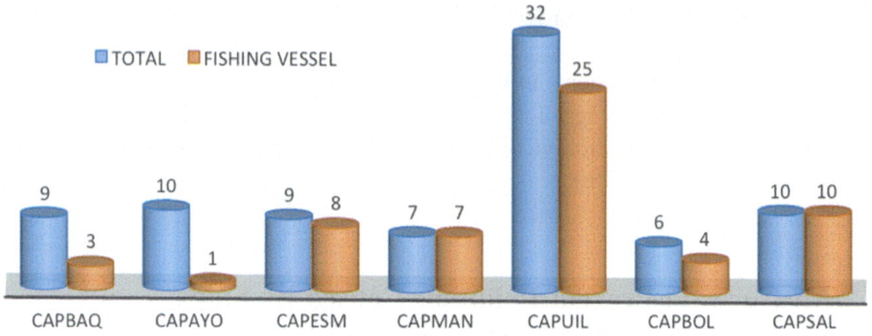

Fig. 3 Number of accidents that resulted in sinking by port captaincy

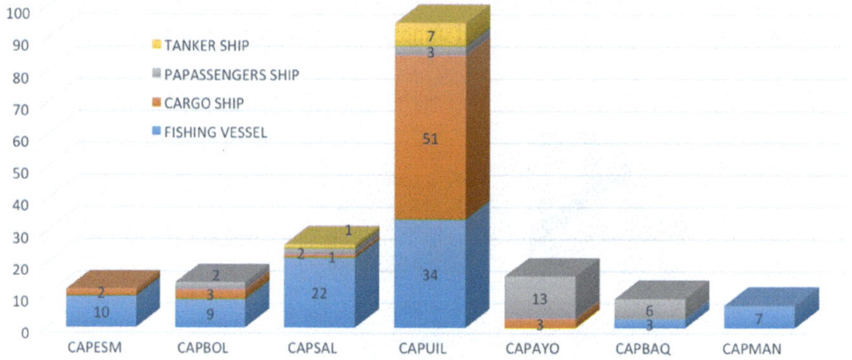

Fig. 4 Number of accidents by type of ship for each port captaincy

or effluents that are harmful for the sea environment. Marpol 73/78 specifies the monitoring, inspections, and supervision for the issuing of international certificates and grants the designation of special areas particularly sensitive for countries that show ecosystem vulnerability [9]. Ecuador was granted said particularly sensitive area designation for the Galapagos Islands in the year 2005 [10].

Marpol 73/78, in its annex VI, establishes the requirements for the regulation of atmospheric pollution released by ships, including the release of substances that deplete ozone, such as nitrogen oxides (NOx), sulfur oxides (SOx), volatile organic compounds, and onboard incineration. It also establishes dispositions for reception facilities for waste generated by the exhaust cleaning systems, incinerators, fuel quality, drilling platforms, and offshore platforms and for the establishment of SOx Release Control Zones [11].

In 2011, mandatory energy-efficiency technical and operational measures were stipulated for the reduction of greenhouse effect gases by ships [2] and in 2015 new directives for the convention appeared due to an increase in the size of the sulfur emissions in the controlled areas (SECA). This zone will cover the international waters of the Republic of Ireland in 2020, culminating with the inclusion of the whole of Western Europe [11].

2.2 *Reception Facilities and Garbage Disposal*

Marpol 73/78 demands the construction of reception facilities in the ports to unload pollutants coming from ships: hydrocarbons, chemicals, sewage, ballast water, and so on [12]. Through the reusing and recycling of materials the production of garbage onboard the ships should be decreased (3 kg–6 lbs/person/day). The IMO establishes specifications for garbage incinerating plants of over 4000 kW per unit [13].

2.3 *Role of the Ecuadorian Maritime Authority*

The current Ecuadorian maritime authority is the Subsecretariat for Ports and Fluvial and Maritime Transportation, whose role is to control the compliance of the current regulations for ships and ports [14].

2.4 *Training*

The most frequent cause of accidents is human error; the use of simulators for the training of the crew is relevant, performing tasks that are very similar to reality. In the Global Maritime University of Sweden, an organism of the United Nations specialized in graduating and investigating level maritime education, including IMO conventions [15], Ecuadorians have been trained, but not in sufficient numbers.

2.5 *New Building*

New tankers count with double hull, separated ballast tanks, slop tanks, and pollution prevention equipment (bilge separator, alarms, automatic detention device, etc.). This is a product of the regulations introduced by the Marpol convention and of the most recent oil spills.

2.6 Routes and Identification and Report

The ships must follow the navigation charts; if the traffic density is high, the traffic separation device and the use of electronic charts must be applied. The authorities should identify and report before the possibility of pollution or other illegal activities performed by ships.

2.7 Establishment of a Specialized Body and Raising Maritime Awareness

The establishment of a specialized regional body is relevant with the aim of coordinating, controlling, and giving scientific and technical assistance after international law violations, such as the European Maritime Safety Agency (ESMA) which is based in Lisbon, Portugal [16]. As such, the participation and education of all the participants in all the areas of the maritime community involved in the prevention of maritime pollution is of great importance.

3 Methodology

At a global level, compliance with the IMO regulations has reduced the number of maritime disasters, especially the loss of ships, loss of lives, and oil spills, despite an increase of the fleet that has occurred through the years. In the statistic for 2016, given by the International Federation of Pollution of Tanker Owners [6], it is evident that the number of oil spills has decreased in the last two decades; see Fig. 5.

It is worth noting that if the IMO conventions were fully observed in Ecuador, the number of accidents, of the previous figures, would decrease, in line with the global tendency.

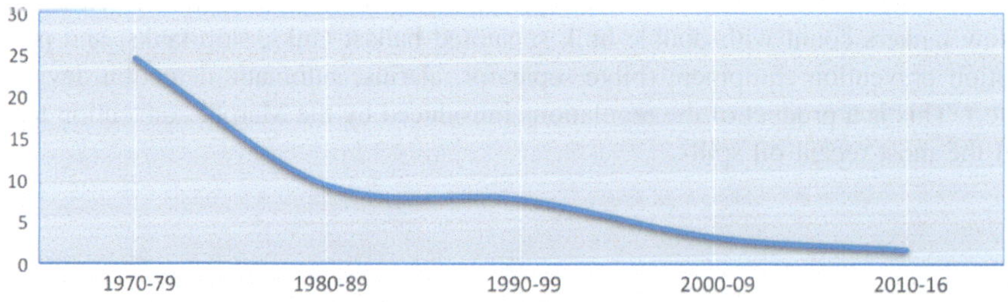

Fig. 5 Average amount of big oill spills per year (>700 tons)

Table 1 Population of maritime managers

Maritime managers	Population	Sample
Naval force educational centers	20	20
Ecuadorian universities that form third-level professionals in the maritime field	4	4
Marine gencies affliated to the Ecuador maritime chamber corporation	41	27
Naval engineers college	31	27
Total	96	78

With the purpose of showing the relevance given to IMO conventions inside the Ecuadorian maritime community, the poll technical instrument was employed, targeted to a sample of 78 managers of organizations and maritime educative institutions; see Table 1 [5].

Next, the five questions that composed the poll instrument were applied to maritime personnel. As shown in each of the graphs, the opinions of each maritime manager are tabled by their categories: totally, in large quantity, moderately, very little, and nothing.

Question 1: Do you consider convenient to continuously train about maritime conventions? In Fig. 6 the sample criterion is categorized based on the poll.

Question 2: Do you consider relevant to build reception facilities in the Ecuadorian ports? In Fig. 7 the sample criterion is categorized based on the poll.

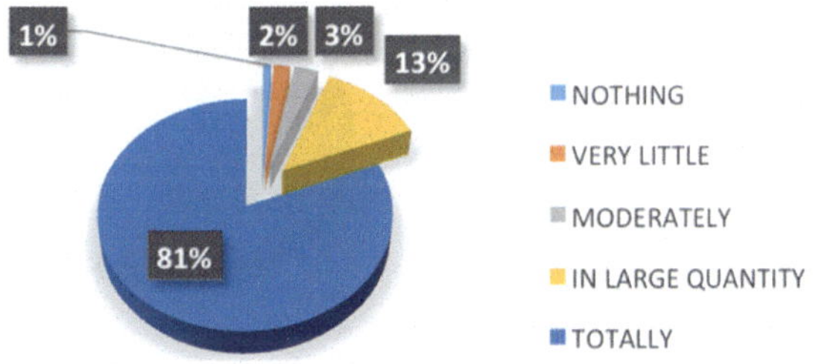

Fig. 6 Continuous training in international and national regulations

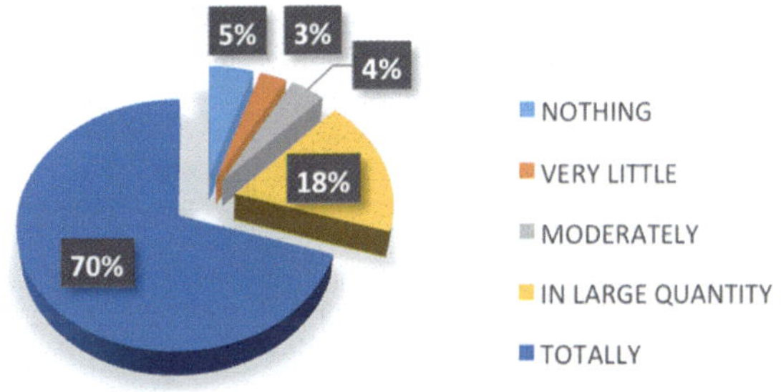

Fig. 7 Building of reception facilities in Ecuadorian ports

Question 3: Do you believe Ecuador is prepared to solve the problem of oil spillage? In Fig. 8 the sample criterion is categorized based on the poll.

Question 4: Do you consider necessary to increase the fines by illegal discharges of pollutants by ships in Ecuadorian waters? In Fig. 9 the sample criterion is categorized based on the poll.

Question 5: Do you believe it is necessary to raise an ecological awareness in the maritime and coastal communities of our country? In Fig. 10 the sample criterion is categorized based on the poll.

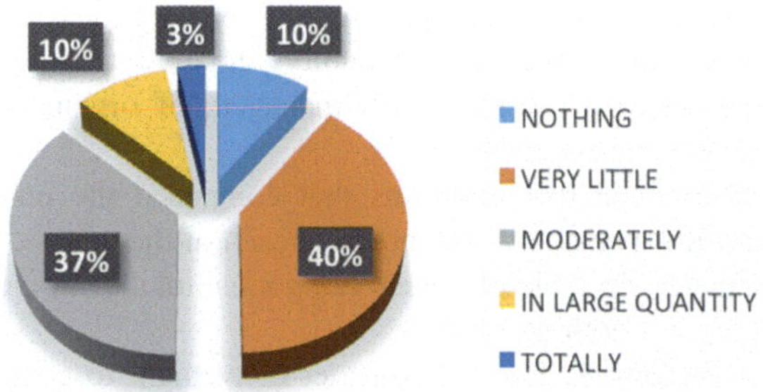

Fig. 8 Ecuador prepared to solve an oil spill

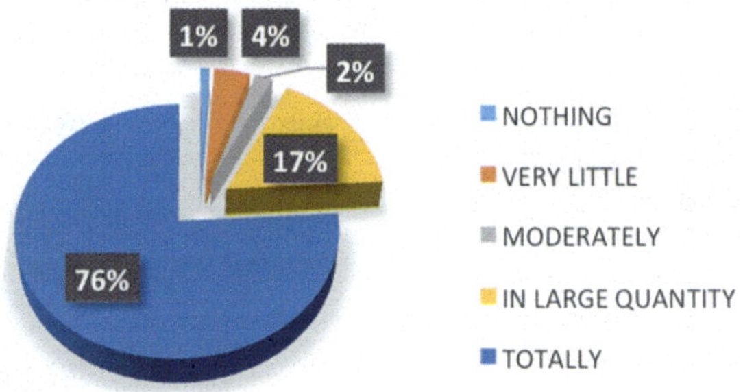

Fig. 9 Increase of fines for illegal discharge of pollutants from ships

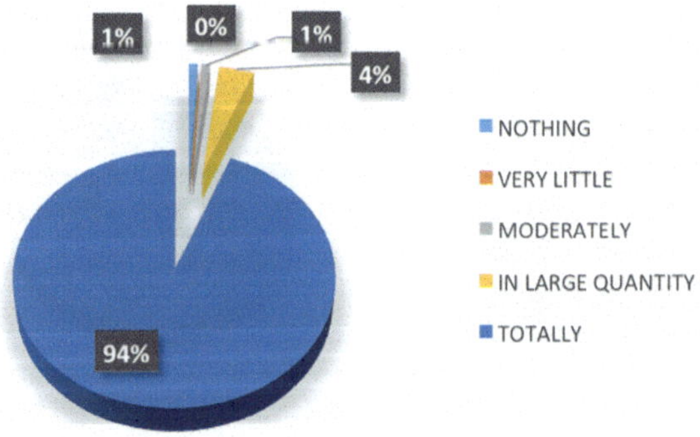

Fig. 10 Creation of ecological awareness in maritime and coastal communities

4 Discussion of Results

The global statistic of Fig. 5 indicates that the introduction of new IMO regulations and their compliance, along with the normalization of ships to existing laws, has allowed the gradual diminishing of the loss of ships, lives, and oil spills, despite the increase of tonnage of the world fleet.

In Fig. 6, it is revealed that the Ecuadorian maritime community is conscious that the technological changes and ship accident reports continuously modify IMO conventions, thus continuous training is required. It is verified that for the community, the IMO conventions are relevant and mandatory, as it also is in Fig. 7, where the need for the building of reception facilities in ports to fulfill the acquired compromise is revealed.

In Fig. 8, the maritime community recognizes that Ecuador is not ready to solve an oil spillage. According to Fig. 9, the requirement for the increase of fines for illegal pollutant discharges from ships is proven. In Fig. 10, the relevance of raising maritime ecological awareness in the Ecuadorian coastal and maritime communities is verified through the application of communication programs, action plans, and campaigns that promote an environmental culture.

5 Conclusions

Ecuador is not sufficiently prepared to prevent the pollution caused by ships. It is concluded that if the Ecuadorian maritime authority, through its inspectors, continuously and mandatorily verifies the observance of the IMO regulations in Ecuadorian ships and ports, the amounts of the national statistics from Figs. 1 through 4 will be reduced in the future, as has happened with world statistics.

In this study no access was obtained to the information about the inspections in the ships, however, the importance was shown of maritime authority in the prevention of pollution caused by ships, and in the verification of the observance of the IMO regulations performed by captaincy supervisors in ships and ports. If both activities were to be coordinated by the maritime authority, a positive effect will occur regarding the reduction of pollution.

In this paper it was sought to learn, through a poll, the compliance and importance of the Marpol convention by maritime managers. The results obtained in Fig. 6 confirm what is exposed from the theoretical point of view: the updating of knowledge of IMO instruments through continuous training would stop the human error and, thus, pollution. As such, the national maritime laws should be reformed periodically and workshops organized about the new requirements and implications of the Marpol conventions, national and regional contingency plans, promoting the exchange of knowledge and contributing to the spread of technology and resources among countries. The Ecuadorian maritime community can be supported with scholarships for students with high academic performance to participate in investigations abroad.

With the purpose of sharing the real importance of prevention of pollution by ships, it is necessary to raise maritime awareness through communication programs, action plans, and campaigns that help promote an environmental culture, as there were participants in the poll that did not answer because of their lack of interest or knowledge.

References

1. Kasmin, S.: Enforcing ship-based marine pollution for cleaner sea in the Strait of Malacca. Environ. Asia **3**, 61–65 (2010)
2. OMI Homepage.: http://www.imo.org/es/About/Conventions/ListOfConventions/Paginas/International-Convention-for-the-Prevention-of-Pollution-from-Ships-%28MARPOL%29.aspx. Accessed 2017/09/17
3. OMI Homepage.: http://www.imo.org/es/OurWork/Paginas/Home.aspx. Accessed 2017/09/17
4. OMI Homepage.: http://www.imo.org/es/About/Conventions/StatusOfConventions/Paginas/Default.aspx. Accessed 2017/09/17
5. Mendieta, N.: Training in maritime in public and private organizations in continuing education and proposed safety of a module of prevention of accidents and illicit actions at sea. Master's thesis, University of Guayaquil, Guayaquil, Ecuador (2012)
6. ITOPF Homepage.: https://www.itopf.com/knowledge-resources/data-statistics/statistics/. Accessed 2017/09/17
7. Leifer, I., et al.: State of the art satellite and airborne marine oil spill remote sensing: application to the BP Deepwater Horizon oil spill. Remote Sens. Environ. **124**, 185–209 (2012)
8. Valverde, R.: Effects of oil spills in marine ecosystems. La Bici **2**(2), 15 (2002)
9. International Maritime Organization: International Convention for the Prevention of Pollution from ships, 1973 and Protocol 1978. MARPOL 73/78. Consolidated Edition. International Maritime Organization, London, United Kingdom (2011)
10. Cordova, E.: Safety on vessels of cabotage that operate in the Galapagos Archipelago. B.A. thesis, University of the Armed Special Forces, ESPE, Quito, Ecuador (2014)
11. WIKIPEDIA, Homepage.: https://en.wikipedia.org/wiki/MARPOL_73/78. Accessed 2017/09/19
12. Torres, N., Berguño, J.: Marine Debris: global and regional impacts. In: Annals of Patagonia institute, 137–146 (2011)
13. International Maritime Organization: Imo Marine Environment Protection Committee completes 66th session. IMO NEWS The magazine of the International Maritime Organization **2**, 10–12 (2014)
14. Ministry of Transport and Public Works Homepage, http://www.obraspublicas.gob.ec/puertos-y-transporte-maritimo-y-fluvial/. Accessed 2017/09/17
15. World Maritime University Homepage.: https://www.wmu.se/. Accessed 2017/09/18
16. Balan-Rusu, M., Pircalabescu, N.: Maritime Safety and Prevention of Pollution Caused by Ships in the European Union. J. Danubian Studies Res. **2**(1), 113–119 (2012)

Analysis of the Possible Risk of Capsizing in a 45 m Tuna-Fishing Vessel That Has Been Elongated in the Middle Section

Oscar E. Viteri and Franklin Dominguez Ruiz

Abstract The increased distance of tuna fishing grounds necessitates the fishing industry to seek the best use of existing fishing vessels. This project begins with a 36 m sardine vessel and converts it into a tuna-fishing vessel with an insulated storage space. Once the new main dimensions have been calculated, an intact stability analysis is conducted according to the International Maritime Organization's criteria for this type of vessel. Subsequently, a simulation to evaluate the hazard criteria of a parametric roll is performed. Hereby, the matching period of roll and period of encounter, the areas of instability p and q, and Dunwoody's critical height are evaluated. Then, a comprehensive calculation is made by applying the methodology proposed by Themelis and Spryrou (Probabilistic Assessment of Ship Stability. Transactions—SNAME, 2007, [15]) as well as by adding a water shipping component to the general roll equation. Finally, to analyze the possible risk of loss (capsizing) based on a confidence interval of at least 86.5%, corresponding to a significant $[H_s]$, an analysis of the critical wave height at each node, as proposed by Dunwoody (Roll of a ship in Astern Seas—Metacentric height spectra, 1989, [4]), is carried out.

Keywords Capsizing · Parametric roll · Water on deck

1 Introduction

The analysis of the possible risk of capsizing, as explored in this paper, focuses on establishing a set of vulnerability criteria that enable an easy application for tuna-fishing vessels. In this case, the risk criteria are applied to a case study of a 36 m sardine vessel that has been lengthened to 45 m.

O. E. Viteri (✉) · F. Dominguez Ruiz
Escuela Superior Politécnica del Litoral, Guayaquil, Ecuador
e-mail: oeviteri@espol.edu.ec

F. Dominguez Ruiz
e-mail: jodoming@espol.edu.ec

A. Vega Sáenz et al. (eds.), *Proceedings of the 25th Pan-American Conference of Naval Engineering—COPINAVAL*, https://doi.org/10.1007/978-3-319-89812-4_35

Table 1 Primary vessel data

Length, L	(m)	45.10
Length between PP, LPP	(m)	38.48
Beam (B)	(m)	7.92
Depth (P)	(m)	4.75
Draught (H)	(m)	4.38
Displacement, Δ	(Ton)	803.30
Block coefficient, C_B	(–)	0.56
Ship speed, V_b	(Knots)	10.50
GM_T in calm water	(m)	0.57
Period of roll, T_{roll} to calm	(s)	0836
GM_L	(m)	34.32
Pitch period, T_{pitch}	(s)	9.91

The first part of this study carries out the dimensional studies and the respective optimization based on a genetic algorithm to define the main dimensions listed in Table 1 [1].

The case study consists of the following analysis process.

- Noting the navigation route, both departure and return
- Identifying equidistant nodes from the port to the fishing area, and vice versa, using weather forecasts
- Making oceanographic calculations of spectral densities and solving RAO to estimate the average heights, $H_{1/3}$, $H_{1/10}$, $H_{1/100}$

The risk criteria are applied as follows.

1. Criterion for possible matches [2]
2. Criterion for areas of instability (p, q) [3]
3. Critical height criterion [4, 5].

1.1 State of the Art

The number of capsizing incidents has given rise to much research on this type of disaster. The International Maritime Organization (IMO) [6, 18] has identified the following phenomena as factors in capsizing.

1. Pure loss of stability
2. Parametric roll
3. Surf-riding/broaching
4. Dead ship
5. Excessive accelerations

The IMO further provides captain recommendations on how to avoid the resonant movements that can lead to capsizing.

This research draws upon the related studies by Belenky [7], Dunwoody [4, 5] Garvaky [8], Himeno [9], and ABS [10] and takes their findings into consideration. These works have contributed to the mathematical description of the parametric resonance phenomenon. The present study aims to establish the hazard criteria for parametric resonance in such a way that critical weather conditions leading to roll resonance can be identified in the design stage as well as during ship operations.

1.2 Literature Review

Probability of Wave Occurrence. Short-term irregular sea conditions can be described using the spectral density function. The sea spectra are in line with statistical distribution. Rayleigh's [11] function is used to calculate the probable density of a particular wave.

$$p(H_i) = \frac{2 \cdot H_i}{\bar{H}^2} \mathrm{e}^{-2H_i/\bar{H}^2} \tag{1}$$

Roll Damping. Roll damping [$\boldsymbol{B_{44e}}$] is an essential characteristic in the parametric balance phenomenon. Roll damping is heavily affected by viscous effects and ship speed.

A model with a unique degree of freedom that describes the movement of a roll of a vessel is shown in Eq. (2):

$$A_\varphi \cdot \ddot{\varphi} + B_\varphi \cdot \dot{\varphi} + C_\varphi \cdot \varphi = M_\varphi(\omega \cdot t) \tag{2}$$

The term $[B_{44e}]$ can be obtained by a combination of the linear and nonlinear components, using Eq. (3), which was proposed by the ITTC [12].

$$B_{44e} = B_{\varphi e} = B_{\varphi 1} + \frac{8}{3\pi} \omega_e \cdot \varphi_a \cdot B_{\varphi 2} \tag{3}$$

Wave Spectrum, DNV Formulation. Steady sea conditions are characterized by a set of environmental parameters: significant wave height H_s, spectral peak period T_p, bathymetry, and wind direction. DNV's sea spectrum formulation has been used, taking the meteorological data of operation as a base. The peak increase factor $[\gamma]$ used by DNV is determined by the significant wave height and the period modal [13], and is given in Eq. (4).

$$S_{\mathrm{DNV}\xi}(\omega_w) = \alpha \cdot \omega_w^{-5} \cdot \exp(-\beta \cdot \omega_w^{-4}) \cdot \gamma^{\exp\left(\frac{-1}{2\cdot\sigma^2}\cdot\left(\frac{\omega}{\omega_p}-1\right)^2\right)} \tag{4}$$

2 Risk Criteria

To analyze the risk of capsizing, the four main criteria of wave height, namely average waves H_{prom} and significant waves $H_{1/3}$, $H_{1/10}$, and $H_{1/100}$, are important. To analyze the resonance in the roll, the following criteria are proposed.

2.1 *Criterion for Possible Matches*

The natural roll at each of the following conditions is equal to:

$$T_\varphi \approx \frac{2}{n} \cdot T_e, \tag{5}$$

where n = 1.2, …

The length of wave, λ_ω, should be within a range of: (0.8–1.2 L).

2.2 *Criterion for Areas of Instability* (p, q)*—Damped*

A second risk criterion is proposed to evaluate the dimensionless parameters p and q in the graph "Ince–Strutt–damped," which is a graphic solution of the differential equation by Mathieu [3]. To obtain a real approximation of p and q, shipping of water at the height of the metacentric calculation (GM_t) must be taken into consideration.

The parameters p and q are part of Mathieu's Eq. (6), defined in Eq. (7) as

$$\frac{d^2\varphi}{d\tau^2} + \mu\frac{d\varphi}{d\tau} + (p + q \cdot \cos(\tau)) \cdot \varphi = 0 \tag{6}$$

$$\begin{aligned} p &= \left(\frac{\omega_m}{\omega_e}\right)^2 - \mu^2 \\ q &= \left(\frac{\omega_a}{\omega_e}\right)^2 \end{aligned} \tag{7}$$

$$\omega_m = \sqrt{\frac{\Delta \cdot \mathrm{GM}_m}{I_{xx} + A_{44}}} \quad \omega_a = \sqrt{\frac{\Delta \cdot \mathrm{GM}_a}{I_{xx} + A_{44}}} \quad \mu = \frac{B_{44e}}{(I_{xx} + A_{44}) \cdot \omega_e}$$

$$\mathrm{GM}_m = 0.5 \cdot (\mathrm{GM}_{\max} + \mathrm{GM}_{\min}) \quad \mathrm{GM}_a = 0.5 \cdot (\mathrm{GM}_{\max} - \mathrm{GM}_{\min})$$

where I_{xx} is inertia, A_{44} is added inertia, and GM_m and GM_a, are the mean value and the range of variation of GM in waves, respectively.

2.3 *Dunwoody's Critical Height Criterion*

This criterion evaluates "the critical level of wave height" according to Dunwoody's procedure [4]. This approach proposes that the fluctuations of GM_t can be expressed as a function of the spectral density $s_{e\xi}$, the frequency of encounter ω_e, and the response ($\mathrm{GM}a/\zeta_a$) evaluated in the frequency of encounter.

From the spectrum of response formula, the critical height of a wave 2*ζ (Eq. 8) can be obtained, thus fulfilling the condition of reducing the dimensionless damping $\Delta\xi$ (Eq. 9).

$$S_{\mathrm{GM}_t} = \left(\frac{\mathrm{GM}_a}{\zeta_a}\right)^2 \cdot S_{e\xi}(\omega_\varphi) \Rightarrow \zeta_a = \mathrm{GM}_a \cdot \left(\frac{S_{\mathrm{GM}_t}}{S_{e\xi}(\omega_\varphi)}\right)^{-\frac{1}{2}} \tag{8}$$

$$\frac{B_{\text{total}}}{B_{\text{crit}}} - \Delta\xi \leq 0 \tag{9}$$

$$\Delta\xi = \frac{\pi \cdot g^2 \cdot S_{\mathrm{GM}_t}}{4 \cdot \omega_\varphi^3 \cdot k_{xx}^4}$$

where $B_{\text{total}} = B_{44e,}$ is the damping coefficient, B_{crit} is the critical damping coefficient (Eq. 10), ω_φ is a ship's natural roll frequency, and k_{xx} is the virtual turning radius.

$$B_{\text{crit}} = 2 \cdot \sqrt{(A_{xx} + I_{xx}) \cdot \rho \cdot g \cdot \nabla \cdot \mathrm{GM}} \tag{10}$$

To link the period of significant wave height in each node, Wiegel's [14] relationship is used:

$$H_s = k_1 \cdot \overline{T}^{K_2} \tag{11}$$

Returning DNV's wave spectrum (Eq. 4), it is possible to leave the significant wave height spectrum and find the height of the critical wave.

3 Case Study

3.1 Route

The route Manta–fishing grounds–Manta was chosen. It is described by nodes, as shown in Fig. 1. The full length of the journey was estimated at 1218 nautical miles with an average speed of 10.5 knots. To simplify calculations, only the maximum displacement condition has been considered. The weather conditions are displayed in Table 2 for the half length of the journey.

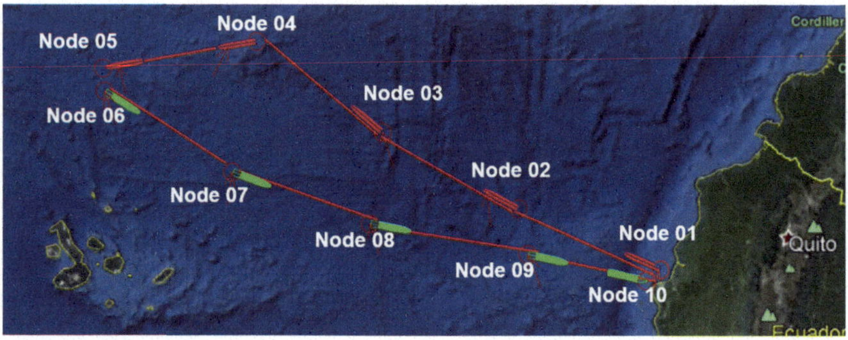

Fig. 1 Outline of the node selections used in this study by Google Earth Pro [15]

Table 2 An example from the port to the fishing grounds, ECMWF data [16]

	Node 01	Node 02	Node 03	Node 04	Node 05
Latitude	0° 46′ 38.27″S	0° 16′ 4.94″N	1° 34′ 8.16″N	3° 15′ 51.56″N	2° 51′ 53.74″N
Length	80° 47′ 6.91″W	83° 16′ 45.49″W	85° 34′ 22.78″W	87° 39′ 24.27″W	90° 36′ 36.76″W
Wave height (m)	1.1	2.2	2.1	2.2	2.4
Wave period (1/s)	9	9	8	9	10
Encounter angle	81	55	32	98	86
Bathymetry (m)	500	3000	3000	2000	2000

4 Results

4.1 Coefficient Calculations of the Roll Equation

This study has obtained the value of added inertia $[A_{44}]$ of 2310.5 [ton*m^2] through the calculation based on the mapping by Lewis [17], and a moment of inertia $[I_{XX}]$ of 8434.2 [ton*m^2]. The values of the damper B_{44e} vary as a function of speed, the period of the wave, the shape of the ship, and the encounter angle of each node.

4.2 Metacentric Height, GM in Waves

A ship's stability in waves has been evaluated from various crest positions along the length and at different heights at each node. We have thus obtained the metacentric minimum and maximum heights of each iteration.

Figure 2 shows the float variation depending on the position of the wave in the hogging and sagging of the ship. A group of waves that continuously generate this effect can eventually cause a ship to capsize.

Figure 3 shows an example calculation of a significant wave height of 2.4 m and varies the length of the wave to determine GM's greatest point of variation in the waves. The wave length selected for this calculation is $\lambda = 1.0\,\mathrm{L}$. This parameter is used by the criterion to determine whether the failure mode can cause instability.

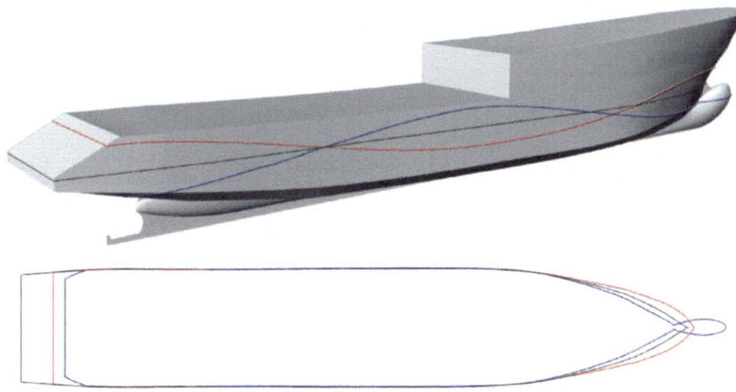

Fig. 2 Waterplane variation depends on the position of the crest, $H_s = 2.2$ m

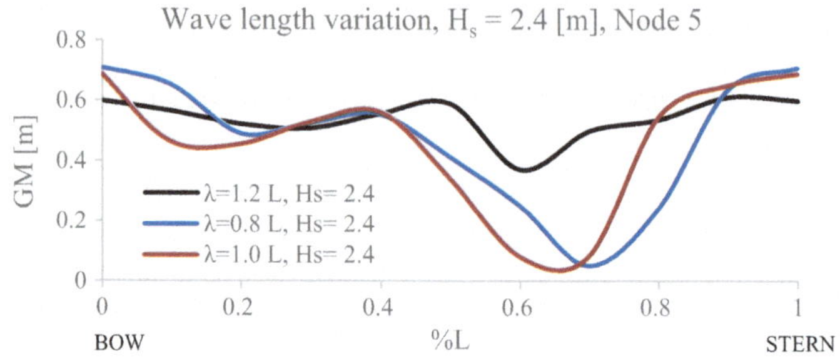

Fig. 3 *GMt* for different positions of the crest along the vessel. $H_s = 2.4$ m

4.3 Possible Criteria Matches (Criterion 1)

Table 3 presents the evaluation of Criterion 1 at corresponding significant heights. Sailing under unstable conditions increases the possibility of capsizing. The area of instability has been marked with the letter "I" and the area of stability has been given the letter "E".

4.4 Criterion for Areas of Instability (p, q)—Damped (Criterion 2)

As a calculation example, Table 4 shows the values obtained for node 4, from the port to the fishing grounds, with $\mu = 0.084$ being the nondimensional damper for this node. Figure 4 shows the obtained values.

Table 3 Summary of Criterion 1 results

	Λ—Deep water (m)		Te (s)	Tφ (s)		Criterion I
	Value	Crit.	Value		Crit.	
Node 01	126.47	E	0957	8.71	E	E
Node 02	126.47	E	11.54	14.59	E	E
Node 03	99.92	E	12.63	13.79	E	E
Node 04	126.47	E	8.25	14.59	E	E
Node 05	156.13	E	10.25	16.30	E	E
Node 06	156.13	E	7.91	16.30	I	**I**
Node 07	156.13	E	9.33	12.34	E	E
Node 08	126.47	E	7.63	13.79	I	**I**
Node 09	56.21	E	4.61	8.70	E	E
Node 10	56.21	E	4.97	8.68	E	E

Bold corresponds to instability of the ship in the correspondent node

Table 4 Results of Criterion 2 at node 4

	$\varphi(^{\circ})$	P_e	$H_{\omega}(\mathrm{m})$	KG fluido(m)	$\mathrm{GM_a(m)}$	GM_m (m)	$\omega_m\left(\frac{\mathrm{rad}}{\mathrm{s}}\right)$	$\omega_a\left(\frac{\mathrm{rad}}{\mathrm{s}}\right)$	$\widehat{\omega_m}$	$\widehat{\omega_a}$	**P**	**q**
ave.	4.62	0.45	1.37	3.54	0.02	0.55	0.64	0.13	0.88	0.18	0.76	0.03
1/3	7.38	0.14	2.19	3.54	0.21	0.40	0.55	0.40	0.75	0.55	0.55	0.3
1/10	9.37	0.04	2.80	3.54	0.29	0.37	0.53	0.47	0.72	0.64	0.51	0.40
1/100	12.31	0.00	3.67	3.54	0.33	0.27	0.45	0.50	0.62	0.68	0.37	0.46

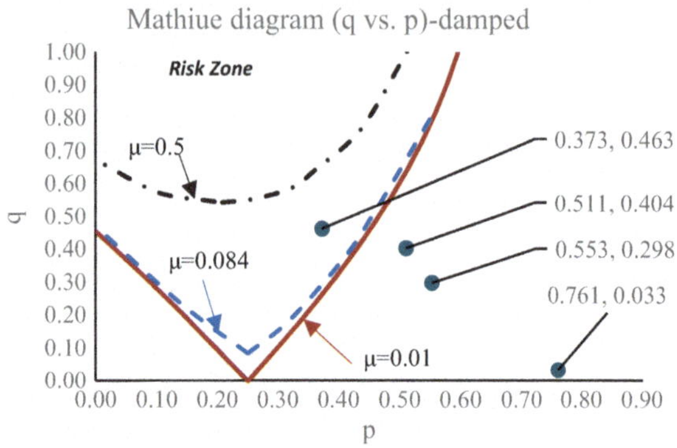

Fig. 4 Ince–Strutt diagram damped with the results of node 4

4.5 *Dunwoody's Critical Height Criterion (Criterion 3)*

Table 5 summarizes the critical height calculations for each node and their respective GM variations, spectrum density, and angle of encounter.

Table 6 shows a comparison of each of the risk criteria for the corresponding H_s of each node.

Table 5 Results of Criterion 3 for each node considering FN = 0.0 (without speed)

Node	H_s–Nodal (m)	Ψ (deg)	GM_a (m)	B_{44e}, FN = 0 (ton*m²/seg)	$H_{crit,\ FN=0}$ (m)	Criterion $_{FN=0}$
1	1.10	81	0.54	380.68	2.52	E
2	2.20	55	0.40	371.47	3.23	E
3	2.10	32	0.41	367.63	4.21	E
4	2.20	98	0.40	387.20	2.22	E
5	2.40	86	0.39	377.14	2.44	E
6	2.20	140	0.40	392.08	1.84	**I**
7	1.90	102	0.43	382.10	2.19	E
8	2.10	118	0.41	394.49	1.98	**I**
9	1.10	125	0.54	441.35	1.82	E
10	0.90	111	0.54	430.66	1.96	E

Bold corresponds to instability of the ship in the correspondent node

Table 6 Results of the three criteria evaluated at each node, with their respective H_s

Node	1	2	3	4	5	6	7	8	9	10
Criterion 1: Possible matches	E	E	E	E	E	**I**	E	**I**	E	E
Criterion 2: Areas of instability (p, q)	E	E	E	E	E	**I**	E	**I**	E	E
Criterion 3: Dunwoody's critical height, FN = 0	E	E	E	E	E	**I**	E	**I**	E	E

Bold corresponds to instability of the ship in the correspondent node

5 Discussion

The study of parametric resonance largely follows the stability calculation based on the influence of waves, the damping ratio calculation, and the possibility of shipping of water, all of which contribute in some way to the risk of capsizing.

The case study found instability at nodes 6 and 8 at their respective significant heights; node 6 is located in the fishing grounding, which forces the vessel to operate at different speeds and different angles of encounter.

6 Conclusions

1. Nodes 6 and 8 are unstable for all of the proposed criteria.
2. The dimensionless values of p and q (damping included and with correcting for speed) tend to enter the zone of instability when the GM_t variation and wave height increase.
3. The results for Criterion 3 took into consideration the damping ratio and did not correct for speed.
4. This study's calculations have been based on significant height, with a probability of wave height exceeding 13.7%.

From the research, Criterion 2 can be found to be at a probability of 86.3%. Criterion 2 shows the areas of instability p and q (damped), allowing the capsizing risk to be estimated for any wave height.

7 Recommendations for Future Work

1. Future studies should take into consideration the effect of the variation of the damping coefficient of roll and perform the analysis for extreme conditions.
2. Future research could also analyze the speed variation and make polar charts depending on the direction and height of the wave, whereby parametric roll risk areas could be clearly defined.

References

1. Dominguez, F., Viteri, O.: Lengthening Optimization Module. ESPOL, Guayaquil (2016)
2. Levadou, M., Guithem, G.: Operational Guidance to Avoid Parametric Roll. Maritime Research Institute, Netherlands (2002)
3. Belenky, V., Bassler, C., Spyrou, J.: Development of Second Generation Intact Stability Criteria. vols. NSWCCD-50-TR, no. 065 (2011)
4. Dunwoody, B.: Roll of a ship in Astern Seas—Metacentric height spectra. J. Ship Res. **33**(3) (1989)
5. Dunwoody, B.: Ship in Astern seas—response to GM fluctuations. J. Ship Res. **33**(4) (1989)
6. IMO: MSC.: 267(65) Código International de estabilidad sin avería. Código IS (2008)
7. Belenky, V., Sevastianov, N.: Stability and Safety of Ship: Risk of Capsizing. SNAME, Jersey City (2007)
8. Garkavyy, V.: Deterministic chaos in the task of roll motion ship with a small freeboard. HADMAR: Int. Symp. Hydro and Aerodyn. Mar. Eng. **2**(49), (1991)
9. Himeno, Y.: Prediction of Ship Roll Damping. Report of Department of Naval Architecture anfd Marine Engineering. No. 239. University of Michigan (1981)
10. ABS: Assessment of parametric roll resonance in the design of container carriers (2004)
11. Bhattacharyya, R.: Dynamics of marine vehicles. John Wiley & Sons, Inc (1978)
12. International Towing Tank Conference: Numerical estimation of roll damping (2011)
13. DNV: Modelling and analysis of marine operations. DNV-RP-H103 (2011)
14. Wiegel, R.: Oceanographic Engineering. Prentice Hall, Englewoods Cliffs (1964)
15. Themelis, N., Spryrou, J.: Probabilistic Assessment of Ship Stability. Transactions—SNAME (2007)
16. Wind, P.: Forecast, 11 July 2017. [Online]. Available: https://www.predictwind.com/
17. Lewandowski, E.: The dynamics of Marine Craft. In: Advance Series on Ocean Engineering. Maneuvering and Seakeeping, Washington DC (2004)
18. Juana Gamo, J: Fenómenos Dinámicos de Estabilidad de un Buque en olas Longitudinales: Resonancia Paramétrica. ETSIN (2015)

Part IV
Marine Corrosion, Legal, and Offshore Technology

Experiences in the Application of Performance Standards for Protective Coatings Intended for Seawater Ballast Tanks

Pedro Martínez Villa

Abstract The present work is related to the acquired experiences during the supervision jobs developed by the author in the ship construction of 10 bulk carriers in the People's Republic of China from 2010 to 2013, specifically in relation to the application of painting in ballast tanks, according to what is established in IMO Resolution MSC215(82), "Performance Standard for Protective Coatings intended to seawater ballast tanks in all types of ships and double-skin spaces of bulk carriers," better known as PSPC (Performance Standard for Protective Coating) for its English initials, adopted in December, 2006. This denomination of PSPC is maintained throughout this paper. The fundamental objective of this work is to sensitize all personnel involved in the ships' construction process in our geographical area and specifically to those who supervise or work in the application of paint in ballast tanks and spaces of double skin in bulk carriers, in the importance of correctly following the steps in the whole process to achieve the required quality.

Keywords Coatings · Protection · Ballast tanks

1 Introduction

Along with the requirements to reduce the pollution by tankers in the marine environment, in 1990, the EE.UU's government introduced the double skin for new building of tankers.

In 1991 classification societies introduced increased supervision and supervision of the condition of paint coatings in ballast tanks. That way, they were classified as Good, Adequate, or Poor, and if classified for this last condition, the tank should be repaired or the ship had to be inspected annually. The IMO (International Maritime Organization) in 1993, 1995, and 1996 kept on with the changes to MARPOL

P. M. Villa (✉)
IPIN Cuba, Lawton, Habana, Cuba
e-mail: pedromv@nauta.cu

A. Vega Sáenz et al. (eds.), *Proceedings of the 25th Pan-American Conference of Naval Engineering—COPINAVAL*, https://doi.org/10.1007/978-3-319-89812-4_36

(International Convention for the Prevention of Pollution by Ships) and SOLAS (Safety of Life At Sea) on these aspects (Figs. 1, 2, 3, 4, 5, 6, 7 and 8).

In 1998 SOLAS introduced the requirement that the ballast's tanks should be covered with suitable paint coatings and issued guidelines for their selection and maintenance. These guidelines of 1998 have been finally replaced by the PSPC, because of the united work of IMO and IACS (International Association Classification's Societies), along with shipbuilders, developing this standard for the protective coatings in ballast tanks.

From the points established in this standard, both shipyards and paint manufacturers were given the task of preparing the conditions in their respective entities for the successful application of the same and as natural for each of them, it means a change in how it was carried out up to the moment. So, for the shipyard, it means a longer time for construction and slower deliveries, with less available options as to shop primes, paint systems for ballast tanks and dry film thickness (DFT), and also the need to train workers and to have more approved personnel, according to FROSIO or NACE, to carry out inspections.

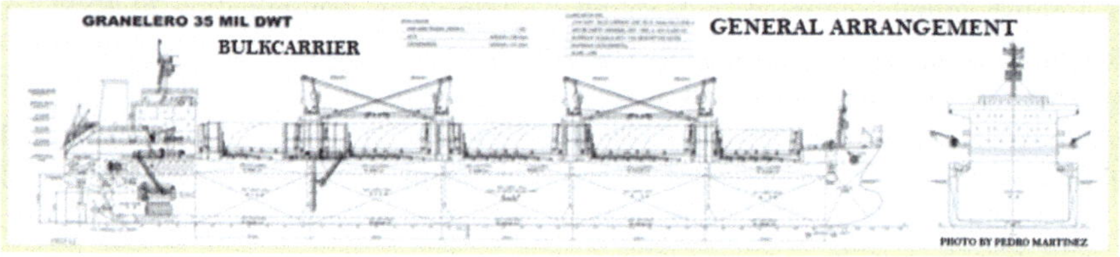

Fig. 1 General arrangement

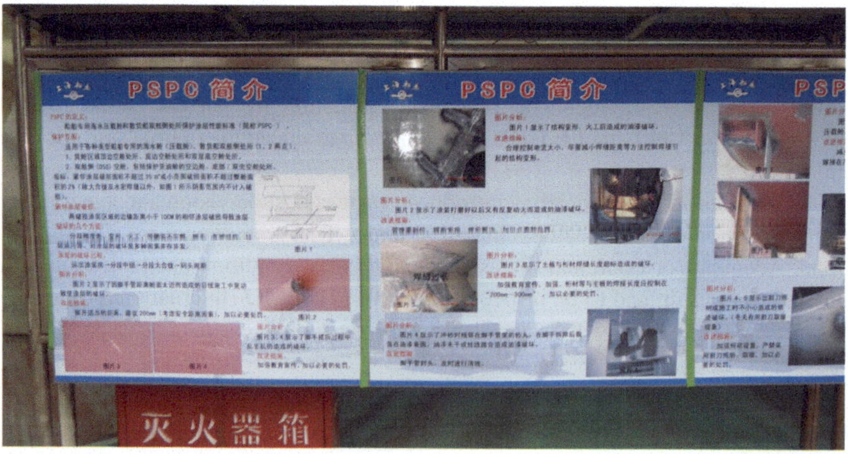

Fig. 2 Graphic circulation in Shanghai Shipyard about PSPC application

Fig. 3 Rounded edges (R-2 mm), according to the standard, before assembling the block sections

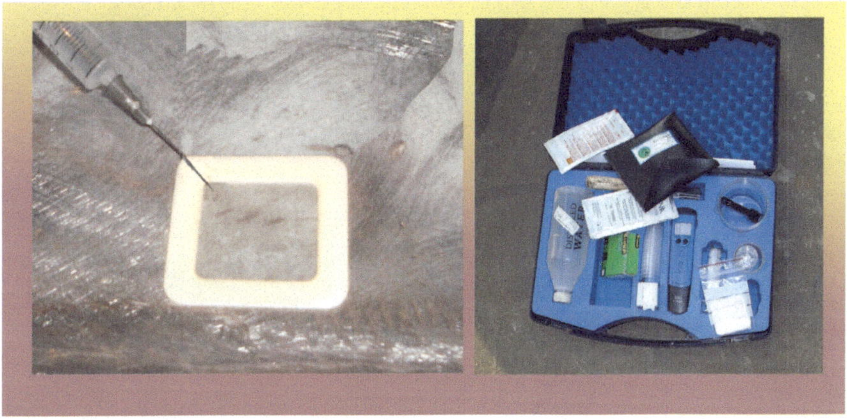

Fig. 4 Salt test in the surface to be painted

Fig. 5 Covering areas for gritblasting and paint processes and outdoors areas for finally painting works

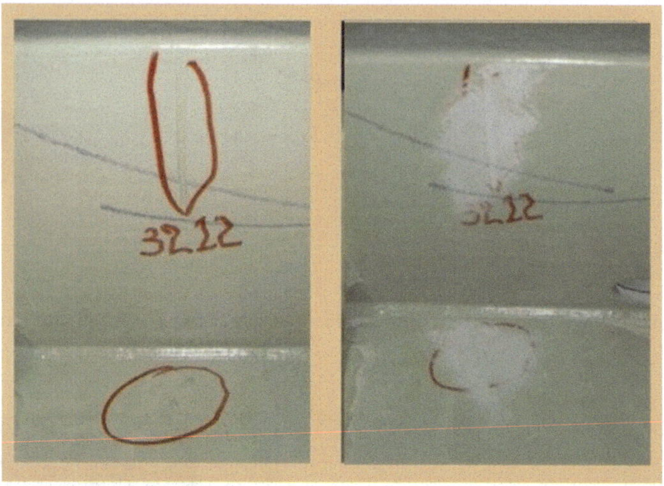

Fig. 6 Remarks in paint's inspections and some parts of its corrections

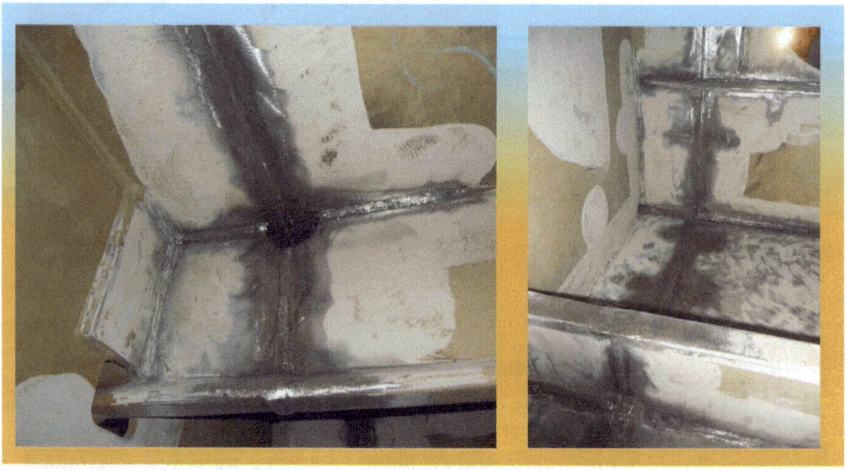

Fig. 7 Secondary surface preparation of affected areas in block's joints

For the shipyards, the level of required documentation on the selection, application, and inspection of the paints for ballast tanks is significantly higher and the new buildings must fulfill the specifications of PSPC. Equally, they should complete the requirements for all the data of the inspection (from the naked steel to the final coat) and be available and auditable, as well as be into the CTF (Coating's Technical File) whose preparation is a responsibility of the shipyard.

For the paint manufacturers, the application of this standard means a bigger need of proofs and documentation for the necessary products, with larger requirements

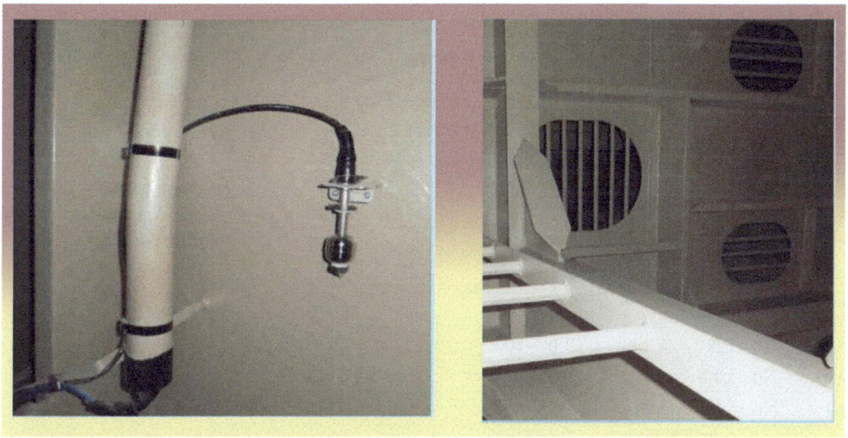

Fig. 8 Termination of the PSPC process in the ballast tanks

and costs to get certificates for ballast tank systems at independent laboratories and the need to modify technical record cards according to the PSPC. Also, they have the responsibility for approval of the stages of the coating works and participation in the making of CTF as support to the shipyard and for the training of more personnel according to FROSIO or NACE to carry out the inspections.

Although for the shipowners the application of this standard had meant an increase of the costs in the stage of construction in order to comply with supervisory criteria of the PSPC, the benefits are significant, having ships of higher quality with minor costs of maintenance and a better retention of the value of investment. It also implies, for the shipowners, a need for better documentation of the maintenance process to the paints of ballast tanks and to maintain a CTF on board.

For classification societies, the adoption of this standard meant additional preparation to the class surveyors in the stages of the ship's building and the responsibility of approving all the documentation deriving from its application.

According to the above, we want to show the experiences acquired during the application of paint in ballast tanks and double skin space, according to that established in Resolution MSC215(82), during 10 bulk carrier construction projects in the Shanghai shipyard in the People's Republic of China, from 2010 to 2013.

2 Project Details—Particulars of Built Ships

This project began on September 15, 2010, in Shanghai's shipyard; it has been one of the most important projects in the last few years for the Cuban Maritime Fleet, specifically for the Nordstrand Maritime Trading Company. This first ship, of a series of 10, was delivered on August 10, 2011; the rest (nine ships) were delivered successively, and the last one was delivered on September 5, 2013. Just as

Table 1 Ballast tank arrangement

No.	Ballast tank no.	Location (frame) [(Cdn) no.]	Area (m^2)	In addition to the ballast tanks (total area more than 46,000 m^2), the PSPC treatment was given to the 4 double skin spaces distributed in two by side, so that the general area added was more than 50,000 m^2
1	No. 1 Port Side (PS)	175–208	2655	
2	No. 1 Starboard (S)	175–208	2655	
3	No. 2 PS	139–175	4196	
4	No. 2 S	139–175	4196	
5	No. 3 PS	103–139	4292	
6	No. 3 S	103–139	4292	
7	No. 4 PS	67–103	4079	
8	No. 4 S	67–103	4079	
9	No. 5 PS	35–67	3148	
10	No. 5 S	35–67	3148	
11	No. 1 (TST) PS	175–208	1240	
12	No. 1 (TST) S	175–208	1240	
13	No. 2 (TST) PS	139–175	1035	
14	No. 2 (TST) S	139–175	1035	
15	Fore peak	208–225	3246	
16	After peak	−5 to 12	1926	
		Total area	46,462	

established in the resolution, the new buildings that had signed contracts after July, 2008 or keel laid after January, 2009, should be based on the application of the PSPC, and for the ballast tanks in ships more than 500 GT and for the double skin spaces or in bulk carriers of 150 m and over in length. Observing the characteristics of those ships, indicated below, the cause of the application of the PSPC during the construction of those ships is perfectly understood.

Bulk carrier particulars: Length overall—179.90 m, Length between perp.—171.50 m, Total Beam—28.40 m, Draft (max)—15.00 m, and DWT-35000 t. Lloyds Register Class. The ballast's tank arrangement is shown in Table 1.

3 Preparation of the Shipowner's and Shipyard's Supervision Teams

Inside the supervisors' group on the shipowner's part (structural part of the project), three were trained in PSPC (the author of this work included), with the aim of enabling them in this innovative aspect for guaranteeing correct supervision of the application of the paint in all the ballast tanks and double skin spaces indicated previously. Sven Mortensen, from Marine Training Services of Lloyd's Register, who imparted the course, issued an appropriate certificate, indicating the received preparation.

At the same time, the supervisors' practical preparation of the shipowner's and shipyard's quality inspectors was accomplished during the first months of construction of the first ship of the series, directed by Mortensen, in areas where the paint was applied or inside the corresponding blocks.

As this process was carried out for the first time at the Shanghai shipyard for several projects at the same time, the shipyard management decided to create good conditions to guarantee the quality of work, which included the necessary equipment purchase of ventilation equipment and dust extractors, equipment for humidity control, the measurement and control instruments for all stages of the coating process, and so on. Equally, the quality inspectors were issued the certificates as established in the resolution in NACE Coating Inspector Level 2 or in FROSIO Inspector Level III and workers were chosen among those having experience and were enabled conveniently; the shipyard included the graphic circulation of this action and its characteristics.

The shipyard elaborated a procedure specifically for the realization of the inspections and the methods of repair of the defects in the application of paint during all phases of construction, including the construction of blocks and the fitting-up phase during the slip assembly of the ship.

As experience about what has been previously stated about the action of this shipyard as to the initial preparation for the application of the PSPC and its effectiveness, we recommend to those shipyards that start up in these aspects, first to do an analysis of the equipment that they possess and to check if it is adequate according to the requirements of the standard. We can summarize the following for the stage of preparation.

- Acquisition of suitable equipment for the gritblasting works, priming, for maintenance of the conditions of work and of control instruments.
- The technical staff's theoretic and practical preparation, including the certification in NACE Coating Inspector Level 2 or in FROSIO Inspector Level III and of the workers who take part in the whole process. To take into account that the worker carrying out this task is, in the end, the one who guarantees the quality process and the final result will depend on his instruction and training.
- To carry out the procedures of the shipyard for application, the inspections, and for the repair of imperfections. It is important for the shipyard to have these procedures (which may be in only one procedure) to be in harmony with the resolution MSC215(82).
- To sign an agreement between the parts involved in the whole PSPC's process (shipyard–shipowner and paint manufacturer) with the revision of the classification society in order to determine the responsibilities of each in this process (tripartite agreement).

4 Construction Process and PSPC's Application—Standards

The construction form of these bulk carriers was carried out from the initial construction of sections with all the structural elements corresponding to the inside of the ballast tanks, edges properly rounded, according to that specified by the standard.

Afterwards, the blocks were built and once constructed and approved by the structural part, passed to the areas destined for gritblasting application, that is, the primary preparation of surfaces. Those requirements are: Sa 2½, with profiles between 30 and 75 μ and blasting shall not be carried out if the relative humidity is above 85%, or the surface temperature of steel is less than 3 °C above the dewpoint.

During the gritblasting inspections, among other defects, were those that could exist in welds and the actions to be followed were recommended and, finally, the cleanliness of the steel surface and the ruggedness profile was verified before the application of the first coat, according to the paint manufacturer's recommendations, in this case JOTUN paints, as established in the Product Description, Type Approval Certificate, and Technical Data Sheet.

The degree of salinity on the surface was controlled (sodium chloride) to verify if it was within the limit of what is permissible according to the standard which was 50 mg/m^2.

Once the gritblasting process was approved, the blocks were transferred to another shed for the application of paints. Those blocks that were a part of the ballast's tanks or double skin space completed their PSPC's treatment in areas bordering these sheds.

About the standards, I want to indicate that these are established to assist in defining procedures and results, relative to the status of the surfaces, to the selection of methods and the manner of realization, and to the quality of the final result. So, the standards determine the base on which the control tasks can take effect in order to guarantee that all the parts involved understand the requirements in the same way.

All the paint inspectors use various standards, either those internationally recognized, which every inspector should know, or national associations' standards, that every inspector working in that country should know, or the standards of the shipyard, which all paint inspectors working in that shipyard should know.

During the supervisory stage at Shanghai's shipyard sometimes I found different interpretations of the same standard and on other occasions others not included in the specifications were used, which, of course, were rejected. The version or versions to which the specification of the paint refers should be known as recommendations before starting work and agreement reached on any variation that can appear.

Reverting to the theme of the treatment of blocks in the outdoor areas, the majority of PSPC paint inspections were carried out here, in order that the double space and ballast tanks, were carried to the assembly area, already certificated. The

Table 2 Scheme of paint for ballast's tanks

Name of paint/manufacturer	NDFT	Color	Application method
Primer W53-02/Lanyan	$20^{+/-}$ mic	Grey	Automatic machine
1st coat JOTACOTE BT ALU/Jotun	160 mic	Red	Airless
Stripe coat JOTACOTE BT BEI./Jotun		Beige	Brush and roller
Stripe coat JOTACOTE BT ALU/Jotun		Red	Brush and roller
2nd coat JOTACOTE BT BEI./Jotun	160 mic	Beige	Airless

following scheme of paint was established in this project with the correspondent nominal thickness of dry film (NDFT) according to what it is stipulated by the standard (Table 2).

As inspector I determined if safety was guaranteed by the shipyard in order to proceed to the inspections, with a correct positioning of the scaffolding and access ladders, and in turn we never came to these places without suitable clothing; we also used protective masks, gloves, and kneecaps when needed.

I should indicate that the utilization of gloves was more necessary during grit-blasting inspections but they were used only as a measure of safety in places determined for the paint inspection, but not at the inspection site because one of the main tools of a paint inspector is the sense of touch, along with vision and common sense.

On the other hand the inspector should have his DFT meter, relative humidity meter, lantern, telescopic mirrors to inspect scallops and openings that are areas difficult to revise with the naked eye, photographic camera, and chalk without fat for gritblasting inspection, and permanent marker for the paint inspections. Although the development of electronic instruments and computers advances quickly it should never be forgotten that such instruments can only supplement, not substitute, the careful observations and common sense, as we pointed out previously, along with the planning and the registration. It should not be forgotten that these instruments should be adjusted and well used, and they are, just in this way, valuable documentation tools.

During the development of the intermediate inspections (before stripe coating or second coat), I verified the presence of dust (visual, by touch, or with a white cloth), because if present this would cause a bad adherence of the coating and its later detachment. The zones cleaned insufficiently were to be washed with clean compressed air and in closed spaces, by means of cleanliness by aspiration.

Also abnormalities (pulverized, orange skin, remove from hanging, unpainted zones, pores, etc.) in the formation of the coat affecting the aspect and the coating's protective properties, were corrected. I, as inspector, should determine, if the aesthetic aspect was important, that with excessive extension of the pulverized or orange skin, and if the affected zones were scratched, sandpapered, and after eliminating the dust, the unpainted zones and pores were painted.

Generally, during these inspections at the blocks I, as inspector, advanced in the learning process and all the necessary measures to avoid the same errors in successive works were taken. Once the course of action of paint relative to the PSPC at all the blocks that were applicable was approved, these moved on to the process of slip assembly in other installments in the existing dry docks.

In the course of the slip assembly in the unions of blocks, where affectations of the previously applied paint were present it was necessary to apply the corresponding mechanical cleaning of welds and other affected areas (St 3); the above should have been in agreement with the resolution, as Butts St 3 or better or Sa 2½ where practicable; small damages up to 2% of total area: St 3; where there are contiguous damages over 25 m^2 or over 2% of the total area of the tank, Sa 2½ should be applied. All the necessary measures were taken to avoid damages in the paint of the tanks; for this reason fire blankets were placed.

The inspections of the ballast tanks and double skin space in the assembly slip, already included all related matters with the correct conclusion of the whole area and in addition to the pipe supports and the pipes themselves, the anode supports, the ladders, covers, and the like, that is, all those metallic miscellanies that were inside the tanks and treated with the PSPC's same scheme.

To achieve a suitable quality in PSPC's treatment the following aspects should be taken into account.

- Avoiding the application of paint without optimal conditions of humidity, ventilation, cleanliness of the surfaces, and the correct drying of previous coats.
- Control of the quality of the shipyard should guarantee that at all times during complete application of the paint the parameters of relative humidity and dew point rank below the established limit (generally 85% for the HR). The shipyard's supervisor should monitor these measurements and ask for repetitions and estimates necessary during the process.
- Not to proceed to the areas recently painted until they are totally dry and the optimal conditions of ventilation and illumination are guaranteed. The shipyard is to ensure that these aspects are complied with and not carry out the inspections without these guarantees.
- To have records of all the deficiencies found and the application of the solutions.

Finally PSPC's treatment of each of the ballast tanks and the double skin space were approved. After the conclusion of all of them the CTF was checked and the PSPC procedure was concluded in the ship, approved by Lloyd's Register Classification's Society.

5 Conclusions

The important aspects to be taken into account for the application of the Performance Standard of the Protective Coatings of the Tanks intended to ballast (PSPC) are the initial preparation of the shipyard to accomplish this process regarding equipment and the technical staff's preparation to guarantee the correct application of the mentioned standard. In addition, it is important to achieve a positive interrelation between the supervisory staff of the shipyard, the shipowner supervisor, and the supervisor of the paint manufacturer, starting from the pact in the mentioned tripartite agreement and executing their lists within the process. All supervisors should have an ample visual field and know the critical zones in each inspection, the disposition and the necessary firmness supported by the monitoring of the supervisory tasks, and possess ample knowledge of the paint's applications as well as have good training for the specific work to be developed.

The experiences shown in this work can be applied and taken into account at the constructions that are occurring right now or are beginning to increase in our geographic area's construction shipyards.

The Obligation of Seaworthiness: Shipowner and Charterer

Stephen Girvin

Abstract The obligation to provide a seaworthy ship is basic in the carriage of goods by sea and this is no the less the case where the contract of carriage is between a shipowner and a charterer or between a charterer and subcharterer. As the concept of seaworthiness is not usually defined in standard form for charterparties, the meaning has to be ascertained from cases decided under the common law. In charterparties, whether time, voyage, or bareboat, it is normal for the seaworthiness obligation to be laid down in express wording, but often describing the standard required as one of due diligence. Alternatively, such a due diligence standard is imported into the charterparty by means of a paramount clause, bringing into the charterparty the terms of the Hague (or Hague–Visby) Rules.

Keywords Seaworthiness · Common law · Charterparties

1 Introduction

The seaworthiness obligation is basic in every contract for the carriage of goods by sea, including charterparties,[1] and for this reason there is an implied undertaking on the part of the shipowner,[2] in the absence of an express contractual

[1]A substantially longer and more detailed paper may be found at NUS Centre for Maritime Law Working Paper 17/11, https://law.nus.edu.sg/cml/wps.html, last accessed 2018/1/28. The standard references are: Eder, B.: Scrutton on Charterparties and Bills of Lading. 23rd edn. Sweet & Maxwell, London, art 68; Treitel. G.; Reynolds F.M.B.: Carver on Bills of Lading. 4th edn. Sweet & Maxwell, London, para 9-013; Bennett, H.N.: Carver on Charterparties. Sweet & Maxwell, London, para 3-072.
[2]This arises by virtue of it acting as shipowner: Lyon v Mells (1804) 5 East 428, 437.

S. Girvin (✉)
Centre for Maritime Law, National University of Singapore, 469G Bukit Timah Road, Singapore 259776, Singapore
e-mail: sdgirvin@nus.edu.sg

A. Vega Sáenz et al. (eds.), *Proceedings of the 25th Pan-American Conference of Naval Engineering—COPINAVAL*, https://doi.org/10.1007/978-3-319-89812-4_37

undertaking or the mandatory application of the Hague (or Hague–Visby) Rules,[3] to provide a seaworthy vessel:

> … Where there is no agreement to the contrary, the shipowner is, by the nature of the contract, impliedly and necessarily held to warrant that the ship is good, and is in a condition to perform the voyage then about to be undertaken, or, in ordinary language, is seaworthy, that is, fit to meet and undergo the perils of the sea and other incidental risks to which she must of necessity be exposed …[4]

As most contracts of carriage by sea, including charterparties, do not provide a definition of "seaworthiness," this must be interpreted in accordance with common law principles, where much of the groundwork has been laid.

2 Meaning of Seaworthiness at Common Law

What is meant by the shipowner's obligation to provide a seaworthy ship has been referred to in a substantial number of cases,[5] among them the *Charter party of the Cheritie* (1531).[6]

2.1 Nature of the Seaworthiness Obligation

2.1.1 Structural Fitness: Loading Stage

Every voyage at sea has an antecedent phase, the loading stage. It is established that, during the loading, the vessel must be fit to receive the cargo and to encounter the ordinary perils of the loading stage. In *McFadden v Blue Star Line*, Channell J explained that

> [T]he warranty is that at the time the goods are put on board she is fit to receive them and to encounter the ordinary perils that are likely to arise during the loading stage; but that there is no continuing warranty after the goods are once on board that the ship shall continue fit to hold the goods during that stage and until she is ready to go to sea …[7]

[3]As, for example, in the Carriage of Goods by Sea Act 1971, c 21, s 1(2).

[4]Kopitoff v Wilson (1876) 1 QBD 377, 380 (Field J). See also Lyon v Mells (1804) 5 East 428, 437; Steel v State Line Steamship Co (1877) 3 App Cas 72, 77; 84; 88.

[5]Many cases also arise on policies of marine insurance and older cases treat the concept as being one and the same as cases on the carriage of goods by sea: J & E Kish v Charles Taylor Sons & Co [1912] AC 604, 611. Cf, however, The Bunga Seroja [1999] 1 Lloyd's Rep 512, [18]; Carver on Charterparties (n 2), para 3-080.

[6]Marsden, R.G.: Select Pleas in the Court of Admiralty. vol 1. Selden Society, London, 35.

[7][1905] 1 KB 697, 704–705. See also A E Reed & Co Ltd. v Page, Son & East Ltd. [1927] 1 KB 743, 756.

As Channell J indicates, once the loading stage is complete, this obligation comes to an end; the vessel must forthwith be seaworthy for the next stage, usually the voyage or some intermediate stage before the voyage.

2.1.2 Structural Fitness

The seaworthiness obligation extends to the structural fitness of the vessel for the intended voyage.[8] Thus in *Steel v State Line Steamship Co*[9] a vessel's orlop deck port hole was insufficiently fastened and water entered through the port, damaging a cargo of wheat. The House of Lords unanimously held that there was an implied obligation to tender a seaworthy vessel, remitting the case to the lower court to determine whether unseaworthiness had caused the loss.[10] That court subsequently found that the vessel had been proved to be unseaworthy.[11] Other incidences of structural unfitness have been held to include defective masts and sails,[12] a leaking hull,[13] a defective screw shaft,[14] leaking rivets[15] or bolts,[16] leaking hatch covers,[17] fractured shell plating,[18] and having corroded bottom plates.[19]

2.1.3 Equipment

Closely linked to the vessel's physical structure is her equipment, including her engines.[20] This requirement has, however, also been held to include having properly functioning boilers,[21] the provision of adequate coal[22] and

[8]Kopitoff v Wilson (1876) 1 QBD 377, 380; Steel v State Line Steamship Co (1877) 3 App Cas 72, 77; 84; 88.

[9](1877) 3 App Cas 72.

[10]See particularly at 90–91 (Lord Blackburn).

[11]Steel & Craig v State Line Steamship Co (1878) 5 R 622, 623.

[12]Namby v Joseph & Seagar (1890) 9 NZLR 227.

[13]Cohn v Davidson (1877) 2 QBD 455.

[14]The Glenfruin (1885) 10 PD 103.

[15]The Christel Vinnen [1924] P 208; Charles Brown & Co v Nitrate Producers Steamship Co (1937) 58 Ll LR 188.

[16]Spillers Milling & Associated Industries Ltd v The Bryntawe (1928) 32 Ll L Rep 155.

[17]The Gundulic [1981] 2 Lloyd's Rep 418.

[18]The Toledo [1995] 1 Lloyd's Rep 40.

[19]The Torenia [1983] 2 Lloyd's Rep 210.

[20]The Antigoni [1991] 1 Lloyd's Rep 209; MT Cape Bonny Tankschiffarts GmbH & Co KG v Ping An Property & Casualty Insurance Co of China Ltd [2017] EWHC 3036 (Comm); [2018] 1 Lloyd's Rep 356, [118].

[21]The Tatjana [1911] AC 194.

[22]Thin v Richards & Co [1892] 2 QB 141; The Vortigern [1899] P 140.

bunkers,[23] lubricating oil,[24] and having provisions[25] and necessaries for the voyage, including medicines[26] and dunnage.[27] Also included is having procedures in place to check on refrigerated containers,[28] having navigational aids in good working order,[29] and up-to-date charts.[30]

2.1.4 Cargoworthiness

To be seaworthy, a vessel must also be cargoworthy, as explained in *Stanton v Richardson*[31]: "It seems to me that the obligation of the shipowner is to supply a ship that is seaworthy in relation to the cargo which he has undertaken to carry."[32] In this case, a vessel was engaged to carry a cargo of sugar in bags but when wet sugar was loaded this gave off such a quantity of molasses that the vessel was rendered unseaworthy. In the case of cargo that needs to be refrigerated, the equipment must be adequate[33] and, where a vessel is contracted to carry live animals, the vessel must be free of disease.[34] Likewise, a cargo that cannot be offloaded because of an infestation of insects would also render the vessel unseaworthy.[35]

The carriage of sophisticated cargoes by sea has necessitated the taking of measures to protect the crew and prevent marine pollution; these measures include the International Maritime Dangerous Goods (IMDG) Code,[36] the International Maritime Solid Bulk Cargoes Code (IMSBC Code),[37] and the International Bulk

---

[23]Northumbrian Shipping Co Ltd v E Timm & Son [1939] AC 397; E B Aaby's Rederi A/S v Union of India (The Evje No 2) [1976] 2 Lloyd's Rep 714.

[24]The Kriti Rex [1996] 2 Lloyd's Rep 171, 185.

[25]The Wilhelm (1866) 14 LT 636.

[26]Woolf v Claggett (1800) 3 Esp 257.

[27]The Marathon (1879) 4 Asp MLC 75.

[28]JP Klausen & Co AS v Mediterranean Shipping Co SA [2013] EWHC 3254 (Comm).

[29]See, eg, Edmund Weil Inc v American West African Line Inc (1945) 147 F 2d 363 (2nd Cir).

[30]Cf The Torepo [2002] 2 Lloyd's Rep 535 (although the discrepancy in the charts was not a material defect in the chart portfolio and the disparity was not causative).

[31](1872) LR 7 CP 421, affirmed (1874) LR 9 CP 390; Tattersall v The National Steamship Co Ltd (1884) 12 QBD 297, 300; AE Reed & Co Ltd v Page Son & East Ltd [1927] 1 KB 743, 754.

[32]At 435 (Brett J).

[33]See Owners of Cargo on the Maori King v Hughes [1895] 2 QB 550; JP Klausen & Co AS v Mediterranean Shipping Co SA [2013] EWHC 3254 (Comm).

[34]See Tattersall v National Steamship Co Ltd (1884) 12 QBD 297; Sleigh v Tyser [1900] 2 QB 333.

[35]The Good Friend [1984] 2 Lloyd's Rep 586, 592; Ciampa v British India Steam Navigation Co Ltd [1915] 2 KB 774, 780.

[36]See www.imo.org/en/OurWork/Safety/Cargoes/DangerousGoods/Pages/default.aspx, last accessed 2017/12/1.

[37]See www.imo.org/en/OurWork/Safety/Cargoes/CargoesInBulk/Pages/default.aspx, last accessed 2017/12/1.

Chemical Code (IBC Code),[38] adherence to which is mandatory under SOLAS 1974[39] and MARPOL 73/78.[40] The obligation to take care to make the vessel seaworthy does not, however, mean that the ship must be immune from the negligence of her crew.[41] In *The Kapitan Sakharov*[42] the court held that under deck stowage of tank containers of isopentane, a flammable liquid,[43] clearly contravened the IMDG Code and rendered the *Kapitan Sakharov* unseaworthy[44] but that the shipowner, even exercising reasonable skill and care, could not have detected the presence of that cargo and had, therefore, exercised due diligence.[45]

2.1.5 Manning

A vessel must also have on board sufficient crew for the voyage.[46] What amounts, however, to "sufficiency" will be a question of fact in each case and, in modern circumstances, will be affected by the requirements of SOLAS 1974, which requires contracting governments[47] to take measures to ensure that all ships are "sufficiently and efficiently managed,[48] in accordance with the IMO Principles of Minimum Safe Manning.[49]

This requirement further extends to the competence of the vessel's crew, including her master,[50] although there is no requirement of perfection.[51]

---

[38]See www.imo.org/en/OurWork/Safety/Cargoes/CargoesInBulk/Pages/IBC-Code.aspx, last accessed 2017/12/1.

[39]The International Convention for the Safety of Life at Sea, Ch VII, reg 3 (carriage of dangerous goods in packaged form); reg 7–5 (carriage of dangerous goods in solid form in bulk).

[40]The International Convention for the Prevention of Pollution from Ships, Annex II (control of noxious liquid substances in bulk); Annex III (prevention of pollution by harmful substances carried by sea in packaged form).

[41]See The Aconcagua [2009] EWHC 1880 (Comm); [2010] 1 Lloyd's Rep 1, [366].

[42][2000] 2 Lloyd's Rep 255.

[43]IMDG Code, class 3.

[44][2000] 2 Lloyd's Rep 255, 266.

[45]At 273 (for the purposes of Art III, r 1 of the Hague Rules, incorporated in the contracts of carriage.

[46]See Hongkong Fir Shipping Co Ltd v Kawasaki Kisen Kaisha Ltd [1962] 2 QB 26, 34.

[47]See The Merchant Shipping (Standards of Training, Certification and Watchkeeping) Regulations 2015, SI 2015/ 782, reg 46.

[48]See ch 5, reg 14.1.

[49]Resolution A.1047(27) (20 December 2011). For consideration in the context of autonomous ships, see Carey, L.: All hands off deck? The legal barriers to autonomous ships. NUS Law Working Paper No 2017/011, available at https://law.nus.edu.sg/cml/wps.html, last accessed 2017/12/1.

[50]Moore v Lunn (1923) 15 Ll L Rep 155, 156 (master and chief engineer "habitual drunkards"); Standard Oil Co of New York v Clan Line Steamers Ltd [1924] AC 100, 120–121.

[51]Rio Tinto Co Ltd v The Seed Shipping Co Ltd (1926) 24 Ll LRep 316; The Jute Express [1991] 2 Lloyd's Rep 55.

In *Hongkong Fir Shipping Co Ltd v Kawasaki Kisen Kaisha Ltd*,[52] the court found that though certain of the vessel's machinery was in a reasonably good condition "… by reason of its age, it needed to be maintained by an experienced, competent, careful and adequate engine room staff." Salmon said that:

> … Would a reasonably prudent owner, knowing the relevant facts, have allowed this vessel to put to sea with this engine room staff? … I have no doubt that the true answer to this question is "No" … as the engines were very old it was necessary to engage an engine room staff "of exceptional ability, experience and dependability."[53]

Competence can also be affected by personality, an aspect of competence that has become much more important as crew numbers on board have diminished.[54] Current law is also affected by enhanced international regulation on the competency of the crew, as laid down in the STCW[55] Convention 1978 (as amended), chapters II and III of which lay down mandatory minimum requirements, respectively, for the certification of the master and deck department and engine department of ships and these are reflected in the relevant national law[56] of contracting states.[57] A further factor is the Maritime Labour Convention 2006,[58] which lays down minimum requirements and conditions of seafarer's employment, and which must also be expected to impact on the scope of the shipowner's obligation to provide a seaworthy vessel.

2.1.6 Relevant Documentation

A wide range of certificates, papers, and documents are today required onboard ship[59] and the absence of such documentation can amount to a breach of the shipowner's obligation to provide a seaworthy ship. Thus, the vessel will be unseaworthy if the necessary health certificate from a port health authority[60]

---

[52][1962] 2 QB 26.

[53]At 34. See also The Makedonia [1962] 1 Lloyd's Rep 316; The Eurasian Dream [2002] EWHC 118 (Comm); [2002] 1 Lloyd's Rep 719.

[54]The Makedonia [1962] 1 Lloyd's Rep 316, 335 (Hewson J).

[55]The International Convention on Standards of Training, Certification and Watchkeeping for Seafarers (amended in 1995, 1997, and 2010).

[56]See The Merchant Shipping (Standards of Training, Certification and Watchkeeping) Regulations 2015, SI 2015/782, pt 2.

[57]See www.imo.org/en/About/Conventions/StatusOfConventions/Documents/Status%20-%202017.docx.pdf, last accessed 2012/12/1.

[58]In force from 20 August 2013. See the statutory instruments passed pursuant to the Merchant Shipping Act 1995, c 21, ss 85–86, during the period 2010–2014.

[59]See, eg, List of Certificates and Documents Required to be Carried on Board Ships (19 July 2017) FAL.2/Circ.131 MEPC.1/Circ.873 MSC.1/Circ.1586 LEG.2/Circ.3, downloadable from www.imo.org/en/Publications/SupplementsAndCDs/Documents/Certificatesonboardships.pdf, accessed 2017/11/30. Different flag administrations also lay down detailed requirements.

[60]See Levy v Costerton (1816) 4 Camp 389; Ciampa v British India Steam Navigation Co Ltd [1915] 2 KB 774.

or consular manifest[61] is not available. In *The Madeleine*, the ship did not possess a deratisation certificate[62] or an exemption certificate and the court held that the charterers had the right to cancel.[63] Other cases have, however, held that a vessel will not be unseaworthy where not carrying a deck certificate of clearance,[64] or an ITF Blue Card,[65] or where RightShip approval has not been obtained.[66]

The standard required has clearly evolved in response to technological and legal change.[67] Indeed, the avalanche of regulation of international shipping has also imposed greater legal requirements on shipowners and this will impact on the potential scope of the seaworthiness obligation, although there may be some relief in the future, given the current momentum towards the issuing of e-certificates.[68] That said, a vessel will likely be unseaworthy if not ISM compliant,[69] carrying a valid ISSC certificate required under the ISPS Code,[70] or an International Sewage Pollution Prevention (ISPP) certificate, required under MARPOL 73/78.[71]

Some charterparty standard clauses make specific provision for the evolving scope of this requirement. In *The Elli and the Frixos*[72] the owners warranted that the vessel complied "with all applicable convention, laws, regulations and ordinances of any international, national, state or local government entity ... including ... MARPOL 1973/1978 as amended and extended ... ." The tankers did not have double-bottom tanks but, before the end of the charterparties, MARPOL Regulation 13H required oil tankers of 5000 tons deadweight and above to be fitted with double bottoms. The court held that the shipowners were in breach because they had failed to obtain exemption under changes to MARPOL, affecting the vessel's cargo-carrying capacity, and that the warranty applied both on and after delivery and expressly referred to MARPOL as amended and extended.[73]

---

[61]Dutton v Powles (1862) 2 B & S 191.

[62]A certificate confirming that the ship is free of rats.

[63]The Madeleine [1967] 2 Lloyd's Rep 224, 241.

[64]Wilson v Rankin (1865) LR 1 QB 162.

[65]The Derby [1985] 2 Lloyd's Rep 325.

[66]The Silver Constellation [2008] EWHC 1904 (Comm); [2008] 2 Lloyd's Rep 440.

[67]See, eg, Martin v Southwark (1903) 24 S Ct 1, 3.

[68]See, eg, Tan, W.Z.: Denmark, Norway and Singapore port authorities ink pact on E-Certificates. Lloyd's List, 2017/4/25.

[69]The International Safety Management Code (the ISM Code), mandatory under SOLAS 1974, ch IX (as amended): The Eurasian Dream [2002] EWHC 118 (Comm); [2002] 1 Lloyd's Rep 719, 739. It is a breach of the carrier's marine insurance cover not to have such certification: see, eg, International Hull Clauses (1/11/03), cl 13.1.4 and 13.1.5.

[70]An International Ship Security Certificate is issued under the International Ship and Port Facility Security Code (the ISPS Code), mandatory under SOLAS 1974, ch XI-2. See Girvin, S.: Commercial Implications of the ISPS Code, 330 Marlus 307–355 (2005).

[71]See Annex IV. Cf The Rewa [2012] EWCA Civ 153; [2012] 1 Lloyd's Rep 510.

[72]The Elli and the Frixos [2008] EWCA Civ 584; [2008] 2 Lloyd's Rep 119.

[73]At [24]; [26].

2.2 Seaworthiness Relative but not Absolute

The obligation to provide a seaworthy vessel is not demanded in abstract terms,[74] being relative to the nature of the ship,[75] the particular voyage,[76] the time of the year,[77] the stages of that voyage,[78] the cargoes the shipowner has contracted to carry,[79] and the relevant standards for the carrying of cargoes at the applicable time.[80] The standard required is not an accident-free ship, nor an obligation to provide a ship or gear that might withstand all conceivable hazards.[81] A temporary defect or one that is trivial and can be remedied will not be enough to render the vessel unseaworthy to encounter the perils of the voyage.[82] However, a defect that is inaccessible and invisible will not ordinarily render the vessel unseaworthy.[83]

Though relative, at common law the obligation is an unconditional one: the shipowner will be absolutely liable, irrespective of fault, for any breach of the undertaking:

> [The] warranty is an absolute warranty; that is to say, if the ship is in fact unfit at the time when the warranty begins, it does not matter that its unfitness is due to some latent defect which the shipowner does not know of, and it is no excuse for the existence of such a defect that he used his best endeavours to make the ship as good as it could be made.[84]

In *Steel v State Line Steamship Co*, Lord Blackburn described the obligation as amounting to an undertaking "not merely that they [the owners] should do their best to make the ship fit, but that the ship should really be fit."[85] In some circumstances, it may be possible for the shipowner to contract out of this absolute obligation, such as by including the wording "… unseaworthiness or unfitness of the vessel at

---

[74]It has been said that "there is no positive condition of the vessel recognized by the law to satisfy the warranty of seaworthiness": Knill v Hooper (1857) 2 H & N 277, 283.

[75]Lord Esher MR has stated that the vessel must be "… in a condition to bear all the ordinary vicissitudes of the voyage …": Thin v Richards & Co [1892] 2 QB 141, 143.

[76]The Fjord Wind [2000] 2 Lloyd's Rep 191, [18].

[77]FC Bradley & Sons Ltd v Federal Steam Navigation Co Ltd (1926) 24 Ll LR 446, 458; Charles Brown & Co v Nitrate Producers Steamship Co (1937) 58 Ll LR 188, 190.

[78]The Vortigern [1899] P 140.

[79]See Stanton v Richardson (1875) 3 Asp MLC 23 (HL); Tattersall v The National Steamship Co Ltd (1884) 12 QBD 297, 300.

[80]See The Good Friend [1984] 2 Lloyd's Rep 586, 592.

[81]The Portland Trader [1963] 2 Lloyd's Rep 278, 280–281 (Dist Ct, Oregon).

[82]Steel v State Line Steamship Co (1877) 3 App Cas 72. See also London Arbitration 2/2005, (2005) 659 LMLN 1.

[83]At 90–91. Cf, however, Gilroy, Sons & Co v W R Price & Co [1893] AC 56, 64; The Schwan [1909] AC 450, 464.

[84]In McFadden Brothers & Co v Blue Star Line Ltd [1905] 1 KB 697, 703 (Channell J).

[85](1877) 3 App Cas 72, 86.

commencement of or before or at any time during the voyage … always excepted."[86] If, however, the language of such a clause is ambiguous, the shipowner will not be able to rely on it.[87] Indeed, the courts have held that, to exclude seaworthiness, "the words used must be express, pertinent, and apposite."[88]

2.3 *Test for Unseaworthiness*

The test for determining seaworthiness is objective, as stated in *McFadden v Blue Star Line*:

> A vessel must have that degree of fitness which an ordinary careful and prudent owner would require his vessel to have at the commencement of her voyage, having regard to all the probable circumstances of it … If the defect existed, the question to be put is: Would a prudent owner have required that it should be made good before sending his ship to sea had he known of it? If he would, the ship was not seaworthy within the meaning of the undertaking.[89]

Although the standards of seaworthiness may rise with more sophisticated knowledge, for example, in ship building and navigation,[90] perfection is not required: "You do not test it by absolute perfection or by absolute guarantee of successful carriage. It has to be looked at realistically … ."[91] On the other hand, it has been suggested that:

> … a prudent owner might well require even a trivial defect to be made good before sending his vessel to sea if, even in a remote contingency, that defect might jeopardise the safety of the vessel or its cargo, upon the basis that every defect, however small, that might do so, must, as a matter of prudence, be corrected before the vessel puts to sea.[92]

2.4 *Time When Obligation Attaches*

The absolute obligation of seaworthiness at common law attaches at two points: firstly, at the commencement of loading, when the ship must be fit to receive her cargo and fit as a ship for the ordinary perils of lying afloat in harbor while

[86]Wiener & Co v Wilsons & Furness-Leyland Line Ltd (1910) 11 Asp MLC 413; Snia Societa di Navigazione Industria e Commercio v Suzuki & Co (1923) 17 Ll L Rep 78, 86.

[87]Nelson Line (Liverpool) Ltd v James Nelson & Sons Ltd [1908] AC 16, 20.

[88]Sleigh v Tyser [1900] 2 QB 333, 337–338 (Bigham J). See also (The Socol 3 [2010] EWHC 777 (Comm); [2010] 2 Lloyd's Rep 221.

[89][1905] 1 KB 697, 706 (Channell J). See also The Derby [1985] 2 Lloyd's Rep 325, 332.

[90]Burges v Wickham (1863) 3 B & S 669, 693.

[91]MDC Ltd v NV Zeevaart Maatschappij Beursstraat [1962] 1 Lloyd's Rep 180, 186 (McNair J).

[92]The Arianna [1987] 2 Lloyd's Rep 376, 389.

receiving her cargo[93] and secondly, on sailing when the ship must be fit in design, structure, condition, and equipment to encounter the ordinary perils of the voyage.[94] At common law the obligation that attaches at the commencement of the voyage is not a continuing one, but only attaches the time of sailing.[95] The requirements have been construed as:

> … An absolute warranty that the carrying vessel must, at the time sailing with the goods, have that degree of fitness as regards both the safety of the ship and also the safe carriage of the cargo in the ship which an ordinary careful and prudent owner would require his vessel to have at the commencement of the voyage, having regard to the probable circumstances of that voyage and its nature.[96]

3 Voyage Charterparties

3.1 Express Clauses

Most voyage charterparties contain an express clause setting out the obligation of seaworthiness on the shipowner. In the well-known Gencon 1994 form, the first part of clause 2 provides that:

> The Owners are to be responsible for loss of or damage to the goods or for delay in delivery of the goods only in case the loss, damage or delay has been caused by personal want of due diligence on the part of the Owners or their Manager to make the Vessel in all respects seaworthy and to secure that she is properly manned, equipped and supplied, or by the personal act or default of the Owners or their Manager.[97]

The clause has been described as "an exceptions clause operating in favour of the owners."[98] After establishing the scope of the owners' responsibility for loss, damage, or delay in delivery of the goods, the clause, much like the Asbatankvoy clause, provides that such responsibility only falls on the shipowner following a personal want of due diligence for making the vessel "in all respects seaworthy" and securing that "she is properly manned, equipped and supplied."

[93]There is no implied obligation that the ship must be seaworthy on the approach voyage to the port: The Nizeti [1960] 1 Lloyd's Rep 132, 137.

[94]The Rona (1884) 51 LT 28, 30–31.

[95]McFadden v Blue Star Line [1905] 1 KB 697, 703.

[96]Virginia Carolina Chemical Co v Norfolk & North American Steam Shipping Co [1912] 1 KB 229, 243–244 (Kennedy J).

[97]Lines 15–21. See The Fjord Wind [2000] 2 Lloyd's Rep 191.

[98]London Arbitration 7/2000, (2000) 539 LMLN 3.

3.2 The Doctrine of Stages

In many instances there may be different stages in a voyage, some of which occur naturally, as, for example, when a vessel has to sail along a river to reach the high seas.[99] In *The Vortigern*, A. L. Smith stated that:

> The only way in which this warranty can be complied with is for the shipowners to extend the existing warranty to the commencement of each stage, and I can see no reason why such a warranty should not be implied, and I have no difficulty in making the implication, for it is the only way in which the clear intention of the parties can be carried out, and the undoubted and admitted warranty complied with.[100]

In this type of case[101] the common law requires that the vessel must be seaworthy at the beginning of each stage.[102] There is some controversy as to whether there is an implied warranty at common law during the period before the commencement of loading[103] but it is clear that this will be required where the parties expressly so provide.[104] In the case of bunkering stops, the vessel must take on sufficient fuel to reach "a particular convenient or usual bunkering port on the way,"[105] also having regard to the ordinary incidents of navigation on the voyage at the time of year in question.[106]

4 Time Charterparties

4.1 Express Clauses

At common law, the seaworthiness obligation in time charterparties attaches at the commencement of hiring.[107] However, as is the case with voyage charterparties, standard form time charterparties always contain an express seaworthiness

[99]Northumbrian Shipping Co Ltd v E Timm & Son Ltd [1939] AC 397, 403–404.

[100][1899] P 140, 155; McFadden Brothers & Co v Blue Star Line Ltd [1905] 1 KB 697, 704.

[101]Possibly also in liner trades where a ship calls at ports in advertised sequence. See Maxine Footwear Co Ltd v Canadian Government Merchant Marine Ltd [1959] AC 589 (PC), 604 (Lord Somervell): "The doctrine of stages had its anomalies and some important matters were never elucidated by authority".

[102]Thin v Richards & Co [1892] 2 QB 141.

[103]The Nizeti [1960] 1 Lloyd's Rep 132, 140.

[104]New York & Cuba Mail Steamship Co v Eriksen & Christensen (1922) 10 Ll L Rep 772.

[105]Thin v Richards & Co [1892] 2 QB 141, 143.

[106]Walford de Baerdemaecker & Co v Galindez Bros (1897) 2 Com Cas 137.

[107]Giertsen v Turnbull & Co 1908 SC 1101, 1110. It has been suggested that the shipowner's obligation is to exercise due diligence before the start of each voyage (London Arbitration 24/1989, (1989) 259 LMLN 4) but this is yet to be confirmed in any reported case.

clause.[108] Though regarded as outdated, the still heavily utilized New York Produce Exchange form, NYPE 1946, provides in the Preamble that

> Vessel on her delivery to be ready to receive cargo with clean-swept holds and tight, staunch, strong and in every way fitted for the service, having water ballast, winches and donkey boiler with sufficient steam, or if not equipped with donkey boiler, then other power sufficient to run all the winches at one and the same time (and with full complement of officers, seamen, engineers and firemen for a vessel of her tonnage) ...[109]

Several features of this clause should be noted. The first is that, as at common law, the obligation applies "on delivery." The second is that the wording embraces the main requirements identified at common law, namely, provision of a cargoworthy vessel ("ready to receive cargo with clean-swept holds"),[110] a vessel that is physically able to withstand the perils of the sea ("tight, staunch, strong and in every way fitted for the service"),[111] and one properly equipped ("having ballast water, winches ..." etc.),[112] and properly manned ("with full complement of officers, seamen, engineers and firemen for a vessel of her tonnage").[113] The wording also emphasizes that the vessel must be "in every way fitted for the service."[114]

The NYPE 93 clause is updated in a number of respects. Although the obligation also attaches "on her delivery,"[115] this is no longer in the Preamble to the charterparty but in a "Delivery" clause requiring the vessel to be "tight, staunch and strong and in every way fitted for *ordinary cargo* service."[116] Finally, albeit in a different clause, the NYPE 93 form requires the vessel to have a "full complement of officers and crew."[117] As with NYPE 1946, the master is not expressly included, but this has been addressed in the new NYPE 2015 form, which provides that

> [T]he vessel on delivery shall be seaworthy and in every way fit to be employed for the intended service, having water ballast and with sufficient power to operate all cargo handling gear simultaneously, and, with full complement of Master, officers and ratings who meet the Standards for Training, Certification and Watchkeeping for Seafarers (STCW) requirements for a vessel of her tonnage.[118]

---

[108] The Madeleine [1967] 2 Lloyd's Rep 224; The Derby [1984] 1 Lloyd's Rep 635.

[109] Lines 21–24.

[110] Line 22. "Readiness" here embraces the requirement that the vessel is "completely ready in all her holds so as to afford the merchant complete control of every portion of the ship available for cargo": Groves, Maclean & Co v Volkart Bros (1884) Cab & El 309, 311 (Lopes J).

[111] Ibid. The vessel must be "fit to commence her chartered enterprise, which consists of loading and sailing when loaded": New York & Cuba Mail Steamship Co v Eriksen & Christensen (1922) 10 Ll L Rep 772, 773 (Greer J).

[112] Lines 22–24.

[113] Line 24.

[114] An express warranty of seaworthiness: The Derby [1985] 2 Lloyd's Rep 325, 331; 333; The Arianna [1987] 2 Lloyd's Rep 376, 389–390.

[115] Line 33.

[116] Lines 33–35. Emphasis supplied. See The Arianna [1987] 2 Lloyd's Rep 376.

[117] Cl 6, line 82.

[118] Cl 2(b), lines 40–44.

The clause specifically states that the vessel on delivery "shall be seaworthy,"[119] echoing the first paragraph of Article III, r 1 of the Hague (and Hague–Visby) Rules, and requires the vessel to be "fit to be employed for the intended service."[120] This replaces previous wording that the vessel is to be "fitted for ordinary cargo service" and may be a response to a comment of Webster in *The Arianna* that

> … although it seems probable to me that there is rarely any practical difference between seaworthiness and fittedness, this case, it seems, may well be one in which there is, if only in theory, a difference … And it seems to me that, although in many cases seaworthiness and fittedness for service have been treated as synonymous … they are not necessarily the same thing.[121]

Then, highlighting the importance of ballasting in shipboard operations, the vessel is required to have "water ballast," as well as "sufficient power to operate the cargo handling gear simultaneously." The latter is important in those cases where the vessel's own gear is used for loading and discharging cargo. The clause also requires a "full complement of Master, officers and ratings," as laid down by the STCW, updating the provision in the earlier NYPE 93 form.

4.2 Continuing Obligation of Seaworthiness

At common law, the shipowner is obliged to maintain the vessel in a seaworthy state, provided that it has a reasonable opportunity of doing so, and as long as this does not cause unreasonable delay or expense to the various interests involved.[122] However, if cargo is damaged by a leak and the master negligently omits to take sufficient steps to stop the leak, causing further damage to the cargo, the shipowner will not be liable if the original and continuing source of the damage is covered by an exception.[123] Thus, in *The Rona*,[124] after leaving her moorings in New York, a vessel stranded and, with initial tug assistance, was able to proceed, although taking in water. Her master chose not to undertake any repairs and proceeded to London, encountering heavy weather en route. The claimant's cargo of flour was damaged by seawater that had come through the vessel's deck and the court held that the master was negligent in not effecting repairs and that the shipowner was liable. Thus, prudent action by the master, in such circumstances, might include putting

---

[119]Line 40.

[120]Lines 40–41.

[121][1987] 2 Lloyd's Rep 376, 389–390.

[122]Worms v Storey (1855) 11 Exch 427, 430; The Rona (1884) 5 Asp MLC 259, 261–262.

[123]See, eg, The Cressington [1891] P 152 (exception of perils of the sea and negligence of the master applied).

[124](1884) 5 Asp MLC 259.

into a port of refuge for repairs[125] or having on board the vessel pumps capable of coping with moisture given off by the loaded cargo.[126]

Putting this common law obligation to one side, a substantial majority of time charterparties[127] also contain wording placing an ongoing obligation on the shipowner to provide a seaworthy vessel. In *Tyndale Steam Shipping Co Ltd v Anglo-Soviet Shipping Co Ltd*, this wording was interpreted as follows.

> … [T]his … does not constitute an absolute engagement or warranty that the shipowners will succeed in so maintaining her whatever perils or causes may intervene to cause her to be inefficient … The engagement of the shipowners is this: that if accidents happen or events arise to cause the ship to be inefficient or the winches to be ineffective and out of action, they will take all reasonable and proper steps that reasonable men could take to put them back again.[128]

Among the other well-known forms, NYPE (1946) obliges the shipowner to "keep" the vessel in the same state as on delivery, and both NYPE 93 and NYPE 2015 provide that the Owners "… shall … keep her in a thoroughly efficient state in hull, machinery and equipment for and during the service."[129]

5 Demise Charterparties

The most widely utilized demise (or bareboat) charterparty standard form, Barecon 1989, contains an express seaworthiness obligation requiring "… due diligence [that] the Vessel [is] seaworthy and in every respect ready in hull, machinery and equipment … . The Vessel shall be properly documented at time of delivery."[130]

This clause was considered in *The Eye-Spy.*[131] The bareboat charterer alleged that the failure of the *Eye-Spy's* starboard stern tube assembly (SSTA), was caused by the failure to exercise due diligence. The court confirmed that the defect in the SSTA was a latent defect[132] and that due diligence required the shipowner to consult an expert repairer and have the vessel slipped to pull the shafts so the

125 Phelps, James & Co v Hill [1891] 1 QB 605; J & E Kish v Charles Taylor Sons & Co [1912] AC 604. The vessel must not leave the port of refuge until again seaworthy: Worms v Storey (1855) 11 Exch 427.

126 Stanton v Richardson (1875) 3 Asp MLC 23 (HL), 24.

127 Also some voyage charterparties: Scott v Foley, Aikman & Co (1899) 5 Com Cas 53.

128 (1936) 54 Ll L Rep 341, 344–345 (Lord Roche); Snia Societa di Navigazione Industria e Commercio v Suzuki & Co (1924) 17 Ll L Rep 78, 88.

129 Cl 6, lines 81–82 (NYPE 93); cl 6(a), lines 94–96 (NYPE 2015). This form of wording is almost exactly the same as in the Baltime (1939) form.

130 Cl 2, lines 11–14.

131 Delaware North Marine Experience Pty Ltd v The Ship "Eye-Spy" [2017] FCA 708; [2017] 2 Lloyd's Rep 463.

132 At [256].

bearing in the stern seal could be properly examined.[133] The failure by the shipowner to take this step, being on notice about a problem that had not been properly rectified, was a breach of its principal seaworthiness obligation under the charterparty.[134]

In the amended Barecon 2001 form the clause has been updated,[135] containing a separate subclause specifying that the vessel must be "properly documented on delivery in accordance with the laws of the flag State ... and the requirements of the classification society ... ."[136] As will be evident from the wording in Barecon 89 and Barecon 2001, the obligation must be exercised "before and at the time of delivery," an intention that the obligation is owed also during the period before the commencement of loading.[137] The required standard is due diligence.

The recently issued Barecon 2017[138] makes a number of important changes to the previous Barecon regime. Thus, in place of the wording in the older forms, the new clause provides that the shipowner must deliver the vessel "in a seaworthy condition and in every respect ready for service."[139] It is submitted that the effect of this is broader than under the Barecon 2001 form. The subclause on documentation is largely the same[140] as in the Barecon 2001 form. The main change, however, is that the required standard of seaworthiness is no longer one of due diligence. Accordingly, the standard is an absolute one, unless modified by the wording of the clause paramount.[141]

6 Due Diligence and Seaworthiness

6.1 Seaworthiness Under the Hague (and Hague–Visby) Rules

As noted earlier, the standard of seaworthiness in many charterparty forms is one of due diligence.[142] This derives from the obligation in Article III, r 1 of the Hague

[133]At [264].

[134]Ibid [ie under cl 2, lines 10–13, of Barecon 89 (now cl 3 of Barecon 2001)].

[135]See cl 3(a), lines 18–21.

[136]Cl 3(b), lines 26–29.

[137]See New York & Cuba Mail Steamship Co v Eriksen & Christensen (1922) 10 Ll L Rep 772, 773.

[138]Issued on 12 December 2017. See "New Barecon 2017 calculates future trends", https://www.bimco.org/news/press-releases/barecon, last accessed 2017/12/12.

[139]Cl 3(a), line 28. Cf the wording in the NYPE 2015 clause.

[140]Cl 3(b), lines 34–35.

[141]See cl 12(a), lines 140–143.

[142]See Clarke, M.: Aspects of the Hague Rules. Martinus Nijhoff, The Hague, 203; Chacón, V.: The Due Diligence in Maritime Transportation in the Technological Era. Springer, Cham.

(and Hague–Visby) Rules. There are several distinctive features of this provision. The first is the period of application, "before and at the beginning of the voyage," which covers the period "from at least the beginning of the loading until the vessel starts on her voyage."[143] If, however, the vessel is unseaworthy owing to some earlier breach of due diligence, the shipowner will be liable on the ground of actual or imputed knowledge of the defects or failure to use due diligence continuing to the date relevant to the particular contract of carriage.[144] Once the vessel starts on the voyage, the obligation no longer applies[145] and the period of coverage, unlike that at common law, is not broken by the doctrine of stages.

The second feature of the provision relates to the extent of the obligation, which, as already noted, is one of "due diligence." The common law implied absolute obligation is abolished.[146] Due diligence has been interpreted by the courts as "indistinguishable from an obligation to exercise reasonable care,"[147] and "lack of due diligence is negligence … ."[148] The obligation is, however, an overriding one[149] and is not delegable to servants or agents: "the shipowners' obligation of due diligence demands due diligence in the work of repair by whomsoever it may be done."[150] If particular responsibilities are delegated to independent contractors or surveyors and such persons are negligent, the shipowner remains liable,[151] it being no defense that reliable experts were engaged or that the shipowner lacked the necessary expertise to check their work.[152] However, the shipowner will not be responsible until the vessel comes under its "orbit,"[153] or its ownership, possession, or control.[154] Thus, if a new vessel is commissioned or a vessel is chartered or purchased from another person,[155] the shipowner will not be liable for existing defects rendering the vessel unseaworthy, unless these were reasonably

143 Maxine Footwear Co Ltd v Canadian Government Merchant Marine Ltd [1959] AC 589 (PC), 603; The mv Devon [2012] EWHC 3747 (Comm), [43].

144 W Angliss & Co (Australia) Pty Ltd v P & O Steam Navigation Co [1927] 2 KB 456, 463; The Kriti Rex [1996] 2 Lloyd's Rep 171, 185.

145 Leesh River Tea Co Ltd v British India Steam Navigation Co Ltd [1967] 2 QB 250, 274–275 (upholding [1966] 1 Lloyd's Rep 450, 457).

146 See, eg, Carriage of Goods by Sea Act 1971, c 19, s 3.

147 The Eurasian Dream [2002] EWHC 118 (Comm); [2002] 1 Lloyd's Rep 719, [155].

148 Ibid.

149 Maxine Footwear Co Ltd v Canadian Government Merchant Marine Ltd [1959] AC 589 (PC), 602–603.

150 The Muncaster Castle [1961] AC 807, 844 (overruling [1959] 1 QB 74; [1960] 1 QB 536 (CA)).

151 The Fjord Wind [2000] 2 Lloyd's Rep 191, 199.

152 This would not extend further to responsibility for manufacturers, exporters, or shippers: The Kapitan Sakharov [2000] 2 Lloyd's Rep 255, 272.

153 The Muncaster Castle [1961] AC 807, 867.

154 The Happy Ranger [2006] EWHC 122 (Comm); [2006] 1 Lloyd's Rep 649, [37].

155 W Angliss and Co (Australia) Pty Ltd v Peninsular and Oriental Steam Navigation Co [1927] 2 KB 456, 461.

discoverable by the exercise of due diligence at the time of takeover.[156] If, however, the defect could have been apparent on a reasonable inspection of the vessel at the time that the vessel was taken over, the shipowner cannot rely for protection even on the certificate of a surveyor or any other classification society.[157] Thus, where a shipowner failed to appreciate that there had been inadequate proof testing of certain crane hooks by a classification society, for whose failings it was responsible, it was held not to have acted with due diligence.[158]

The third feature of the provision is the wording in (a), (b), and (c). This is understood as embracing each of the distinct elements of seaworthiness recognized at common law and does not have an extended or unnatural meaning.[159] However, the provision is then explicit in requiring due diligence as to manning, equipment, and supply of the vessel. Underlying cargoworthiness, the "holds, refrigerating and cool chambers, and all other parts of the ship in which goods are carried, fit and safe for their reception, carriage and preservation."

6.2 Paramount Clauses

Neither the Hague nor the Hague–Visby Rules apply to charterparties.[160] If the parties wish to apply the Rules to a charterparty[161] this must be effected by means of an incorporating clause, a "paramount clause" or "clause paramount."[162] The effect of such a clause was considered in one of the leading cases, *The Saxon Star*,[163] which arose out of a consecutive voyage charterparty for a tanker. The shipowner selected and appointed incompetent engine-room staff, the vessel breaking down on the first voyage to her loading port. The charterer lost the services of the vessel for an extended period but the shipowner claimed that its obligation was limited to the exercise of due diligence.[164] The House of Lords found for the shipowner, Viscount Simonds explaining that the parties had agreed to "... an obligation to use due diligence in place of the absolute obligation ... ."[165]

---

[156]Ibid.

[157]The Happy Ranger [2006] EWHC 122 (Comm); [2006] 1 Lloyd's Rep 649.

[158]At [62].

[159]The Aquacharm [1982] 1 WLR 119; The Benlawers [1989] 2 Lloyd's Rep 51, 60.

[160]See Article V. This ensures that "the shipowner retains absolute freedom to conclude charter parties on the terms he wishes and that he can insert whatever clauses he likes, as in the past ...": Carver on Bills of Lading (n 2), para 9–310.

[161]Examples are: Asbatankvoy, cl 20(b)(i); NYPE 1946, cl 24; NYPE 93, cl 31(a), lines 318-328; NYPE 2015, cl 33(a), lines 526–538.

[162]For fuller consideration, see Carver on Charterparties (n 2), para 5-009.

[163][1959] AC 133.

[164]As contained in the US Carriage of Goods by Sea Act 1936.

[165][1959] AC 133, 154; The Satya Kailash and Oceanic Amity [1984] 1 Lloyd's Rep 588, 593–594; The Eternity [2008] EWHC 2480 (Comm); [2009] 1 Lloyd's Rep 107, [19].

The same approach has been applied to trip time charterparties[166] but whether it will also apply to other charterparties, including time charterparties,[167] is yet to be resolved.[168]

7 Conclusion

Most aspects of the seaworthiness obligation have been thoroughly tested in the courts over several hundred years. In the charterparty context much of this law continues to apply and, as this paper has shown, many of the seaworthiness cases today require consideration of express clauses in the standard forms, demonstrating the continuing vitality and centrality of the concept. From the shipowners' perspective, the weight of regulation has been heavy, particularly since the establishment of the International Maritime Organisation,[169] and continues unabated. This is continuing to have some impact on the seaworthiness obligation, particularly as many charterparty standard forms are revised. Indeed, the notable trend is for these later revisions of the standard forms to embrace more specific detailed seaworthiness requirements. Not all the difficulties have been resolved, however. In particular, the real impact of new challenges, such as autonomous ships[170] and the so-called fourth industrial revolution and "smart shipping," are still to be felt.[171]

[166]The Aliakmon Progress [1978] 2 Lloyd's Rep 499, 501.

[167]The Hermosa [1980] 1 Lloyd's Rep 638, 647–648.

[168]At 648; London Arbitration 24/1989, (1989) 259 LMLN 4.

[169]See www.imo.org, last accessed 2017/12/7.

[170]Carey (n 50).

[171]See, eg, Clayton, R.: Outlook 2018: Prologue of a new era. Lloyd's List, 2017/11/29.

Seaworthiness and Major Accidents at Sea—An International Perspective

Sarah Fiona Gahlen

Abstract This study gives an overview of the international understanding of "seaworthiness," both in contracts of carriage and in charterparties. The obligation to provide a seaworthy ship is of great importance in all contracts of carriage that are governed by the Hague or the Hague–Visby Rules, and presents a duty that cannot be derogated from. The study considers a number of aspects, namely the required condition of the vessel, the necessary standard of care in ensuring the seaworthiness of the vessel, the point in time in which seaworthiness is assessed, and the question if fault or negligence of third parties can be attributed to the person in charge of the vessel. Precedents on these aspects from different countries are compared in order to assess whether there is international uniformity in the understanding of the term. In each case, a comparison is drawn between the provisions of the HVR and the law applicable to charterparties, which is usually governed by standard contracts rather than mandatory international law.

Keywords Seaworthiness · Contracts of carriage · Charterparties

1 Definitions of "Seaworthiness"

The idea that a vessel should by all means be in an appropriate state, technically and in other respects, for the voyage that she is about to undertake is certainly a very old one, and was reflected in the maritime law of many legal systems before finding its way into international conventions.

It appears logical that the person in charge of the vessel—the owner, or in the case of a contract of carriage, the carrier—ought to assume responsibility for the state of the vessel towards his contractual partners, who are not in control of the ship. The duty to provide for a "seaworthy" vessel thus naturally became an

S. F. Gahlen (✉)
Lebuhn & Puchta, Am Sandtorpark 2, 20457 Hamburg, Germany
e-mail: sarah.gahlen@lebuhn.de

A. Vega Sáenz et al. (eds.), *Proceedings of the 25th Pan-American Conference of Naval Engineering—COPINAVAL*, https://doi.org/10.1007/978-3-319-89812-4_38

obligation of shipowners and carriers. Still, it is a very simple term to describe an obligation with many facets.

As shown below, the question of whether a vessel is seaworthy is sometimes difficult to assess. It is in many cases a question of facts and considered by the courts on a case-by-case basis. There is no recognized international definition.

As a matter of common law, "seaworthiness" has been defined to "express that the ship should be in a condition to encounter whatever perils of the sea a ship of that kind, and loaded in that way, may be fairly expected to encounter" [1]. In US maritime law, a vessel is considered as seaworthy if she is "reasonably fit to carry the goods which she undertakes to transport" [2]. Similarly, the German Supreme Court defined seaworthiness as the "ability of the vessel to withstand the usual perils of the sea on the intended voyage" [3].

A lack of seaworthiness can have many causes; apart from mere technical shortcomings causing nautical unseaworthiness, "commercial" or "administrative" unseaworthiness can also occur. This describes a situation in which the ship is technically sound, but unlikely to complete the voyage successfully for other reasons, for example, in view of expected measures by port states taken due to a lacking documentation or even in view of a threatened arrest [4]. This study focuses on seaworthiness in connection with major accidents and therefore mainly considers technical and maintenance defaults.

2 Seaworthiness and the Circumstances of the Voyage

It should not be forgotten that it is also in the carrier's or the owner's best interest to maintain the vessel in a seaworthy condition. Cases in which the carrier or the owner consciously neglect their duties are rare. On the other hand, they will also be interested in keeping the cost of maintenance and equipment proportionate: a vessel can never be outfitted for every possible risk.

As can be seen from the definitions quoted above, a ship can generally only be seaworthy with regard to the intended voyage. The vessel needs to be equipped and outfitted for the voyage planned; there is no need to cater for risks that will not or are unlikely to arise on this trip. With regard to this matter, it is often submitted that the requirements of seaworthiness would thus vary, for example, according to the cargo [5], the geographic region of the voyage, and the weather conditions to be expected.

The German Supreme court has, however, emphasized that the actual weather forecast would not be a criterion to take into account, because this could change in a matter of days or even hours [6]. Accordingly, the assessment would need to be made with regard to the usual perils of the voyage and thus, the weather conditions that can be expected on this trip at the given time of the year.

Establishing seaworthiness requires an assessment of whether a particular peril can be expected on a given voyage. Difficulties in this respect might arise with regard to cases in which the voyage becomes, all of a sudden, more dangerous, or where it becomes obvious that the voyage will be more dangerous than initially expected.

3 Seaworthiness as an Obligation to Exercise Due Care

At common law, the obligation to provide a seaworthy vessel is an implied obligation, and is objectively assessed: the vessel needs to be fit for the purpose it is employed for, irrespective of whether the owner or the carrier knew or could have known of any defects [7]. Still, it has been asserted that even this warranty does, however, not extend "to latent defects that could not by any reasonable diligence or skill be discovered" [8], so that there are obviously possibilities of exemption even where the obligation is thought to be an absolute one.

3.1 Contracts of Carriage

From an international perspective, the importance of the term of "seaworthiness" in contracts of carriage is of course due to the provisions of the Hague Rules of 1924, which provide for internationally uniform rules for contracts of carriage covered by a bill of lading, and which, as regards liability, cannot be derogated from. The Hague Rules have been amended by the so-called Visby Protocol in 1967, and the consolidated version is now commonly known as the Hague–Visby Rules (HVR). The provisions discussed in this study have been left unchanged by the Visby Protocol, so that the findings hold true both for contracting states to the Hague Rules and to the HVR.

Considering the time the HVR have been in force for and the fact that most of the countries ensuring the greater part of the world's maritime trade are contracting states, the rules present a sound basis of internationally uniform rules on carrier liability. In contrast to the common law position, the HVR provide for a subjective obligation of the carrier with regard to the vessel's seaworthiness. Art. III (1) (a) HVR reads: "The carrier shall be bound before and at the beginning of the voyage to exercise due diligence to: (a) Make the ship seaworthy."

3.1.1 Standard of "Due Diligence"

Thus, the carrier is only bound to exercise due diligence to render the vessel seaworthy; if he did exercise due diligence, he complied with this obligation, irrespective of whether the vessel was objectively seaworthy. In conclusion, the carrier is not liable for any defects that escaped his attention despite the exercise of due care. This, of course, begs the question for the standard of "due care" required.

The question has very extensively been treated in all the jurisdictions under scrutiny here. A court or tribunal, when deciding whether the measures taken by a carrier were good enough to fulfill a standard of "due diligence" with regard to the specific voyage will always need to consider the specific facts of the case. Still, an overwhelming number of precedents confirm that the standard is that of all reasonable measures that would have been taken by a reasonable shipowner at the relevant time. As a matter of English law, it is well established that the duty to exercise due care to ensure the vessel's seaworthiness is not absolute, and that it is assessed in relation to the knowledge at the time and the standards applicable then [9]. Similarly, the German Supreme Court has held that a carrier cannot be expected to take additional measures in stowing a bulk cargo if the potential danger connected with this cargo was not yet fully known in engineering practice [10].

Consequently, any regulatory frameworks that oblige shipowners or carriers to take security or other measures are of great importance to assess the standard of due diligence. A shipowner or carrier not complying with safety regulations is very likely to fall short of this standard [11]. There is even one English case suggesting that the failure to comply with the standard set out in regulations such as the ISM code might be indicative of a lack of good care even if the regulation has not yet formally entered into force for the type of vessel in question [12].

It can, of course, be argued that every damage has a cause, and that every defect is detectable or foreseeable with the right equipment or the right tests. Consequently, it appears that there are comparatively few cases where a carrier was found to have exercised due diligence, but the damage was not detectable. This was the case both in an English [13] and a Belgian [14] decision, in which the respective vessel's breakdown was in each case caused by fatigue damages in machinery. These could have been detected if further examinations had been made, but which would have required dismantling. During the visual inspections that had been carried out and that had in both cases indisputedly been in line with usual practice, the defect had passed unnoticed. It was held in both cases that it was not unreasonable for the carrier to rely on the visual examination.

The question was also dealt with in the recent German decision regarding the catastrophic accident of the *MOL Comfort* [15]. The vessel literally broke in two on a voyage from Hong Kong to Hamburg. The exact cause of the excessive strain on the hull was not established with absolute certainty, a construction defect, faults in the stacking of containers, and overcharging might all have contributed to it. The court, however, considered the vessel as unseaworthy because of the fact that measurements of the vessel's draft in Singapore had evidenced a significant bending in the hull. Irrespective of any other cause, the court held that the carrier did not comply with his obligation to exercise due care, because the master had failed to take action with regard to this bending.

With a view to the difficulty of assessing if the carrier exercised due diligence or not, it may in practice be very important to establish which party has to bear the burden of proof regarding seaworthiness or the lack of it. The HVR, in this respect, achieve a considerable degree of uniformity by giving a clear answer to this question in Art. IV (1) HVR. According to this provision, the cargo interests wanting to invoke unseaworthiness have to show that the vessel was unfit for purpose. The carrier can then show that he still exercised due diligence to make the ship seaworthy, that is, that the defect escaped his attention despite the exercise of good care. If there are any doubts whether the unseaworthiness actually caused the damage, it is for the cargo interests to prove this [16].

3.1.2 Exoneration for "Latent Defects"

The HVR contain exoneration clauses in Art. IV rule 2, exempting the carrier and the ship from liability for damages arising from certain causes. Among these is the exception of "(p) latent defects not discoverable by due diligence." The additional reference to "due diligence" of course begs the question as to the interplay between the standard of due diligence required in Art. III (1) (a) HVR and the exemption in Art. IV (2) (p). In view of the fact that Art. IV (2) (p) HVR does not refer to specific points in time at all, it can be considered that the exemption clause applies throughout the entire voyage, whereas the obligation to exercise due care to ensure the ship's seaworthiness in any event only arises before and at the beginning of the voyage [17]. It also appears that Art. IV (2) (p) HVR can apply to the rare cases in which there is a latent defect that causes a damage but cannot be considered to affect the vessel's seaworthiness.

In summary, it does not seem that Art. IV (2) (p) HVR could have a wide scope of application after all or would by any national jurisdiction be considered to affect the carrier's obligation under Art. III (1) (a) HVR. For instance, English [18] courts have explicitly stated that seaworthiness is an "overriding" obligation, so that in any event, neither of the exemptions in Art. IV (2) can be invoked if the carrier was found to have provided an unseaworthy ship. This view is certainly supported by the wording of the HVR, considering that Art. III (1) HVR, unlike Art. III (2) HVR, is not made subject to the exemption provisions of Art. IV HVR. It can be concluded that the carrier can invoke the exemption clauses when he fails to comply with the obligation to carry the goods (Art. III (2) HVR), but not when he fails to provide a seaworthy ship.

3.2 Charterparties

In contrast to the law of carriage of goods, the law on charterparties is not governed by international conventions such as the HVR. Subject to the provisions of the law governing the charterparty, the parties are free to contract as they like.

In common law, seaworthiness of the chartered vessel will be considered as an implied warranty of the owner [19], even if the scope of this warranty might not be entirely clear for a given charterparty. The German law on maritime trade, as thoroughly revised in 2013, now provides in §560 of the Commercial Code (*Handelsgesetzbuch*) that in a time charterparty, the owner needs to maintain the ship in the condition that is contractually agreed on, in particular, that it is seaworthy. Any contractual agreement between the parties will, however, prevail over this general provision.

From an international perspective, a considerable degree of conformity is achieved by the use of standard contracts and terms. Almost all standard charterparties contain clauses regarding the condition of the ship upon delivery. In contrast to a contract of carriage, in which the circumstances of the specific voyage will generally be clearly foreseeable, namely time charterparties can be concluded for several years and with worldwide trading limits. Clauses dealing with the vessel's condition and its ability to undertake voyages must be worded and interpreted with a view to this. For instance, the burden of proving unseaworthiness might in some contracts be on the charterer. Furthermore, the obligation to provide a seaworthy vessel could be considered as a continuing obligation, especially in time charterparties, in view of the fact that there will be more than only one loading port (also see Sect. 4 below).

Where the provisions of the HVR are incorporated into a charterparty, by way of a "Clause Paramount" or in any other way, the HVR provisions on the vessel's seaworthiness cannot be contracted out of, of course.

4 Seaworthiness and the Beginning of the Voyage

As a matter of fact, most legal systems only require a carrier to ensure the vessel's seaworthiness at the beginning of a voyage. This idea may appear outdated in modern times. In the past, vessels were expected to encounter unexpected perils on their voyage, with very little help or assistance available under way, so that it appeared fair not to hold a carrier or shipowner liable for unseaworthiness that only developed during the voyage, when the vessel had left the sphere of her owners' control.

4.1 Contracts of Carriage

According to Art. III (2) of the HVR, the carrier is under a duty to exercise due care to make the vessel seaworthy "before and at the beginning of the voyage."

The pertinent point in time may sometimes be hard to determine or even lead to arbitrary results. In the case of *Maxine Footwear Co. Ltd v. Canadian Government Merchant Marine Ltd* [20], a vessel caught fire in port, during the loading operations, when frozen pipes were thawed with an acetylene torch. The vessel was completely destroyed by fire and in fact never sailed, and it was held that the carrier had failed to comply with his duty to exercise due diligence "before and at the beginning of the voyage," because the ship was unseaworthy due to the fire damage. Had the fire been detected later, after the vessel had set sail, the decision would have been different.

On the other hand, it has been confirmed by the American courts that the vessel's seaworthiness actually needs to be assessed in the pertinent port, and that it is not sufficient to rely on inspections and findings made during an earlier voyage [21], not even if the call in the first part of the voyage was a very short one.

Difficulties arise where a vessel is scheduled to call in different ports. In this way, it could have been questionable if the carrier in the German case of the "*MOL Comfort*" was under a duty to act on findings made during an intermediary call in Singapore, when the vessel had actually sailed from Hong Kong and the bills of lading were issued for a voyage from Hong Kong to Hamburg. Neither the court in the first [22] nor in the second instance, however, considered this point, which may appear surprising. The German Supreme Court had in 1972 [23] still held that damage which occurred during a bunkering call in Bremerhaven did not cause a vessel's unseaworthiness "at the beginning of the voyage" when the bills of lading were issued for a voyage from Hamburg.

Courts in various jurisdictions have dealt with the issue of defects that are present in the beginning of the voyage but could later be discovered and then amended. It has been explicitly held in American [24] and German law [25] that a ship is not unseaworthy if it is to be expected that the matter will, in the ordinary course of the ship's operation, be fixed before any danger can materialize. There must, however, be an actual probability for this to happen; even if the means to mend the defect are on board, there is no likelihood if the master has given to understand that he did not intend to use them [26].

Even under a contract of carriage, a ship might always be ordered to a different port during the voyage. The question of whether a new voyage then begins in the moment that the new order reaches the ship has explicitly been decided by the higher German courts. It was held that a new order can only present the beginning of a new voyage if the ship is in port in the moment the order is given [27]. The court argued that even in modern times, the ship can only be thoroughly inspected and defects be repaired in port. The judgment was rendered with regard to the German DTV standard clauses for hull insurance that require seaworthiness "at the moment the ship is sent to sea," but of course, the reasoning can also apply to seaworthiness "before and at the beginning of the voyage" according to the HVR.

Under the wording of the HVR, the carrier's obligation is clearly restricted to the time "before and at the beginning of the voyage." As can be seen from the German case of the "*MOL Comfort*," there might in more recent times be a tendency to be generous with the interpretation of where the "beginning of the voyage" is to be located.

4.2 Charterparties

Under the terms of most standard charterparties, the duty to exercise due care to provide a seaworthy ship is less clearly a temporary one. Obviously, a chartered vessel, namely under a time charter, will often undertake more than one voyage and call in several ports.

Under English law, the "doctrine of stages" was made use of to deal with this problem. According to the doctrine, a vessel would pass through several stages during a trip, and the shipowner or carrier is under a duty to ensure the vessel's seaworthiness at the beginning of each stage. The doctrine of stages was held to be inapplicable when the obligation to make the ship seaworthy only attached to the beginning of the voyage under the HVR but could even then still apply to the carrier's ancillary duty to bunker at the appropriate time, that is, at the beginning of each stage requiring bunkering [28]. It seems that the stages of a charterparty are determined according to the vessel's foreseeable actual needs and most probably with a view to the intended calls in port.

Early general maritime law in the United States frequently considered seaworthiness as a matter of marine insurance law, and in this context considered that the shipowner's duty to maintain the vessel seaworthy applied throughout the voyage, even though any deficiencies could also be dealt with before the relevant stages [29].

Modern standard charterparties, of course, deal with this issue by providing for clauses as to the condition of the ship and its trading limits. As a more recent codification, the German law on maritime trade makes the owner's duty to maintain the vessel in a seaworthy condition a permanent obligation, which applies—in the absence of contractual stipulations to the contrary—"for the duration of the time charter contract" (§560 of the Commercial Code).

5 Seaworthiness as a Nondelegable Obligation

As shown above, shipowners and carriers have the duty to provide a seaworthy ship, or to exercise due care in providing a seaworthy ship, regardless of fault or negligence; if they failed to comply with this standard, it does not matter whether they acted negligently in doing so.

In many cases damages, however, occur because of actions or decisions taken by agents or servants of the carrier or shipowner, or even independent contractors. This begs the question as to the role of these persons when it comes to assessing seaworthiness.

5.1 Contracts of Carriage

According to the wording of Art. III (2) HVR, the "carrier shall be bound [...] to exercise due diligence" to make the vessel seaworthy. Hence, there are many precedents and far-reaching agreement to the effect that the carrier cannot escape liability if the duty is delegated to a third party. The English [30], American [31], and German courts [32] all agree that if the carrier delegates the duty to other persons, he then needs to show that these other persons also exercised due diligence when carrying out the task on his behalf.

As neither of the exoneration reasons of Art. IV (2) will be available to the carrier if the vessel's unseaworthiness is established, the carrier cannot rely on the exonerations of Art. IV (2) HVR, namely the exceptions of nautical mistake of the crew or fire. In this respect, the decision in *Maxine Footwear Co. Ltd v. Canadian Government Merchant Marine Ltd.* (see above, Sect. 4.2) appears very strange, because the court seems to hold that the unseaworthiness was caused by the fire itself, and not by the acts of the carrier's crew or contractors when working with open fire.

Whereas a carrier might still be in a position to show that his crew or his agents exercised due diligence, considering that he exercises control over their work, the matter becomes more difficult with regard to truly independent contractors, namely shipyards or suppliers. In the English case of *Eridania SpA v Oetker (The Fjord Wind)* [33], the vessel had had a history of crankpin failures, the exact cause of which was unknown even to the manufacturers of the engine. The case was a charterparty case, but the pertinent clause of the contract was considered to create a nondelegable duty. The shipowner could show that he had made considerable efforts to detect the reason for the failures, including by addressing the matter with the manufacturers. Still, he could not show that the manufacturers, in turn, had exercised due diligence in investigating the reason for the failures.

In view of this, it is interesting to touch briefly upon the role of classification societies. Whereas many parties in the maritime trade rely on classification certificates for practicable reasons, it does, however, not seem that a carrier would be entitled to refer only to the classification society's findings. In accordance with the principle that the shipowner's duty is nondelegable and that the exercise of due diligence by his servants or contractors must also be shown it has, for instance, been held that it does not suffice if tests have been carried out to the satisfaction of the classification society's surveyor [34] or a classification certificate has been delivered [35] if there is evidence that the defect could have been detected by a more thorough inspection. According to some precedents in national law, the certification can indicate that the carrier has exercised due diligence [36] or even create a presumption to this end [37], but it appears that the carrier will still need to show that the classification exercised due diligence in turn.

5.2 Charterparties

In contrast to Art. III (2) HVR, standard charterparties do not necessarily provide that the shipowner would be under an nondelegable duty to provide a seaworthy ship.

Charterparty clauses can delegate both the duty and the liability for a breach of the duty to third parties, so that the shipowner will not be responsible for the negligence of those persons, to the extent that he himself did not act negligently. It is by no means uncommon for charterparties only to provide for liability for the "personal" act, default, or omission of the owners.

6 Conclusion

Considering that the assessment of seaworthiness is almost always a question of fact and requires a case-by-case consideration by the courts, the present study on some aspects of the principle of seaworthiness in shipping law reveals the importance of internationally uniform law.

Even if the HVR do not take account of the developments in technology and communication, they create a remarkable degree of legal certainty and international uniformity on the issue of seaworthiness. The advantages of international legal standards have repeatedly been recognized by the courts, and there are decisions in which national courts refer to international precedents in their reasoning.

With a view to charterparties, it seems that this function is partly fulfilled by the existence of standard contracts and clauses, but that national law still plays an important role.

References

1. Steel and Craig v. The State Line Steamship Company (1877) 4 R. (H.L.) 103, 105
2. The "Silvia" (1898) 171 U.S. 462
3. German Supreme Court (Bundesgerichtshof), 21.04.1975, II ZR 164/13; 15.10.1979, II ZR 80/77 ("Manny") (both with regard to inland waterway vessels); 20.02.1995, II ZR 60/94
4. Explicitly confirmed to cause unseaworthiness under US law: Morrisey v. SSA & J. Faith, 252 F. Supp. 54 (N.D. Ohio 1965)
5. The French Cour de Cassation for instance held that the handling of a cargo of fuel oil required permanent surveillance, [1951] 2 Gazette du Palais 225 ("Paraskevopoulos")
6. German Supreme Court (Bundesgerichtshof), 20.02.1995, II ZR 60/94
7. Steel and Craig v. The State Line Steamship Company (1877) 4 R. (H.L.) 103, 111: "That is generally expressed by saying that it shall be seaworthy […] not merely that they should do their best to make the ship fit, but that the ship should really be fit"
8. Kopitoff v. Wilson 1 Q.B.D. 377, 383
9. FC Bradley & Sons v Federal Steam Navigation Co. Ltd (1927) 27 Ll. L. Rep. 395, 396
10. German Supreme Court (Bundesgerichtshof), 11.03.1974, II ZR 45/73 ("Neuwardersand")
11. Suggested in Golden Fleece Maritime Inc. v. ST Shipping and Transport (The "Elli" and the "Frixos") [2008] EWCH Civ. 584, even though the case concerned a charterparty and was decided on a different provision
12. Papera Traders Co. Ltd. & Others v. Hyundai Merchant Marine Co. Ltd (The "Eurasian Dream") [2002] EWHC 118 (Comm), para. 143
13. Union of India v NV Reederij Amsterdam (The "Amstelslot") [1963] 2 Lloyd's Rep. 223
14. Rechtbank van Koophandel te Gent, 21.05.1996, European Transport Law 1996, 688 ("Adriano")
15. Higher Regional Court of Hamburg, 02.03.2017, 6 U 86/16 ("MOL Comfort")

16. American Law: Bernstein Co. v MS "Titania" A.M.C. 2040, 2044 (E.D. Louisana 1955); German Law: German Supreme Court (Bundesgerichtshof) 15.10.1979, II ZR 80/77 ("Manny")
17. Lefebvre, G.: L'obligation de navigabilité et le transport maritime sous connaissement. Les cahiers de droit **31**(1), 81–123, 100 (1990), footnote 69
18. Maxine Footwear Co. Ltd v. Canadian Government Merchant Marine Ltd. [1959] A.C. 589, 603
19. Giertsen v. Turnbull, 1908 S.C. 1101
20. [1959] A.C. 589
21. Union Carbide & Carbon Corp v. The "Walter Raleigh" et al., 109 F. Supp. 781, 792 (S.D. N.Y. 1951)
22. Regional Court of Hamburg (Landgericht), 19.04.2016, 411 HKO 99/14 ("MOL Comfort")
23. German Supreme Court (Bundesgerichtshof), 14.12.1972, II ZR 88/71
24. Middleton & Co. (Canada) Limited v. Ocean D.S.S. Corp., 137 F.2d 619 (2 Cir. 1943)
25. German Supreme Court (Bundesgerichtshof), 14.12.1972, II ZR 88/71
26. German Supreme Court (Bundesgerichtshof), 17.01.1974, II ZR 172/72, regarding a vessel undertaking a voyage in the Baltic in stormy weather with unsecured hatch covers. The means to secure them were on board but the master was found to have consciously decided not to use them, because he trusted in a favourable weather forecast.
27. Higher Regional Court of Hamburg (Oberlandesgericht), 20.06.2006, 6 U 222/05 ("Cap Triunfo")
28. The "Makedonia" [1962] P. 190, 195
29. Dupeyre v. The Western Marine & Fire Insurance Co., 41 (2) Louisiana Reports, 457, 460 (S.C. Louisiana 1843)
30. Riverstone Meat Co. v. Lancashire Shipping Co. (The"Muncaster Castle") [1961] A.C. 807
31. General Motors Corp. v. The "Olancho" et al., 115 F.Supp. 107 (S.D. N.Y. 1953), p. 115
32. German Supreme Court (Bundesgerichtshof), 28.06.1971, II ZR 66/69
33. [2000] C.L.C. 1376, 1388
34. American Linseed Co. v. United States, 40 F.2d 657 (E.D. N.Y 1930)
35. Rechtbank van Koophandel of Antwerp, 18.05.1994, (1995) European Transport Law 616
36. Fireman's Funds Insurance Companies v The "Vignes", 794 F.2d 1552 (11th Cir. 1986)
37. Suggested by French courts: Cour d'Appel de Rennes, 13.06.1985, (1986) DMF 625 ("Gogofrio"); Cour d'Appel de Paris, 19.06.1959, (1960) DMF 86 ("Merkurius"), but the presumption was considered rebutted in both cases.

The Fate of BTEX Diluted in the Macondo Blowout

Rubén A. Rodriguez and Serguei Lonin

Abstract The following paper presents the results of blowout numerical modeling formulated in Eulerian coordinates with a special emphasis on the 3D distribution of diluted benzene, toluene, ethylbenzene, and xylene (BTEX). The Eulerian approach is applied only to the liquid phase of a stratified water column with 3D dynamics. The hydrodynamic model used simulates the spatial–temporal conditions of the Macondo's well accident in the Gulf of Mexico in June 2010. The model's domain is limited by −89.125278 to 88.485833°W and 28.516944 to 27.894167°N. The development of the Lagrangian plume model provides information on the trajectories of oil droplets of different diameters (between 600 and 800 μm) that compose the plume, as well as for conceptualizing an Eulerian model for the fate of the diluted BTEX. The generated output of the model agrees with field measurements by order of magnitude for the benzene concentrations in the affected area of the Macondo´s well. The results of the investigation show that BTEX was dissolved in deep waters and it was part of the main toxic hydrocarbon compounds, not detectable by remote sensors or visual observations.

Keywords Hydrodynamic model · Blowout · Aromatic compounds

1 Introduction

Oil is not merely a term, but actually describes a wide variety of natural substances of different origins, as well as a range of synthetic compounds. Crude oil is naturally occurring oil generated by natural processes including geological and

R. A. Rodriguez (✉)
COTECMAR - Science and Technology Corporation for Naval, Maritime and Riverine Industry Development, Cartagena de Indias, Colombia
e-mail: rrodriguez@cotecmar.com

S. Lonin
Research Group of Oceanology, "Almirante Padilla" Naval Cadet School, Cartagena, Colombia

A. Vega Sáenz et al. (eds.), *Proceedings of the 25th Pan-American Conference of Naval Engineering—COPINAVAL*, https://doi.org/10.1007/978-3-319-89812-4_39

geochemical. A variety of petroleum components, such as aromatic compounds, are consequential from this natural source. Due to their changing compositions, every category of crude oil product has special and singlular properties. These characteristics influence how petroleum will perform when it is spilled and determine its effects on habitats [8].

Crude oil has aromatic compounds, and these are identified as being carcinogens. Minimizing risks for the sea environment needs an understanding of how petroleum spills behave depending on their different components and the ecological effect they produce [8].

Monoaromatic (single-ring) components found in crude oil are regularly denoted as BTEX (benzene, toluene, ethylbenzene, and xylene) [3]. Aromatic compounds include at minimum one benzene ring, and are chemically stable structures. These are persistent in the environment and can have negative effects on organisms. In any oil spill, the aromatic compounds are of particular interest especially for their toxicity grade [12].

Studied spills have confirmed the presence of BTEX inside marine animals, such as in the Exxon Valdez Spill in 1989 [3].

On April 20th, 2010, the Macondo's well discharged 4.1 million barrels of crude, and 2.1 million gallons of dispersants were applied in the Gulf of Mexico after the accident of the Horizon deep platform of perforation [10].

The spill happened at 1.5 km of depth. The event demonstrated the importance of physical, chemical, and biological oceanography for understanding the risks of crude transport and its dispersion in ultradepth water discharges [4].

In deep-water crude blowouts, there are inertial as well as buoyancy-driven stages of the mixture. The time and extent of the spread of each stage depend on the characteristics of the discharge and on environmental parameters such as temperature, salinity, and currents. To simulate the behavior of BTEX realistically, numerical models must be applied to handle the mixture of oil compounds [9].

As water depth increases, exceeding 200 m, the basic dynamics of the crude jet (plume) becomes more complex, principally due to the increase in hydrostatic pressure. This leads to the possibility of gas hydrate solids formation [11]. This paper describes the behavior and fate of BTEX in the catastrophic accident of the Gulf of Mexico blowout, and simulates its oceanographic spatial and temporal conditions.

The ECOMSED hydrodynamic model (three-dimensional hydrodynamic and sediment transport model) [5] is used in this study with an original Eulerian block for soluble substances. The latter is related to the Lagrangian plume of the crude. The ECOMSED has been tested in a variety of applications that are not described in this paper [1].

Verifications of the Eulerian model were not carried out in this investigation due to the lack of in situ data related to the diluted BTEX. Nevertheless, the results of this study demonstrate a good qualitative agreement between modeled benzene concentrations and data brought from technical reports [4].

2 Methodology

The present approach deals with a new model concept for advective and diffusive processes of aromatic compounds in a plume of oil in deep-water blowouts, formulated in Eulerian coordinates.

To simulate the behavior of diluted BTEX in the blowout, a Lagrangian model was implemented for simulating 3D trajectories of the plume. Each particle was defined with a spherical form and characterized by its location.

The domain of the model (35 × 35 km) was spatially adjusted setting the Macondo well in the center of the domain (28°12′ N, 88°48′ W) [14].

During the research, no information with a fine spatial resolution was found regarding the currents, temperature, and salinity fields. It was estimated that the appropriate grid step of the model should be on the order of 500 m. To achieve such a resolution, it was necessary to apply a special algorithm for "spreading" the available thermohaline information throughout the model domain along a previous cold-start simulation. Data for temperature and salinity profiles were found at the National Center for Environmental Information (https://www.nodc.noaa.gov/deepwaterhorizon/).

Toward this aim, temperature T(z) and salinity S(z) data were prepared at standard levels as a single profile (z as the vertical coordinate) to apply the technique of sea-level anomalies assimilation [6]. Satellite altimetry information for absolute dynamic topography (ADT) was required to "reconstruct" the 3D temperature and salinity fields for the above-mentioned data assimilation technique.

Due to the open domain boundaries, a special procedure of volume control was implemented in ECOMSED. This procedure checks the water inflow and outflow in proportion to the integral transport rate in the vertical in each point of the open boundaries and in each time step.

Figure 1a, b show the current fields at 500 and 1000 m of depth as a result of ECOMSED on June 12, 2010.

In the Eulerian model, the volume of cells was defined within a fixed controlled volume around each particle of the Lagrangian plume. New variables were incorporated, such as the discharge time, the mass flow, and the coefficient of mass transference [2] of the BTEX.

The change of mass of the BTEX drops diluted in seawater was expressed by Eq. (1) [13]:

$$\frac{dn}{dt} = -KA(S - C_a) \tag{1}$$

Herein, t is time; n is moles of the diluted compounds; K is the mass transference coefficient for BTEX (m/s), A is the surface area for a drop (m^2), S is solubility of the BTEX (mol/m^3), and C_a is the concentration of the dissolved BTEX (mol/m^3). It was presumed that C_a in (1) is negligible ($C_a << S$).

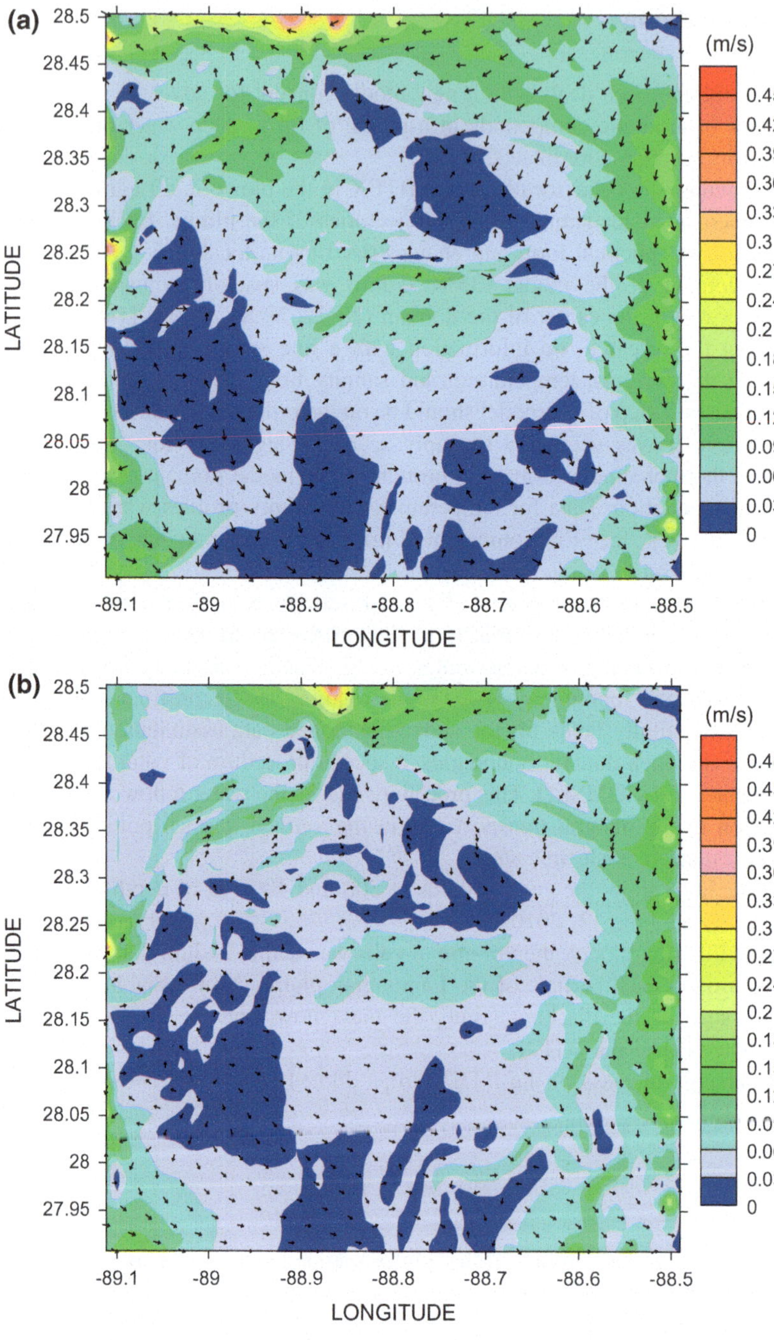

Fig. 1 **a** Currents field, at 500 m depth—ECOMSED. **b** Currents field, at 1000 m depth—ECOMSED

To calculate the mass transference coefficient in the present model, it was necessary to use the concept of effective internal coefficient and heat transference to solid spheres [7]:

$$K = \frac{10D_V}{D_p} \tag{2}$$

Herein, D_v is the diffusion coefficient of BTEX (m^2/s) and D_p is the droplet diameter (m).

Substance transport is affected by factors such as currents, processes of turbulent diffusion, and pollutants' phase change (solubility):

$$\frac{\partial C}{\partial t} + \frac{\partial CU}{\partial x} + \frac{\partial CV}{\partial y} + \frac{\partial CW}{\partial z} = \frac{\partial}{\partial z}\left(K_H \frac{\partial C}{\partial z}\right) + F_c + Q \tag{3}$$

In (3), C is the pollutant concentration, transported in the time t and space (x, y, z) for the respective current components U, V, and W; Q is the internal source of BTEX ($kg/m^3/s$) due its dilution, and F_C represents the horizontal diffusion. Q stands for the accumulated mass of BTEX originated from the Lagrangian plume and is defined by Eqs. (1) and (2).

The open boundary conditions were defined by specifying $C = 0$, under the consideration that the boundary limits are distant enough from the source. The initial condition was specified as C ($t = 0$) = 0. The flux of BTEX on the surface and ocean bottom was assumed to be zero. Furthermore, from the compounds conforming BTEX, benzene was the only substance considered in the results.

3 Results

Coupling between the Lagrangian model of the plume (loss of the substance in the plume) and the Eulerian model [increase of the diluted fraction in water, expressed by Eq. (3)], is addressed to quantify the change of BTEX concentrations and its transfer from the crude into the water column.

Figure 2a–f show the changes of concentration of BTEX (benzene) inside the studied domain for the date of June 12, 2010 at 1200, 1000, 750, 500, 250, and 100 m of depth, respectively.

The distribution of the BTEX concentration had certain spatial changes in 3D. The origin of the oil spill was considered to be in the center of the domain. It was observed that there were no significant changes in the concentration field of benzene at 1000 m of depth.

Nevertheless, an increase of BTEX was observed at 250 m with an orientation towards the southwest.

The BTEX concentration increased inside the plume towards the surface of the sea due to the continuous process of dilution as the oil plume emerged upwards.

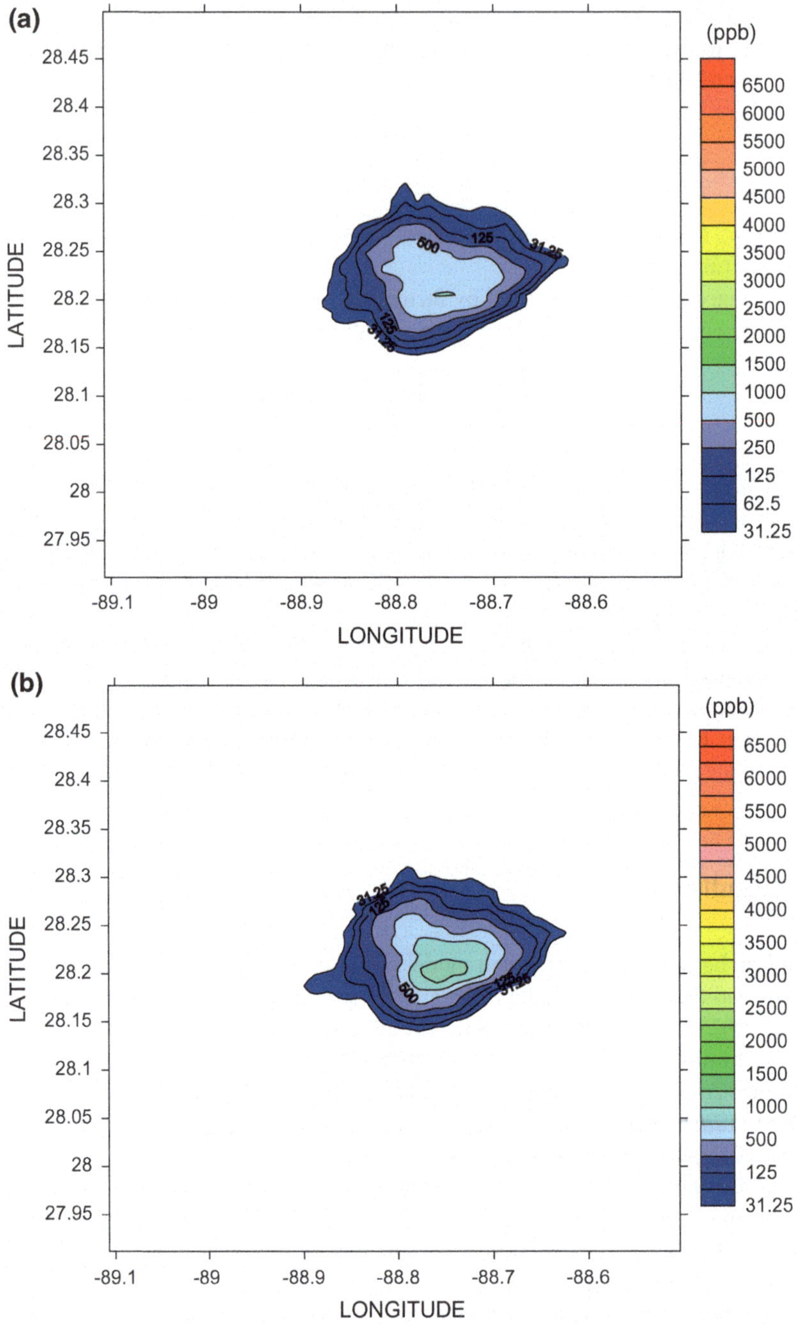

Fig. 2 **a** Concentration field of BTEX at 1200 m of depth. **b** Concentration field of BTEX at 1000 m of depth. **c** Concentration field of BTEX at 750 m of depth. **d** Concentration field of BTEX at 500 m of depth. **e** Concentration field of BTEX at 250 m of depth. **f** Concentracion field of BTEX at 100 m of depth

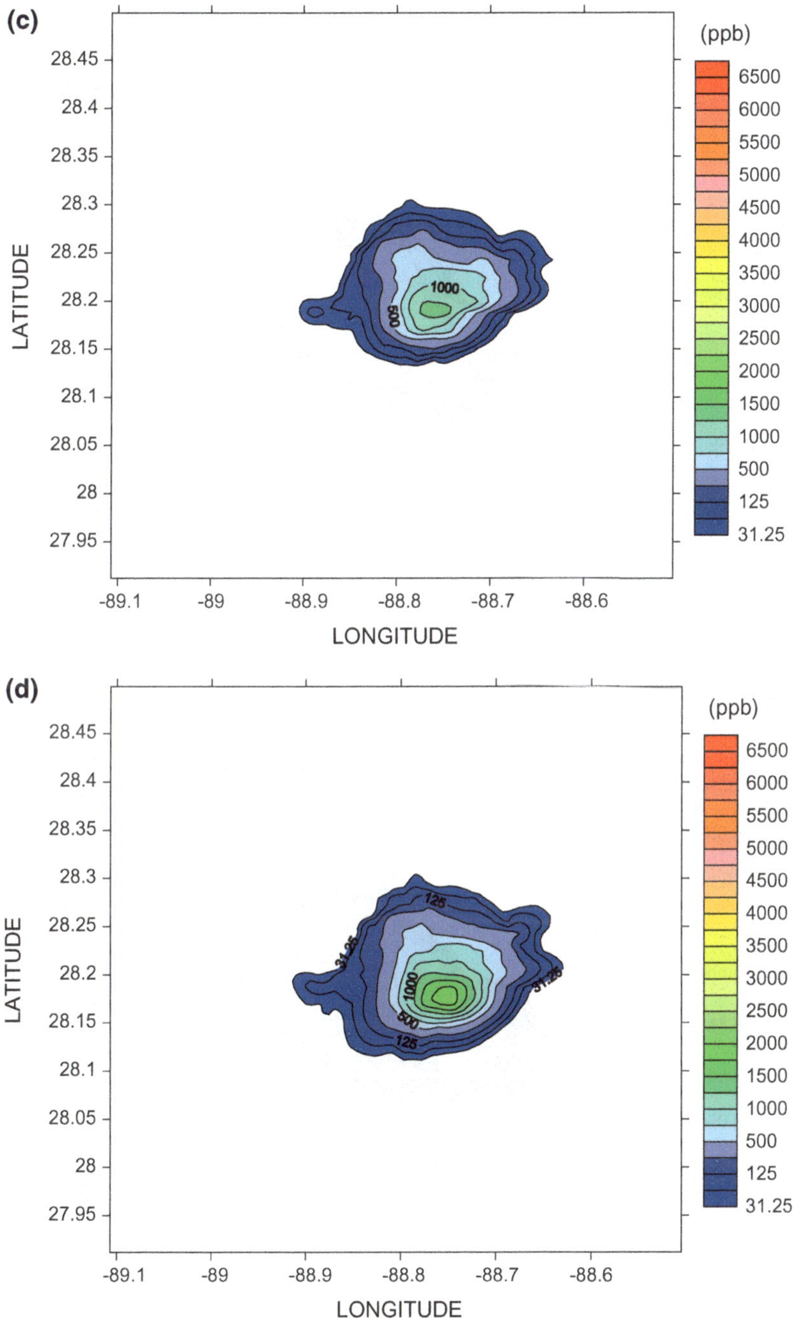

Fig. 2 (continued)

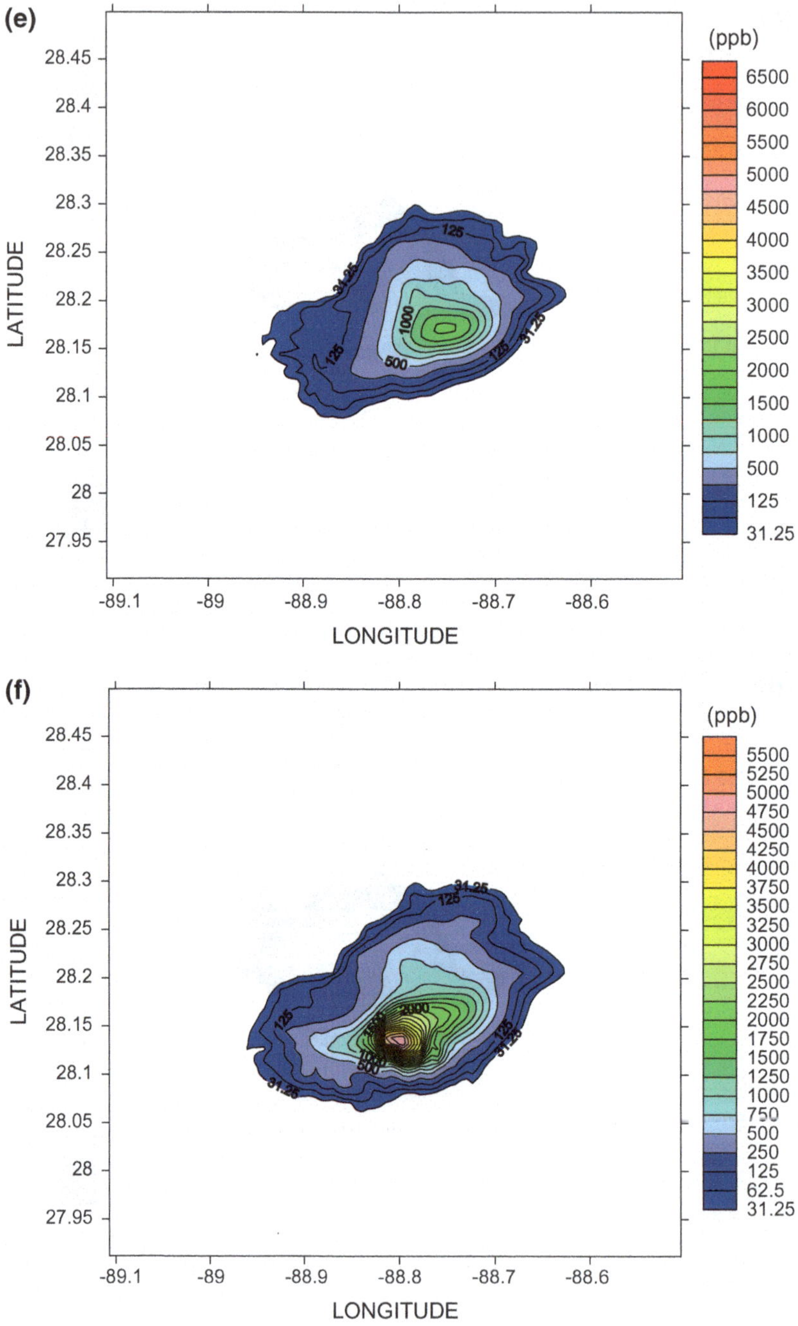

Fig. 2 (continued)

At 100 m of depth, a maximum benzene concentration of 2.500 ppb can be observed, whereas the concentration had an order of magnitude less in levels deeper than 1000 m.

4 Conclusions

The general behavior of diluted BTEX that was part of the oil blowout has been described. The Eulerian model for BTEX composition considered measurable variables such as stratification, depth, physical characteristics of the environment, and the oil composition.

The hydrodynamic model ECOMSED [5] was adapted to simulate the behavior of the soluble BTEX for the case of the Macondo's well accident. The concentration of diluted BTEX changes in seawater and must not be negligible because of its considerable toxicity for marine organisms. Thus, it is recommended that environmental evaluations of offshore activities include the analysis for this type of risk. This knowledge is useful for emergency response plans developed by governmental organizations, and for evaluation and supervision activities of offshore oil explorations. When oil is released in high depths, the model provides a tool for environmental monitoring regarding the dynamics of soluble substances in seawater.

The results of this study show that underwater concentrations of diluted BTEX depend on the oil plume dynamics and that although these components are not remotely measurable, they are significantly important because of their toxicity degree.

References

1. Blumberg, F.A., Mellor, L.G.: A description of a three—dimensional coastal ocean circulation model. Am. Geophys. Union, 1–16 (1987)
2. Branan, C.R.: Practical Solutions for the Chemical Engineers. Mc Graw-Hill, Mexico City (2000)
3. Brenda, E.B., Kimberly, A.K.: Hydrocarbon Residues in Tissues of Sea Otters (Enhydra lutris) Collected from Southeast Alaska. Exxon Valdez Oil Spill State/Federal, Alaska, Anchorage, Estados Unidos (July de 1997)
4. Christopher, M.R., Arey, J.S., Jeffrey, S.S., Sean, P.S., Karin, L.L., Robert, K.N., … Camilli, R.: Composition and fate of gas and oil released to the water column during the Deepwater Horizon oil spill. PNAS, 20229–20234 (2011)
5. HydroQual, I.: Manual: A Primer for ECOMSED. ECOMSED, Nueva Jersey (2007)
6. Lonin, S.A., Anduckia, J.: Assimilation of sea level anomalies in the ocean prediction system of DIMAR. Cartagena: Scientific bulletin 22—CIOH (2004)
7. McCabe, W.L., Smith, J.C., Harriott, P.: Unit Operations in Chemical Engineering. Mc Graw-Hill, Mexico City (2007)

8. NRC, N.R.: Committee on Oil in the Sea: Inputs Fates and Effects. National Academy Press, Washington, DC, GC1085 .O345 2003. Online access: http://www.nap.edu/books/0309084385/html/ (Open book format) (2003)
9. Poojitha, D.Y., Malinda, R.W., Anusha, L.D., Jeffrey, A.D.: How Does Oil and Gas Behave When Released in Deepwater?, pp. 275–285. Elsevier, New York (2012)
10. Sarah, E.A., Brian, W.S., Kim, A.A.: Impact of the Deepwater Horizon oil spill on bioavailable polycyclic aromatic hydrocarbons in Gulf of Mexico coastal waters. Nat. Inst. Health, 1–14 (2012)
11. Spaulding, M.L., Bishnoi, P.R., Anderson, E., Isaji, T.: An Integrated Model for Prediction of Oil Transport from a Deep Water Blowout. Applied Science Associates, Inc, Narragansett, USA (2000)
12. Speight, J.G.: The Chemistry and Technology of Petroleum. Marcel Dekker Inc, New York (1991)
13. Zheng, L., Yapa, P.D.: Modeling gas dissolution in deepwater oil/gas spills. J. Marine Syst., 299–309 (2001)
14. Zhengzhen, Z., Laodong, G., Shiller, A.M., Steven, L.E., Vernon, A.L., Christopher, O.L.: Characterization of oil components from the Deepwater Horizon oil spill in the Gulf of Mexico using fluorescence EEM and PARAFAC technique. Mar. Chem., 10–21 (2013)

Management of the Prevention of Labor Risks in Construction and Repair Shipyards

José Angel Fraguela Formoso, Asunción López-Arranz, María Jesús Rodríguez Guerreiro and Isabel Lamas-Galdo

Abstract Shipbuilding and/or repair activities are considered to be activities of special danger, with high levels of accidents, occupational diseases, dissatisfaction, and psychosocial risk. The risk assessment obliges the employer to manage the necessary preventive activity in order to reduce or control dangerous situations, in accordance with the legislative provisions related to specific risks, taking into account the probability of causing damages and the consequences derived from the exposure of workers to such risks. As in any management system, its implementation requires planning, organization, direction, and control, as well as the development of its corresponding activities, assigning sufficient human, material, and economic resources. The most sensible and effective approach would be the constitution of a proper prevention service, given the danger and variety of the construction and repair work of ships and the need to obtain an immediate response on the control of a certain risk. The implementation of comprehensive management systems for risk prevention, quality control, and environment protection, certification of the shipyard and its auxiliary companies, people and equipment, services, and installed systems, increases competitiveness and adds qualitative values, which will be highly valued by shipowners and shipping companies. This work will be of great help to the management of the shipyards, establishing a model of occupational risk management and defining and developing the activities to be carried out in each of the technical and medical areas.

Keywords Shipyards · Management · Risks

J. A. Fraguela Formoso (✉) · A. López-Arranz · M. J. R. Guerreiro · I. Lamas-Galdo
Universidade da Coruña, A Coruña, Spain
e-mail: jafraguela@udc.es

A. López-Arranz
e-mail: alarranz@udc.es

M. J. R. Guerreiro
e-mail: chus@udc.es

I. Lamas-Galdo
e-mail: isabel.lamas.galdo@udc.es

A. Vega Sáenz et al. (eds.), *Proceedings of the 25th Pan-American Conference of Naval Engineering—COPINAVAL*, https://doi.org/10.1007/978-3-319-89812-4_40

1 Introduction

A shipyard safety and health management system needs to comply with general and specific requirements that allow the promotion of a series of good practices to achieve the objectives of the policy approved by the organization. The evaluation and control of risks is not an easy task and requires important material and human resources. In the implementation and operation of the management system, human resources dedicated exclusively to health and safety at work play a fundamental role.

Its foundation is based on the policies adopted in each country or associations of countries and on existing international standards concerning these management models.

Taking as a reference the European Union, it can be said that following the great reform of the Constitutive Treaties of the European Community through the Single European Act of 1986, the great challenge of the 1980s comprised social policy, promotion and improvement of working conditions, the single market policy, and the free circulation of people, products, services, and capital. These policies, through occupational safety and health directives and product safety directives, converged and complemented each other, increasing the level of risk prevention for the worker.

At the Summit of Heads of State and Government of the Community, held in Dublin in June 1990, the Environmental Policy was launched, accepting the challenge of the 1990s for the European Community, to achieve sustainable development from the ecological point of view. This position considers that the environment depends on our collective actions and that tomorrow's environment depends on our behavior today, not only at the level of the European Union, but also worldwide.

These rulings would not be possible without the European Council Resolution of December 21, 1989 on a global approach to conformity assessment [1], which introduces a new philosophy in Europe: promoting use of European standards in all member states and establishing the procedures that allow evaluating the conformity of the products (accreditation systems, certification, tests, and calibrations).

In terms of social policy, in the European Union, Directive 89/391/EEC [2] was passed decades ago regarding the application of measures to promote the improvement of the safety and health of workers in the workplace. This Directive contains the general legal framework in which the policy of community prevention operates.

It is illustrative to read its Article 6 "General Obligations of Entrepreneurs," which states:

1. Within the framework of his responsibilities, the employer shall take the necessary measures to protect the safety and health of workers, including activities to prevent occupational risks and provision of information and training, as well as the constitution of necessary organization and means.

The employer must ensure that these measures are adapted to take into account changing circumstances and tend to improve existing situations.

2. The employer shall apply the measures provided for in the first paragraph of Sect. 1 in accordance with the following general principles of prevention:

   (a) Avoid risks;
   (b) Evaluate the risks that cannot be avoided;
   (c) Combat the risks at their source;
   (d) Adapting the work to the individual, in particular with regard to the design of work places as well as the choice of work teams and work and production methods, with a view, in particular, to mitigate monotonous and repetitive work and to reduce their effects on health.
   (e) Take into account technical evolution;
   (f) Substitute the dangerous for that which entails little or no danger;
   (g) To plan develop a coherent overall prevention policy which covers technology, organization of work, working conditions, social relations and the influence of the environmental factors in the work place;
   (h) Adopt measures that put collective protection before individual protection;
   (i) Give proper instructions to the workers.

It also indicates that for this the employer must organize prevention and protection services with a sufficient number of workers who have the necessary skills and capacity, sufficient resources, and appropriate time to dedicate to the effort.

In addition, the employer must take into consideration the occupational skills of workers in matters of safety and health to perform the task entrusted, ensure that places with serious and specific risks are only accessible to workers who have received sufficient and adequate information, and anticipate that preventive measures are effective in the face of distractions or reckless behavior by the worker.

These principles of preventive action give an idea of the field of work on which one has to act within the company. But there is more to it than that. Nowadays it is normal that this initial field of prevention and limitation of risks to the workers is extended to the control of damage to flora, fauna, property, and the environment, caused by the company facilities in its material usage, maintenance operations, storage of raw materials, manufactured products, and industrial waste, especially toxic and hazardous waste. All these aspects must be addressed in the implementation of a risk prevention system by the company.

2 Management Models

The occupational safety and health management standards are intended to be an effective tool to achieve the objectives formally expressed by the company's top management. In this sense OHSAS 18001: 2007 [3], in its planning section, indicates that the procedures for hazard identification and risk assessment must take into account, among other things, the infrastructure, equipment, and necessary materials in the workplace. It also says that the organization must establish, implement, and maintain programs to achieve the desired objectives, assigning responsibilities and delegating authority to the various levels of the organization.

The implementation of an occupational safety and health management system requires sufficient material and human resources and adequate skills. In addition, the functions of each of the levels of the preventive organization must be defined [4].

But the standards do not go far enough. Terms such as sufficient human resources, necessary, or suitable leave open a range of possibilities, which we need to specify.

To achieve the health and safety objectives, the workers that make up a prevention service should have an interdisciplinary nature and should be sufficient and adequate for the preventive activities to be developed.

A part of this interdisciplinary team will be composed of the technical team. The other part will be composed of the medical team. We cannot forget that the success or failure of a mission lies in teamwork. In the prevention of occupational risks, the technical part and the medical part must work in close collaboration and be of an adequate number for the needs of the company.

The guidelines and general objectives of business management increasingly include aspects related to integration within the company's management system, prevention management systems, quality management systems, and systems of environmental management. They should be considered as convergent management systems, intertwined and inseparable, and they should be applied to equipment, systems, and facilities [5].

And that is the case because the principles on which they are based are coincidental: they begin with management; they are based fundamentally on preventive action and not on dividing action; they have to be applied in all the phases of the lifecycle of the products and in all stages of the productive processes; they can be measured; they are everyone's task; and they are achieved through training. In addition, none of the fields is a priority, but rather two sides of the same coin and any action to improve working conditions has a favorable effect on the attitude of workers and leads to an improvement of the quality of manufactured products, the image, and the competitiveness of the company, and the conservation of the environment.

The Integral Management System represents the whole of the planning of the activities, the organizational structure, responsibilities, practices, procedures, processes, and necessary resources to develop, implement, carry out, review, and keep updated company policies, to ensure that the working conditions, the quality of the products produced, and the preservation of the environment remain within the levels intended in advance.

The correspondence between the Prevention Management Systems of the OHSAS 18001 standard, the Quality Management Systems of the UNE-EN-ISO 9001 standard [6], and the Environmental Management Systems of the UNE-EN standard—ISO 14001 [7], makes the implementation of an integral system easier and cheaper than the implementation of the three systems separately. Periodic evaluation is easier. Any improvement implemented in one of the areas will have a favorable effect on the others. In addition, it improves the work climate in the company, favors competitiveness, reduces insurance premiums by reaching higher levels of efficiency, reduces the cost of waste treatment, and finally increases the confidence that customers and consumers place in the company.

3 Organizational Infrastructure, Functions, and Specific Authority in Health and Safety

In general, the planning of preventive activities will require an organizational structure and material means that will depend on the size of the company and the risks involved in the activity carried out.

In shipyards, there is a great variety of jobs, many of which must be considered as high-risk activities and their size usually exceeds 250 workers, including shipyard workers and auxiliary companies.

In this case, the shipyard needs its own prevention service, with multidisciplinary human resources, specialized in the management of labor risks, organized at the most demanding level, capable of planning the prevention with a detailed programming of activities according to the objectives that are established.

Details of the standards of prevention management systems, such as the organizational model and the activities to be developed, have not been stated in full. For that reason, we have considered it important to develop a model with highly developed and complex structures. This model describes the functions, authority, and responsibility of people who occupy a relevant position in each of the areas of occupational safety and health. The organizational infrastructure that we present has been working effectively in a large shipbuilding shipyard for decades and we have participated in its planning.

Among workplace pathologies, the most important ones are work accidents and occupational illness, although we must also mention fatigue, premature aging, dissatisfaction, and others.

Technically, we should understand a work accident as any undesired alteration of the development of the productive process or one that breaks the continuity of a work or is capable of producing injuries to workers or damage to equipment or the environment.

With regard to occupational illness, we would have to understand it as an organic or functional alteration in whose evolution three stages can be distinguished: the disruption itself, without symptoms and reversible upon cessation of exposure to the contaminant; the longer term effect, with specific symptoms and not fully reversible upon cessation of exposure to the contaminant; and the disease, with permanent consequences.

The fundamental differences between occupational accidents and occupational diseases are the way in which they occur, the causative agent, and the type of damage produced. Thus the work accident occurs suddenly, by a material agent and produces traumatological injuries and/or damage to equipment, facilities, and buildings. In contrast, occupational disease, according to medical criteria, not legal, will occur after the worker has been exposed for a period of time normally long, to physical, chemical, or biological pollutants or to social and psychological environmental factors, producing organic or functional alterations.

To act on risk situations, it is necessary to perform risk evaluation (hazard identification, risk estimation, and risk assessment) and risk control through the implementation of prevention and protection measures.

The prevention of risks can be defined as the set of scientific knowledge and technological means that, applied to the risk factors, eliminate or control their evolution.

The protection aspect comprises the set of scientific knowledge and technological means that, applied to the situation of occupational risk, prevents or minimizes the injuries to people or material damages.

Both preventive and protective techniques will act on the so-called technical factors (set of elements and facilities that the worker uses to carry out his productive task with all the conditions that produce different situations of risk) and on the human factors (the workers with all their personal conditions, who create different risk situations).

Comparing the techniques and actions, the prevention techniques are more effective than the protection techniques. They are more effective, cost less, and are easier to apply. Operative conception techniques are more effective than correction techniques. These techniques should give priority to the technical factor and then also act on the human factor.

3.1 Preventive Management

The director of the shipyard must appoint a prevention supervisor as the uppermost level responsible for the management of prevention (understood to include planning, organization, direction, and control), according to the policy she decides to develop for the control of risks and the improvement of working conditions. She will be responsible for both technical areas (safety, hygiene, ergonomics, psychosociology) and medical areas (preventive medicine and healthcare medicine).

The prevention supervisor will analyze and update the chosen model of preventive management, in relation to the following areas.

- Management commitment: roles and responsibilities
- Prevention planning
- Prevention bodies
- Information, training, and participation of workers
- Basic preventive activities
- Personal protections
- Statistical control of accident rate

- Accident investigation
- Evaluations of working conditions
- Risk control
- Work rules and procedures
- Preventive maintenance
- Emergency plans.

3.2 Safety at Work and Safety of Facilities

From the point of view of security, as a risk prevention technique that tries to avoid accidents at work, the human factor and the technical factor must be acted upon, which is why its areas of action must include not only safe work procedures (safety at work), but the safety of the equipment, services, and systems according to the detailed instructions established in the industrial regulations (safety of the facilities).

Occupational safety is the area of accident prevention whose action is fundamentally immediate or very short term. That is, its work is carried out by controlling and supporting the production line, so that those situations of risk are corrected at the moment or in the shortest timeframe possible. Therefore, it carries the day-to-day weight in the prevention of accidents, based on the rules and work procedures of the shipyard's internal regime.

The safety of the facilities has its main line of action in the project phase and design of the equipment, services, and systems of the shipyard, for which it will be supported in all those administrative dispositions and norms of obligatory fulfillment, as well as in norms and technical rules of groups or entities of a public or private nature.

Its actions in the medium and long term are developed in the scope of the control exercised over the inspections and periodic reviews that are necessary to set up certain installations and equipment. All of this will be imposed by the regulations of industrial facilities and periodic reviews within the established plans in the shipyard.

The complexity, importance, and enormous losses derived from a fire in the ships and buildings of the shipyard make it necessary to act in a preventive way in the works with risk of fire and explosion and immediately with effective means of protection. In any case, workers should receive information and theoretical and practical firefighting training. If the shipyard is large, a team of professional firefighters will be needed.

The following lines of action are included.

- Sites and work spaces
- Work equipment: Installation and use
- Machines and tools: Installation and use
- Lifting and transport devices and equipment: Installation and use
- Pressure devices: Installation and us.

- Storage and handling of chemical substances and preparations
- Storage and use of combustible and flammable liquids
- Storage and use of flammable gases
- Installation and use of high- and low-voltage electrical installations
- Prevention and protection against fire
- Work specific to the industrial sector in question
- Emergency plans.

3.3 Industrial Hygiene

The American Industrial Hygiene Association (AIHA) defines industrial hygiene as "Science and art devoted to the anticipation, recognition, evaluation, prevention, and control of those environmental factors or stresses arising in or from the workplace which may cause sickness, impaired health and well being, or significant discomfort among workers or among citizens of the community."

From this definition, industrial hygiene can itself be divided into four typical branches, namely theoretical hygiene, field hygiene, analytical hygiene, and operative hygiene. From its relation with occupational medicine, we can assure its totality from the broad and technified work done in this area of prevention.

Without entering into theoretical discussions and along the lines of the other areas, I must say that its work is mainly carried out in the medium and long term, both in the field of recognition, evaluation, and control of risks in existing facilities and processes, and in the design of new facilities. Its documentation of work and its metrology is wide and adequate to cover the variety of contaminants existing in a shipyard.

Its lines of action are:

- Chemical agents (nonliving material): gases, vapors, powder, fibers, fumes, fogs, mists
- Physical agents (energies): noises, vibrations, nonionizing radiation, ionizing radiation, thermal stress, illumination
- Biological agents (living organisms): virus, mushrooms, bacteria.

3.4 Ergonomics

There are many definitions of ergonomics, but we can say simply that it is a set of techniques that aim to adapt the working conditions to the worker. It has an enormous importance in the design of workspaces. To achieve its objectives, its lines of action are focused on:

- The design of the work center: location, relationship between internal and external environments, social facilities
- The design of the workplace: organization of the workplace, anthropometry and geometry of the workplace, ergonomic design of positions using DDS (data design screens), thermal comfort, visual ergonomics, acoustic comfort, quality of the air in the interior spaces
- The study of the work itself: physical load (efforts, postures, repetitive movements), mental load (work demands, individual factors, conditions outside the work environment)
- The study of the individual: human characteristics and behaviors.

3.5 *Applied Psychosociology*

The psychosocial factors that affect the workers of a shipyard are complex and quite varied. In shipyards that embrace a more preventive culture, a great deal of attention is given to these factors and they form an important tool for the reduction of absenteeism in the workplace. Its application and control will be accomplished by taking into account:

- The characteristics of the company: size, location, public image, workplace design, business activity
- The structure of the organization: types of organization (simple structure, departmental structure, others), organizational structure factors (definition of competencies, hierarchical structure, command style, communication and information channels, relationships, personal promotions, work schedules)
- Individual characteristics: cultural conditions, personality, age, motivations, training, attitudes
- The factors of the work itself: work content, workload, autonomy, automation, role in the organization, professional development
- Effects on the worker: physiological, mental, social and labor dysfunctions.

3.6 *Business Medicine*

At present, there are more than enough reasons to go from compulsory company medicine (implemented by law) to progressive company medicine (in line with technological progress and modern concepts of social welfare). Some of these reasons are:

- The functions of the company doctor are fundamentally preventive.
- The recognition, evaluation, and control of existing contaminants in the work environment are carried out by an industrial hygienist.

- The low relevance of classical professional pathology, motivated by the prohibition, substitution, or control of dangerous products in the work environment, or the improvement in working conditions, makes the search and control of nonspecific pathologies acquire more and more attention.
- The increased awareness of the workers brings as a consequence greater exigencies of integral health, proposed and in general agreed upon, by work committees.
- The demands of workers and the achievement of greater performance at work are more than enough reasons to design, agree upon, and implement a preventive program to detect, control periodically, rehabilitate, and treat specific and unspecific pathologies in shipyard workers.

In the preventive program, the following should have a place: occupational toxicology, cardiovascular diseases, sensory deficits, early detection of cancer, vaccinations, diseases derived from overloads, drug addiction, absenteeism control, and occupational epidemiological studies.

Preventive medicine is the area responsible for the planning and control of ordinary periodic recognitions; specific medical examinations; various medical examinations (new hires, return to work, reassignments, etc.); study of assignments; study, consultation, and reports on occupational diseases, in terms of prevention as well as the diagnosing of existing ones; study, treatment, and prevention of drug addiction; identification and therapy of psychosomatic diseases; vaccination campaigns; early detection of prostate and breast cancer; control of sick leave due to occupational diseases; and so on.

Medical care is the arena responsible for the planning and control of emergency areas (emergency room, ambulance, drills, and evacuations, etc.); cure areas (treatment room and its instruments and medication, accident assistance); control of sick leave due to accident; and rehabilitation and physiotherapy areas (rehabilitation and physiotherapy room, appliances and equipment, and rehabilitation programs), among others.

The clinical and toxicological analysis laboratory is the area in charge of the planning and control of clinical analysis, toxicological analysis; sampling of organic fluids for toxicological analysis; study and improvement of analytical techniques; quality control programs; results archives; pharmacy (medication, healing material); microbiology (microbiological analysis of toilets and hygienic services); and programs to control insects and vermin.

3.7 Participation Groups and Union Representative Consultations

It is essential to have regular meetings regarding participation, information, and improvement of working conditions between the management of the company, the various departments of the prevention service and union representatives.

The highest level of representation and improvement body is the Central Committee of Safety and Health Conditions, whose president will be the company director or the person at the highest level that she delegates. In this committee, the most important decisions will be taken concerning the management of risk prevention and the approval of the manual of risk prevention procedures. This committee will depend on the Delegate Commission for Safety and Healthcare Medicine, the Delegate Committee on Hygiene and Preventive Medicine, and the Delegate Committee on Physiology, Ergonomics and Garments.

4 Conclusions

There are great difficulties involved in achieving an adequate organization in any company in a given sector, but it is even more difficult in the construction and repair of ships, considered to have a high risk of accidents and occupational illnesses. In these shipyards the highest level and effectiveness of prevention services depends fundamentally on the interest of the employer and the union pressure at a given moment. Both these aspects are extremely variable among different countries and over time in the same company.

A management system of health and safety at work allows planning, organizing, and controlling the objectives set by the organization, using the resources that the company has prepared for that purpose.

The risk prevention management system must be integrated into the overall management system of the company through the prevention plan. This plan can be understood to be the document that correlates a set of ordered, well-known, and assumed activities necessary to achieve the objectives set by the organization in terms of safety and health.

Risk assessment and the planning of preventive activity are the fundamental instruments of prevention management.

In the company's prevention plan, the model of the organization and the modality chosen will be specified with the allocation of human, technical, and material resources to carry out the established objectives, which will depend mainly on the size of the company and the degree of danger of existing risks.

The procedures used to carry out preventive activities must respond in writing to the questions: What should be done? How will it be done? When will it be done? and Who will do it?

If the organizational structure responds adequately to these approaches, we can achieve the desired objectives; if not, our preventive organization will be unable to achieve the best working and health conditions in our companies.

References

1. Unión Europea.: Resolución del Consejo de 21 de Diciembre de 1989, relativa a un planteamiento global en materia de evaluación de la conformidad. Diario Oficial n° C 010 de 16/01/1990, pp. 1–2 (1990)
2. Unión Europea.: Directiva 89/391/CEE del Consejo, de 12 de junio de 1989, relativa a la aplicación de medidas para promover la mejora de la seguridad y de la salud de los trabajadores en el trabajo (Directiva Marco) Diario Oficial n° L 183 de 29/06/1989, pp. 1–8 (1989)
3. OHSAS 18001:2007.: Sistema de gestión de la seguridad y salud en el trabajo. Requisitos, Ediciones AENOR, Madrid (2007)
4. Formoso, F., Ángel, J.: Implantación de la Calidad en la Gestión de Riesgos. VII Xornadas Galegas Sobre Condicions de Traballo e Saúde, Fundación Caixa Galicia, Ferrol (2000)
5. Formoso, F., Ángel, J.: Sistemas de Gestión de la Prevención. VI Xornadas Galegas Sobre Condicions de Traballo e Saúde. Fundación Caixa Galicia, Ferrol (1998)
6. UNE-EN-ISO 9001:2008.: Sistemas de gestión de la calidad. Requisitos, AENOR, Madrid (2008)
7. UNE-EN-ISO 14001:2004.: Sistemas de gestión ambiental. Requisitos con orientación para su uso, AENOR (2004)

Conclusions

With a tradition that goes back more than 50 years, the Pan-American Institute of Naval, Maritime and Port Engineers (IPIN), an institution formed to promote scientific and technological knowledge in the American continent, chose Panama as a meeting point to celebrate the XXV Congress of Naval, Maritime and Port Engineering (COPINAVAL). There are five decades of experience, of scientific and technological development, of cultural exchange, and of kinship among the peoples, which guaranteed an event of great impact in the maritime sector.

Panama City was the meeting point for more than 150 renowned scientists from countries such as Japan, China, Singapore, Belgium, Spain, Germany, Brazil, Argentina, Uruguay, Chile, Peru, Ecuador, Bolivia, Paraguay, Venezuela, Colombia, Mexico, the United States, and Cuba, who for three days updated the participants in the latest technological and scientific developments and presented their vision of the future of global maritime trade.

Parallel to the COPINAVAL six symposia were organized on strategic issues for the maritime sector: Maritime Education, Marine Corrosion, Offshore and Offshore Operations, Maritime Legal Aspects, Maritime Maintenance, and Maritime Transport. In addition, a Congress of Students of Mechanical Engineering was held. This made it possible for hundreds of university students to have access to an event of a high academic and scientific level that allowed them to update themselves while interacting with other students, professors, and entrepreneurs. This made COPINAVAL one of the most important engineering congresses in the region.

The main scientific contributions of the Congress are presented in this book, which is the product of the great work of the authors, reviewers, and editors and in which those papers evaluated as outstanding are reflected.

This book becomes the first of the series of many that will come in the future, always with a common objective: to meet the demand for new knowledge that our academy, the industry, and students of naval, maritime, and port engineering, continuously have.

A. Vega Sáenz et al. (eds.), *Proceedings of the 25th Pan-American Conference of Naval Engineering—COPINAVAL*, https://doi.org/10.1007/978-3-319-89812-4

The manufacturer's authorised representative in the EU is Springer Nature Customer Service Centre GmbH, Europaplatz 3, 69115 Heidelberg, Germany. If you have any concerns regarding our products, please contact ProductSafety@springernature.com

Printed and bound by CPI Group (UK) Ltd, Croydon, CR0 4YY
05/10/2026
02228141-0014